Introduction to
Population Biology
& Evolution

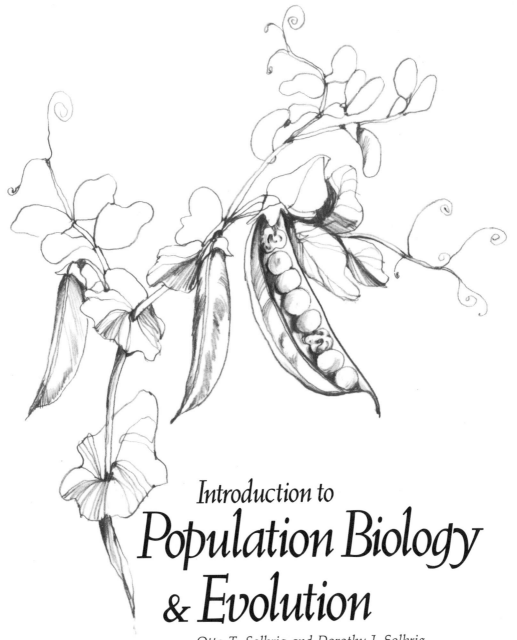

Introduction to
Population Biology
& Evolution

Otto T. Solbrig and Dorothy J. Solbrig
Harvard University

Addison-Wesley Publishing Company
Reading, Massachusetts • *Menlo Park, California*
London • *Amsterdam*
Don Mills, Ontario • *Sydney*

This book is in the
Addison-Wesley Series
in the Life Sciences

Library of Congress Cataloging in Publication Data

Solbrig, Otto Thomas.
    An introduction to population biology and evolution.

    1. Population biology. 2. Evolution. I. Solbrig,
Dorothy, 1945-   joint author. II. Title.
QH352.S66   574.5'24   78-67947
ISBN 0-201-06987-3

ISBN 0–201–06987–3
ABCDEFGHIJ–MA–79

Preface

This book is expected to fill the existing need for an up-to-date, comprehensive textbook in population biology and evolution at the introductory (freshman-sophomore) college level. It is aimed primarily at one-semester introductory courses.

In the present textbook we offer a rigorous treatment of the major aspects of population genetics, population ecology, and evolutionary theory, but we assume no previous knowledge on the part of the reader. We feel that our textbook is also unique in that we start with an explicit discussion of the scientific method and the processes of developing and testing hypotheses, which we apply directly to the theory of evolution. We then refer back to the general theory whenever more narrow and particular hypotheses are presented.

The emphasis is on the presentation and development of ideas, together with the necessary empirical evidence. Throughout the book we have tried to ask four questions in relation to each topic: (1) What is the problem under attack? (2) What is the evidence, the empirical "facts" that are relevant? (3) What hypotheses have been proposed to explain the facts? (4) What experiments or observations exist to discriminate between alternatives? It is our hope that the use of the hypothetico-deductive method will convey to the student the importance of rigorous hypothesis testing and indicate that this approach is as relevant to population biology as it is to molecular biology. Another feature of our book is the use of mathematical models when appropriate. We feel that mathematics, when used judiciously in our field, can impart discipline and insight. But mathematical models can have crippling limitations if they are based on unrealistic assumptions. Consequently, we have explicitly pointed out the assumptions and limitations of each model used. Furthermore, we have derived each formula step by step, so that it is comprehensible even to the student who has had only high school algebra. Finally, we have tried to present as many examples and case studies as space would allow.

Population biology provides the scientific foundation for the study of the diversity of nature, an appreciation of which is necessary to understand and deal with issues of pollution, conservation, and management of natural resources. It is a dynamic field that is moving rapidly. It is our hope that the vitality and importance of population biology have been captured in this book and will be transmitted to its readers.

Many people contributed to the preparation of this book. In first instance, We wish to thank the students at Harvard for whom the book was first outlined. We are indebted to Drs. William Brown of Cornell University, J. Merritt Emlen of Indiana University, William Healey of Holy Cross College, and especially Henry Wilbur of Duke University for their careful reading of the manuscript and many useful comments. We of course take responsibility for any errors that may exist. To Mr. James Funston of Addison-Wesley we are indebted for his support of this project. To the unfailing devotion of Ms. Margaret Cassidy to accuracy and precision this book owes much, and we are truly indebted to her. Last but not least, we wish to thank Mrs. Anita Fahey for secretarial assistance well beyond the call of duty.

Arlington, Massachusetts   O.T.S.
March 1979   D.J.S.

Contents

Part II

Part I

Chapter 1

Evolution, Populations, and Natural Selection

Approximately half a million species of living plants and twice as many species of animals have been found and described to date, and it is estimated that there may be at least one or two million more undescribed species. Furthermore, it is believed that the number of species of plants and animals, now extinct, that once inhabited the earth is five to nine times the number of those living today. The realization a hundred years ago that species are the result of evolution through natural selection in the last three billion years was one of the most significant advances in biology.

SPECIES AND POPULATIONS

We intuitively recognize a *species* as a group of closely similar organisms, such as humans, horses, or carrots. The scientific definition has varied historically (Chapter 10), but one that is often cited today is "a group of morphologically similar organisms of common ancestry that under natural conditions are potentially capable of interbreeding." There are a number of practical problems with this definition, such as ascertaining common ancestry and determining the meaning of "similar" and "potentially capable of interbreeding." We will discuss these points later, but for the moment this definition should give some notion of what is meant by a species.

Individuals within a species do not always live in close proximity. Normally they live in more or less isolated groups called *populations*. In order to study the process of evolution, we look within the species at the population. A population is a community of interbreeding organisms since individuals within a population normally breed with one another. However, occasional individuals may leave the population and breed with individuals of another population. This kind of population, called a breeding population or sometimes a Mendelian population, has to be clearly distinguished from the population studied by the statistician or the demographer. Either of them also defines a population as a group of individuals but one with a different binding property, such as "interaction in space" or "inhabitants of a given geographical locality."

Breeding populations are not necessarily discrete units; often it is hard to draw clear separating lines between them. A good example is the belt of balsam fir, *Abies balsamea,* that extends across northern Canada from Newfoundland to Alberta and from New England to Iowa, sometimes for hundred of miles with no obvious interruption. How can populations be demarcated in this case? Does the impossibility of delimiting the populations invalidate the concept? No. The inability to differentiate distinct breeding populations in certain cases and the overlapping of breeding populations in others does not necessarily mean that they do not exist. The problem of delimiting the breeding population into distinct units may be compared to the problem of delimiting young and old organisms in a nonarbitrary way. Nevertheless, "young" and "old" are useful and valid distinctions.

In organisms that reproduce asexually (such as many bacteria and some plants) there are no breeding populations, because each individual can reproduce independently. However, most asexually reproducing organisms occasionally revert to sexual reproduction. For example, many species of hawthorns, genus *Crataegus*, reproduce by the formation of seeds without recourse to fertilization. However, occasionally some seeds are formed in the normal sexual way by the union of the two gametes, an egg and a sperm. A population of asexually reproducing organisms is a breeding population to the extent that it reproduces sexually.

To understand how evolution works, we must understand processes within the population. The study of these processes is called *population biology*. Its main components are population genetics, which is the study of hereditary processes within the population, and population ecology, which is the study of interactions of members of the population with one another and with their physical and biological environments. Although the population biologist is primarily interested in the evolutionary phenomena that take place within the context of populations, he must be knowledgeable about the kinds of species that result from the evolutionary process. The diversity of living and extinct species, which is the domain of systematics and paleontology, is briefly considered at the end of the book.

ORIGIN OF TERRESTRIAL LIFE

The slow recognition in our day that all living creatures consist entirely of chemicals has had great impact on biology. It used to be thought that "dead" and "living" matter were of different natures, but today we know that this is not so. Since living matter is more complex, more highly organized, than non-living matter, organisms must have evolved from simpler, nonliving components. By tracing the origin of living forms, we gain important glimpses into the very nature of life.

To be sure, our understanding of the origin of life is at present far from complete. Many of our ideas are speculative. They are based on the properties of present-day organisms and on data submitted by physicists, chemists, geologists, and astronomers. At best, then, we have an idea as to how life could have arisen from nonlife, but this provides only some background about the nature of life and a brief glimpse into that nature.

The origin and nature of life have interested humans ever since they became rational beings. One of the earliest explanations brought forward was that life originated by *spontaneous generation* from nonliving matter. Undoubtedly this idea came from the observation that all dead bodies decompose into simpler elements, that everything "reverts to earth," in combination with ignorance about the life of maggots and molds, which seemed to appear from nowhere. When Pasteur in the late nineteenth century demonstrated that spontaneous generation does not take place on earth today, spontaneous generation seemed scientifically untenable, and for a time the idea was abandoned. Spontaneous generation is not possible today because of the presence of living organisms that feed on every energy-rich organic substance available. This fact prevents the transformation of simple organic substances into more complex matter, a very slow process.

But conditions on earth were different at one time. At the dawn of our planet's existence, no organisms were present, and therefore the slowness of the process by which simple chemicals became transformed into more complex ones did not matter. Under these conditions spontaneous generation is believed to have taken place through the slow transformation of atoms and molecules into simple chemical compounds, and of these compounds into more and more complex substances, leading eventually to living organisms. For this sequence of events to occur, certain conditions were necessary, such as temperatures in the narrow range between the freezing and boiling points of water, and the presence of a source of energy, which in this case was solar radiation. Eventually water started accumulating on the earth, and oceans were formed in which various chemical compounds dissolved. The accumulation of water and dissolved chemicals facilitated reactions among the different substances and the formation of still more complex ones.

Many of the compounds broke down into simpler compounds and then into elemental molecules. Thus the concentration of complex substances remained low. However, under certain special conditions, the concentration of chemicals must have increased several fold, since such higher concentrations are necessary for even more complex reactions to take place. The absorption of organic material in soluble colloidal particles, called *coacervates*, has been proposed as an important concentration mechanism. Since each of these coacervates was unique, some were more efficient at obtaining energy and in replicating. Therefore compounds were eventually formed that were capable of producing more of their own kind by a process of duplication. Duplication, or reproduction, is the diagnostic property of life. When a substance that had the property of reproducing its own kind appeared, no matter how simple that substance was, what we call life had originated. From there on, by the process we call evolution through natural selection, the simple organisms became gradually more complex, acquiring the traits and functions that are characteristic of present-day plants and animals.

The nature of all the steps leading to the formation of life is not known. Nevertheless, several plausible ways have been proposed. Since we have no precise way of knowing what the exact conditions on earth were at the time, there is a great deal of conjecture involved. Also, because of the presently changed conditions of our planet and the slowness of the process, it is impossible to verify experimentally the proposed pathways that led to the origin of living matter. All these difficulties notwithstanding, a simple experiment showing the soundness of the fundamental idea was performed by Stanley L. Miller of the University of Chicago in 1953. He circulated in a sealed circuit a mixture of the gases methane, ammonia, hydrogen, and water vapor, which are universally accepted as the gases most likely to have been present in the early atmosphere. At a certain point in the circuit, an electrical discharge simulating lightning was produced (Fig. 1.1). Miller ran the experiment for a week and then analyzed the products. Minute quantities of amino acids—the basic components of proteins and one of the principal chemical organic substances of any living being—had been formed.

One of the more fascinating aspects of this scheme is that once the right conditions were present on the earth, the formation of complex organic substances and life was the inevitable consequence. Each necessary event had a small chance of occurring per unit of time, but total time involved was tremendous. In turn, the position of the earth in the solar system probably made the presence of the right conditions inevitable. But the sun is not a special type of star in the universe; the likes of it are counted by the millions in our galaxy alone and by the billions in the totality of the cosmos. According to the late Harvard astronomer Harlow Shapley, of the approximately 10^{20} stars existing in the universe, 20 percent are essentially identical to our sun in size, luminos-

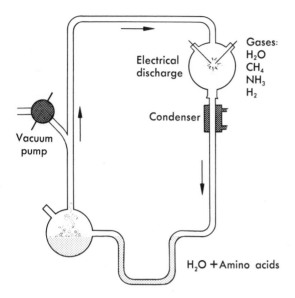

Electrical
discharge

Gases:
H_2O
CH_4
NH_3
H_2

Condenser

Vacuum
pump

H_2O + Amino acids

Figure 1.1

Sketch of the experiment conducted by S.L. Miller. The gases methane, ammonia, hydrogen, and water vapor were circulated for about a week. In each cycle they were subjected to an electrical discharge. Amino acids were recovered from the trap at the bottom. (From O.T. Solbrig, 1966. *Evolution and Systematics*. New York: Macmillan)

ity, and chemistry. If only one percent of these suns have planets, and if only one percent of the planets are in positions relative to their suns similar to the earth's position relative to our sun, some form of life is probable for at least 2×10^{15} planets. If so, life may be a rather common phenomenon after all! In addition, Shapley thought that conditions leading to what we call life are possible also in crusted-over stars—that is, stars with a warm center but a cold outer crust. The reader should clearly understand that "life" is here defined as some kind of self-duplicating system, not necessarily forms of life as found on earth.

THE CONCEPT OF NATURAL SELECTION

The evolution from atoms and molecules to simple and then complex substances and from those to still more complex ones capable of self-duplication is called *chemical evolution* to differentiate it from the evolution of organisms, called *organic evolution*. The difference between these two kinds of evolution lies in the fundamental nature of living matter: the capacity to reproduce itself. In the stages leading to the formation of life, more and more complex substances in ever larger quantities were formed, as more energy was received from the sun and as more chemicals reacted with one another. These substances were capable of growing in a fashion similar to the growth of a crystal, and by accidental breakage they could divide into several units. But that type of growth was not yet reproduction. The moment the property of self-duplication was acquired, that chemical substance—most likely a nucleic acid similar to or identical with the DNA found in chromosomes of present-day plants and animals—could form more of its own at the expense of other and probably simpler compounds. In other words, it could grow and reproduce. When two substances or two strains of the same chemical have the same property, the one that produces the larger number of surviving "offspring" will become more abundant. This is the essence of the process called natural selection, which is an exclusive characteristic of organic evolution.

The concept of natural selection was first proposed by Charles Darwin and Alfred Wallace at the same session of the Linnaean Society of London on July 1, 1858. A year later Darwin (Fig. 1.2) elaborated his ideas in his now famous book *On the Origin of Species by Means of Natural Selection*.

The principle of natural selection is based on three basic observations and two deductions from them. First, Darwin noticed that all organisms have the potential for increasing their numbers in a geometric fashion. That is, all organisms produce more offspring than are needed to replace the parent population. The second observation Darwin made was that the actual number of each kind of organism usually remained about the same from generation to generation. From these two observations he reasoned that a "struggle for existence"

Figure 1.2

Charles Darwin. (Photo copied from original in deCandolle Coll. at Cons. de Bot., Geneva; print at Hunt Institute, Pittsburgh)

must take place. Since more young are produced than can survive, they must be competing with one another for the resources needed for survival. Darwin also observed that no two individuals of any species are alike or, stated another way, that variation is a fact of nature. If variation exists among the adults, it must also be present among the young. When the young compete for resources, more of those with characteristics that aid them in competition (favorable variations) will survive than of those with unfavorable variations. This phenomenon is called *differential mortality*. The surviving population will therefore have a greater proportion of favorable variations than its predecessors had, who in turn had a higher proportion of favorable variations than their ancestors had, and so on. The repetition of this process over time eliminates variations unfavorable in the struggle for existence and selects the favorable ones (Fig. 1.3).

One of the ways in which individuals vary is in the number of offspring they produce. The probability that some offspring of a particular set of parents will survive to reproductive age is greater if they produce more young, other things being equal. Therefore, in a population of individuals similar in other characteristics, those individuals that produce the most offspring will have a greater chance of having some of their offspring survive to adulthood. This process, called *differential reproduction*, together with differential mortality, underlies natural selection (Fig. 1.4).

What are the characteristics that result in an increased rate of survival of the progeny and thus affect the differential mortality? They are many and exceedingly varied, as Darwin realized. Basically, any property that increases the survival probability of the individual and his or her offspring will be favored by natural selection. Any inherited characteristic that makes the individual possessing it more capable of surviving the rigors of the environment is termed adaptive. By this we mean characteristics that will result in circumventing adverse conditions, in obtaining more and better food, in avoiding predators, and so forth. These characteristics will result in survival of a larger number of the young. Adaptive characteristics can be morphological, as is the presence of thick fur and layers of subcutaneous fat in polar bears to withstand cold. They can be physiological, as in apple trees that will not bloom and fruit well unless they have been exposed to a certain amount of cold—an adaptation that ensures the

Figure 1.3

Differential survival. Three hypothetical rabbit pairs differ only in color. Each breeding pair produces ten offspring. The black and white rabbits, being more conspicuous than the gray ones, lose 60 percent and 40 percent, respectively, of their offspring before they can reproduce, whereas the gray rabbits lose only 10 percent. After three generations, the black rabbits, which were 33⅓ percent of the original population, are only 16/250 = 6.4 percent of the population, even though they have increased in number. Similarly, the white rabbits have gone down to 55/250 = 21.6 percent of the population. The gray rabbits, on the other hand, have increased from 33⅓ percent to 180/250 = 72 percent.

▶

Parents	Second generation		Third generation	
	Born	Adult	Born	Adult
	Ten	Four	Twenty	Eight
	Ten	Six	Thirty	Eighteen
	Ten	Nine	Forty-five	Forty

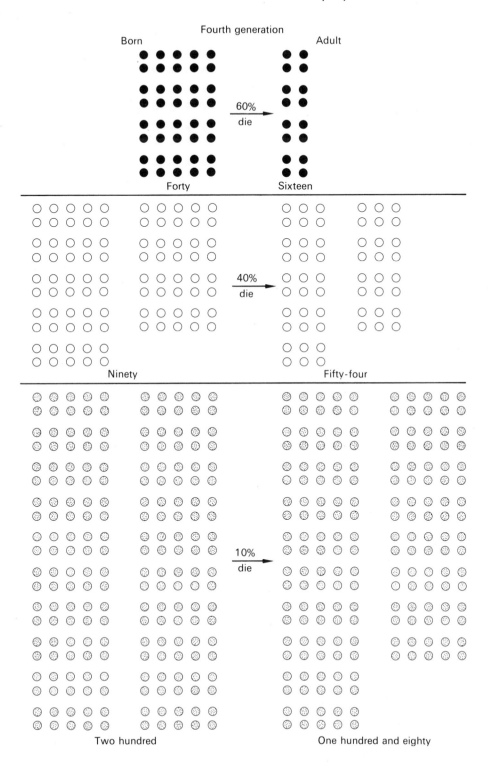

Fourth generation

Born   Adult

60% die

Forty   Sixteen

40% die

Ninety   Fifty-four

10% die

Two hundred   One hundred and eighty

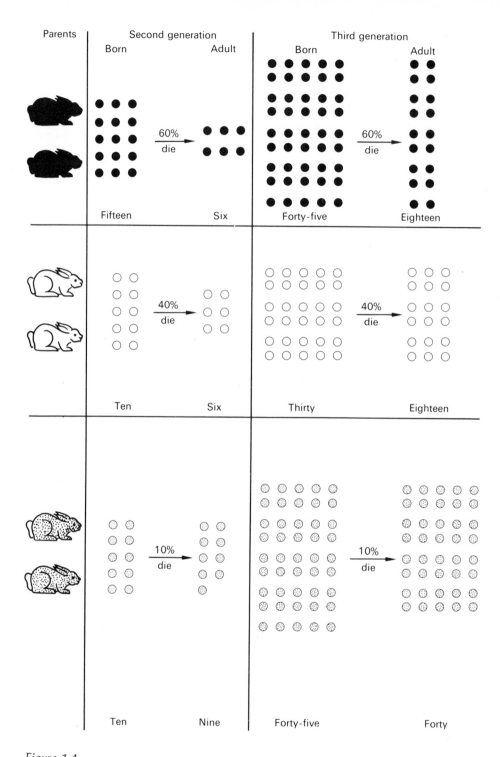

Figure 1.4

Differential reproduction. Suppose the black rabbits of Fig. 1.3 produce 15 offspring at each mating instead of ten while other conditions remain the same. Then, after

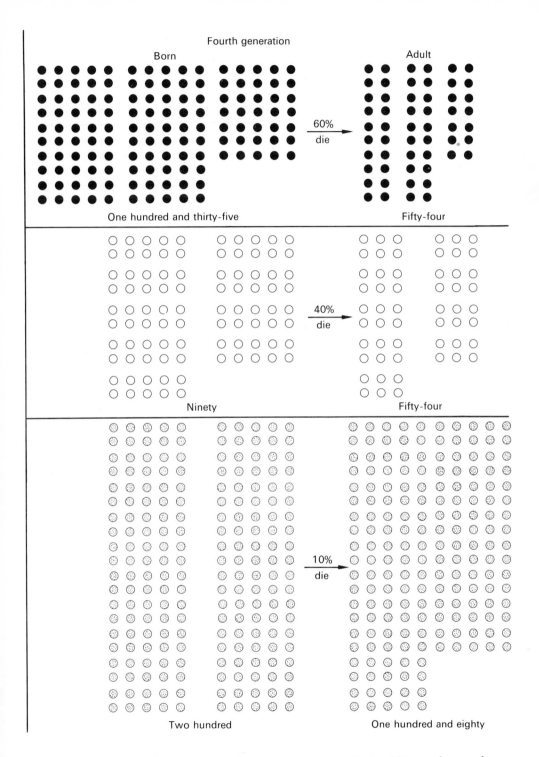

three generations, there will be 54/288 = 18.75 percent black rabbits in the population (as opposed to 6.4 percent in Fig. 1.3), 18.75 percent white rabbits, and 180/288 = 62.5 percent gray rabbits.

plant against blooming in a late fall or early winter "Indian summer." They can be behavioral, as in the complex and highly specific mating rituals of birds that ensure the mating only of two individuals of the same species and thus prevent the formation of ill-adapted hybrids. Finally, a very important adaptive characteristic is the ability to produce the largest number of offspring without sacrificing their ability to survive. All major traits of plants and animals are certainly adaptive or represent expressions of former adaptive traits. It was once believed that every inherited phenotypic characteristic of living organisms, no matter how small, was adaptive, but we think today that this is probably not so.

The weakest point in Darwin's theory of natural selection was the absence of an adequate explanation of the mechanism of inheritance. Since natural selection depends on the inheritance by the offspring of the characteristics of the parents, the mechanism of transmission of these characteristics determines the patterns of natural selection. The theory of inheritance we accept today was first proposed by Gregor Mendel (Fig. 1.5) in 1865 although it was not widely known until the beginning of this century. Mendel in his experiments with garden peas found evidence suggesting that heredity was controlled by particles passed from parent to offspring. Today we call these particles *genes*. The rules governing transmission of genes are discussed in Chapter 4. The idea of particulate inheritance is of great importance in understanding how natural selection works in a population. Its implications are discussed in detail in Chapters 5–9.

EVOLUTION

Evolution can be regarded as changes in any attribute of a population over time. In an ultimate, philosophical sense, this is what evolution means. But for the researcher who is interested in quantifying evolutionary change and studying the mechanisms by which it occurs, this definition is too vague. Certainly not every change results in evolution, and furthermore, the definition gives no idea of the time unit over which evolution occurs.

A very popular definition regards evolution as the change in the frequency of individual genes in a population from generation to generation. This is a very precise definition. Theoretically, the frequency of genes can be measured, and changes in the frequency of each gene over time can be recorded. However, there are some problems with this definition, too. Although gene frequencies can be ascertained in certain cases, at present only a relatively small number of them can be so measured, and there is no way of judging whether these particular genes are representative of the kind of changes that improve survival of the organisms in the population.

A third type of definition tries to cope with the problems posed by the two former ones by stating that evolution is the change in adaptive characteristics in the population and their underlying genes, over time. Although ap-

Figure 1.5

Gregor Mendel. (Photo from
Dörfler, *Botaniker Porträts*,
1907; print at Hunt Institute,
Pittsburgh)

parently more precise, since it specifies that only those changes that improve
the survival of the population should be taken into consideration, it leaves
unresolved what exactly those changes are. Consequently, this definition is
nonoperational and circular.

The diversity of definitions points out how difficult it is to quantify and
measure evolution. Most scientists agree that evolution is change over time,
and that this change has to involve the genetics of the organisms. But evolution
is a very complex process that includes many phenomena. Since many are not
well understood, a foolproof, rigorous definition of evolution is not yet possible. Let us now look at some of the factors important to evolution.

First and foremost is the mechanism of inheritance. The way in which
genes determine the expression of characters and the manner in which genes
are transmitted to the offspring shape the whole evolutionary picture. In turn,
the organization of genes in chromosomes and the behavior of the chromosomes
during cell division affect the mechanisms of inheritance and evolution. Evolution is therefore often specifically defined as changes in the frequency of genes
in a population.

Although evolution can take place without sexuality as known to us, this
has not been the norm. Sexuality is a mechanism that tends to combine the

genetic materials of individuals and to produce new and novel combinations. Its effect is a tremendous increase in variability, the advantages of which are apparently so great that the phenomenon has become almost universal in all plants and animals. Sexuality seems to have developed very early in the evolutionary history of organisms.

Without sexuality and interbreeding, species as we know them today would not exist. But just as important for the evolution, particularly the multiplication, of species has been the development of barriers—geographical, ecological, behavioral, or genetical—to the free exchange of genes. The very simple earliest organisms may have been able to mix their genes with others of the same level of organization, but present-day organisms, with elaborate and complicated developmental pathways, cannot exchange genes with drastically different organisms. When they do, the result of the exchanges is lethality or at best sterility. The selective advantage of barriers that prevent gene exchange in such instances is clearly apparent. When we think and speak of evolution, we refer almost always to the evolution of species. That fact is a clear indication of the important role of sexuality and isolation.

In a slightly different sense, it can also be said that evolution is shaped by the environment. Differential survival is always partly due to capacity to adapt to the environment, particularly the physical environment. Chemical evolution could occur only after our planet changed from the original "ball of fire" to a body where water could accumulate. Organic evolution, which has produced the tremendous number of organisms, is in part a consequence of the adaptation of these organisms to the infinite types of environments found on earth.

THE CONSTRAINTS ON EVOLUTION

Living organisms are complex systems that harvest energy and simple chemical compounds and transform them into an integrated body and into offspring. But energy and materials needed by organisms are not available in unlimited amounts, nor are they distributed evenly on the earth.

There are two sources of available energy for organisms. One is radiant energy in the form of light, present during the day. The other source is chemical energy, available in the form of certain energy-rich inorganic compounds, such as nitrates and sulfates, but primarily in the form of energy-rich compounds furnished by the bodies of living and dead organisms. The same distribution in availability applies to the inorganic molecules needed as building blocks by organisms. They are available in low concentration as ions dissolved in the waters of the seas and of rivers and in the soil water solution, and again, they are available in concentrated form in the bodies of living organisms.

Light energy is harvested by green plants through the process of photosynthesis, and it is transformed into energy-rich carbohydrate molecules. However, plants are able to use less than five percent of the incident radiation; the remainder is reradiated or dissipated in the form of heat. With the chemical energy contained in the sugars produced in photosynthesis, the plant builds its body from the atoms and molecules it gets from the surrounding water (if a water plant) or from the water-soil solution. Harvesting light requires certain specialized structures ("solar panels") we call leaves (or otherwise green stems). They have to be placed so that they do not shade each other; they must be strong enough to resist strong wind currents, rain, and snow; and so on. That is, the harvesting of light constrains the form of the organism. Consequently, in order to "eat" light, a plant has to specialize in its form and function. In turn, other organisms can harvest the chemical energy stored in the leaves of plants (Fig. 1.6). These specialized organisms are the herbivores. To efficiently harvest the energy contained in plants requires in turn specialized morphologies that are very different from those of plants, such as organs to separate the leaves from the rest of the plants (mouth parts) and to extract the energy and the chemical compounds from the tissues (gut). Because plant materials have a lot of compounds that are not digestible, and because their protein and fat content per gram is low (except in some specialized tissues, such as seeds, nectar, or pollen), herbivores have to consume relatively large amounts of plant material to maintain themselves. That requires that they spend a fair amount of their time eating. This characteristic of herbivores allows another kind of specialization, that of carnivores. Carnivores get their energy and materials from herbivores. The bodies of herbivores contain more digestible energy per gram of matter than those of plants. However, carnivores also have to have specialized bodies to be able to catch and overpower herbivores.

The trophic structure, as the different levels of food gathering are called, is then a source of diversity (but not the only one, as we will see further on). It exists because organisms require energy, and the rate at which they gather energy and the way they do so limits them in terms of what they can do and how they are built. In turn, within each trophic level further specialization can and does occur. Herbivores specialize by tissue: flower eaters, wood eaters, leaf eaters. Carnivores specialize in the way they hunt: those that hunt by pursuing their prey, those that sit and wait for the prey to come to them, those that lay traps for the prey. Plants specialize in the efficiency of light harvesting under conditions of high or low light intensity, high or low temperatures, and high or low water availability.

Evolution is then the result of the interplay of many and diverse factors. These factors themselves are subject to change. Early in the history of each lineage of plants and animals, structures or processes developed that have profoundly influenced the evolutionary history (called the phylogeny) of that group. So, for example, a segmented body and an exoskeleton have been major

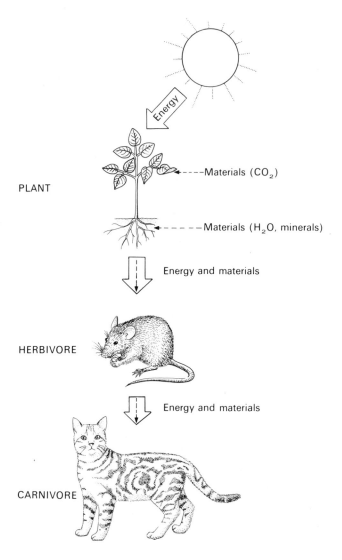

PLANT

————Materials (CO_2)

————Materials (H_2O, minerals)

Energy and materials

HERBIVORE

Energy and materials

CARNIVORE

Figure 1.6

Trophic levels. Green plants harvest energy from the sun and take carbon dioxide from the air and water and nutrients from the soil. Herbivores obtain both energy and nutrients by eating plants, and carnivores get theirs by eating herbivores (or other carnivores).

factors in the success of the insects, but these factors have in turn restricted the size and habits of the members of the class.

It is the multifaceted characteristic of evolutionary processes that makes the study of this subject so fascinating. And our knowledge is far from complete. Much remains to be discovered, much to be learned. The application of new techniques, such as the use of high-speed digital computers, or the application of comparative biochemistry to elucidate developmental pathways or to discover evolutionary relationships at the chemical level is opening grand new vistas in the field.

In the following chapters we will describe in detail some of the major evolutionary factors, analyze their role and relative importance, and investigate their development. We will also try to relate these factors to the field of ecology. Whenever possible, we will attempt to point out how human investigative tools limit as well as help in the acquisition of knowledge about nature. But first it may be useful to dispel some erroneous notions and popular misconceptions about evolution.

WHAT EVOLUTION IS NOT

The misconceptions about evolution are many. Although the reading of this account should dispel most misconceptions, two ideas are so fixed in the popular mind that they deserve to be considered separately. The first is the equation of evolution with struggle. The phrase "survival of the fittest" is Darwin's, but it is clear from his writings that he did not mean it as it is often interpreted. It was the British philosopher Herbert Spencer (1820–1903) who popularized the idea of struggle in his works and his concept of social evolution. In the historical context of the times, the idea of struggle between individuals, from which the winner, the fittest, survives to get the spoils, was appealing to certain groups. The second half of the nineteenth century represented the height of the laissez-faire era, the pinnacle of the British Empire. Spencer's ideas, with apparent backing from the natural sciences, gave a supposed scientific (and, for some, moral) justification to the abuses of the times. In translating organic evolution to the social scene, Spencer misinterpreted the main idea—that is, differential survival and reproduction of the progeny. He replaced it with one element of natural selection, differential mortality due to better adaptation to the environment, and interposed the idea of direct competition as responsible for differential mortality. The ideas as presented are entirely his and have no scientific backing. We must keep always in mind that by the "fittest" Darwin meant the one with the most surviving progeny. That can be and often is a comparatively weak individual.

Almost a hundred years later, the Soviet geneticist Lysenko, in the name of Darwinism and with the official blessing of the Communist party, was to denounce several aspects of modern genetics and cytology and accept the theory of the inheritance of acquired characteristics. In Stalinist Russia, these ideas gave an apparent scientific backing to the efforts of the state to mold the individual into an obedient servant, so that his descendants might become still more docile and obedient. In this case, too, abandonment of scientific accuracy created confusion, and although we are all aware of the fallacies and political motivations of Lysenko, we are not always so much aware of Spencer's mistakes.

Evolution also lacks foresight. At one time it was believed that evolution was "inner-directed" and consequently was purposeful, or goal-directed. This was called the theory of "orthogenesis." Evolutionary change, however, is the result of genetical and ecological circumstances at each particular time and place on the earth. What these phenomena are and how they interact is the substance of this book.

The second important point to understand is that no moral judgment of any sort can be read into evolution. Evolved organisms are not "better" in a moral sense; they are only better-adapted to the environment they occupy compared with their extinct ancestors. We also should remember that every living species, by this very fact, is adapted to the environment it occupies, presumably as effectively as any other living organism, irrespective of the phylogenetic positions of the organisms in question. Evolution is as blind as Justice is supposed to be: Those with the largest surviving progeny will multiply, regardless of how good or bad humans may consider them.

In the coming chapters we will present the background necessary to understand evolution. In Chapter 2 we will discuss the scientific method and in Chapter 3 apply it to the theory of natural selection. The first three chapters of Part II (Chapters 4, 5, and 6) will be dedicated to simple genetic aspects. We will discuss first the basic laws of genetics, then look at the chromosomes, the physical structures that contain the hereditary material, and finish with a detailed presentation of the major factors that account for genetic changes in populations. Chapter 7 deals with the role of chance in evolution and Chapter 8 with reproduction. Chapter 9, which carefully reviews the known patterns of genetic variation in populations of plants and animals to see whether they agree with what has been predicted so far, completes the discussion of basic population genetics.

Part III deals with patterns of evolution at the level of the population, the species, and the community. What the evolutionary units are and how the genetic structure, seen in Part II, is translated into a pattern of individuals, populations, and species constitute the subject of Chapter 10. In the next two chapters we discuss the process of species formation, and Chapter 13 takes up the subject of population growth. In Chapters 14 and 15 we discuss community structure. Chapter 16 is dedicated to a discussion of behavior—what it is, how it affects evolution, to what extent it is inherited. The book finishes with a chapter on the role of paleontology and one on taxonomy.

SUGGESTED FURTHER READING

Anfinsen, C.B. 1965. *The Molecular Basis of Evolution*. New York: Wiley.

Bell, P.R. 1959. *Darwin's Biological Work*. New York: Wiley.

Commoner, B. 1964. DNA and chemistry of inheritance. *American Scientist* 52:365-388.

Darwin, C. 1964. *On the Origin of Species by Means of Natural Selection*. Facsimile of the first edition (1859) with introduction by E. Mayr. Cambridge, Mass.: Harvard University Press.

Dobzhansky, T. 1956. What is an adaptive trait? *Am. Nat.* 90:337–347.

Ghiselin, M.T. 1969. *The Triumph of the Darwinian Method*. Berkeley: University of California Press.

Irvine, W. 1959. *Apes, Angels, and Victorians*. New York: Meridian Books.

Lewontin, R.C. 1970. The units of selection. *Ann. Rev. Ecol. Syst.* 1:1–18.

Miller, S.L., and L.E. Orgel. 1974. *The Origins of Life on the Earth*. Englewood Cliffs, N.J.: Prentice-Hall.

Medvedev, Z.A. 1969. *The Rise and Fall of T.D. Lysenko*. New York: Columbia University Press.

Chapter **2**
The Scientific Method

Throughout history, human beings have observed and attempted to explain nature. But in the last few centuries our knowledge of nature has become many times deeper and broader than in all the previous millennia. The reason is the development of the scientific method, a tool that has provided many valid and useful generalizations. In this chapter we explore the ways by which the scientific method differs from other ways of gaining knowledge.

ACQUISITION OF KNOWLEDGE

Formal logic deals with the possible relations regarding truth or falsity among propositions, no matter what their subject matter. Logic gives the *necessary* conditions for valid inference so as to eliminate false reasoning. However, formal logic is not *sufficient* to establish any material or factual truth in any particular field of knowledge. What, then, are the ways we acquire knowledge?

Most beliefs are based on the acceptance of current attitudes or on our own unquestioned assumptions. Most of us find ourselves challenged at one time or another to support or change our opinions and beliefs. And we do so by various methods. A method is a *procedure* for dealing with a set of problems—in this case, for supporting our beliefs. According to the American philosopher M. Cohen the most frequently used methods are the following four.

1. *The method of tenacity.* This method results from habit or inertia. We defend a statement simply because we have always believed it. Questions of country, race, language, favorite sports team, etc., are most often supported this way. It must be obvious that new knowledge is rarely obtained by this procedure. Nevertheless, most arguments between people, including some of the most heated ones, involve statements that are defended through the use of the method of tenacity.

2. *The method of authority.* Appeal to some higher authority is another method by which humans defend their beliefs. The authorities vary in accordance with the belief of the individual and the subject matter: Marx, the founding fathers, Che Guevara, Mao's little red book, the Bible, the Koran, Amy Vanderbilt, Vince Lombardi, etc., are cited by one or another person as sufficient reason for the correctness of their beliefs in matters of politics, religion, manners, and sports.

There are two distinct situations in which the method of authority is used. An individual who lacks direct factual knowledge can choose to take someone else's word. This is what we do when we consult a dictionary, or a history book for the date of a battle, or a chemistry book for the formula of a compound. Such an appeal to authority is inevitable and reasonable, since we do not have the time to research thoroughly every question. It nevertheless involves the investment of the source with a special authority, even though the individual reserves the right not to accept the opinion of the expert source.

The second case involves the appeal to an authority that has invested itself or has been invested by others with some sort of infallibility. Political, social, and religious questions in particular are subject to this process. The authority, in turn, whether it is a religious or political document or leader, invokes some external force to give sanction to its decisions. Furthermore, the authority divests the individual of the right to modify the opinion of the expert source. The individual who defies such a ban is branded heretic or deviant.

Although the method of authority aims at obtaining unanimity and stability of belief, this goal is not possible, because the "authorities" disagree among themselves, and the number of "heretics" is always substantial.

3. *The method of intuition.* Some values in a culture are so ingrained, and some phenomena so appeal to our senses, that we believe them to be "self-evidently" true. That the earth is flat and that the sun revolves around it, that mold is formed by decaying matter, that professors are always right—these concepts were accepted for a long time as "self-evident." However, many "self-evident" truths have been proved wrong: The earth is round; the sun does not revolve around it; even professors have been found in error. In other

words, our senses are not infallible, and the fact that a proposition is not questioned because it is considered "self-evident" is no guarantee of truth.

4. *The method of science or reflective inquiry.* The methods presented so far try to supply absolute and final answers. However, they are all subject to human error and arbitrariness. Furthermore, they have no built-in way to detect and correct mistakes. The scientific method does not produce final answers, but it provides the best alternative compatible with observation and experimentation.

The scientific method is based on observations of nature, hypotheses formulated to explain those observations, and experimental testing of deductions from those hypotheses. New observations are constantly being compared with old observations and theories. Errors are detected and theories are modified as new information becomes available. Consequently the scientific view of the world is provisional and uncertain. This uncertainty does not deny scientific progress but, rather, makes it mandatory, since misconceptions, which are inevitable, given the fallible nature of humans, are constantly being cleared up. Scientific precepts, like common sense, are opinions, but opinions based on observations rather than arbitrary dicta. Let us now briefly consider what the scientific method is.

THE STRUCTURE OF THE SCIENTIFIC METHOD

Philosophers and logicians of science are not entirely agreed on the structure of the scientific method. Scientists, even the most successful ones, have in general paid little attention to this question, and they usually can give only vague and imprecise views regarding the logical foundations of their work. What follows is a general picture of the scientific method, painted with a broad brush (Fig. 2.1). Interested readers are referred to Bunge's *Scientific Research*, Kuhn's *The Structure of Scientific Revolutions*, Cohen and Nagel's *An Introduction to Logic and the Scientific Method*, Popper's *The Logic of Scientific Discovery*, and Madden's *The Structure of Scientific Thought* for greater insights and further discussion.

The starting point in any scientific inquiry is a realization that we have insufficient knowledge to explain satisfactorily an observed natural phenomenon. For example, why is it that offspring resemble their parents but are not identical to one of them or exactly intermediate between both? The scientist refers to this first stage as "the problem." There are innumerable "problems" that can be investigated. Some are trivial; some are not. Some are inconsequential to the advancement of science or to human welfare; others are tremendously important. The scientist when choosing a problem to research has to decide what its relative importance is, both to the advancement of science and to humanity. Sometimes the problem appears to have very little practical significance, but it eventually proves to be very important. An example is Roent-

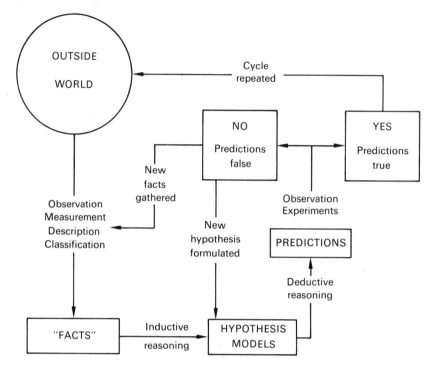

Figure 2.1
One view of how the scientific method operates.

gen's observation that certain types of radiation were not deflected by magnetic fields, which led him to the discovery of X rays, with all its implications.

Once the problem to be investigated has been chosen, it must be formulated very clearly and precisely. The scientist has to state very exactly what the problem is.

The second step in a scientific inquiry is to collect all the facts relating to the problem under investigation. Facts in this context are observations, measurements, specimens, previous ideas, and so on, which are gathered by the investigator from books, from conversations with colleagues, and particularly from his own study of nature. At this stage the complexity of the real world is reduced to a simpler dimension, relevant to the problem under study, and more comprehensible to the scientist.

After the problem has been formulated and all the facts have been assembled, the scientist attempts to find an explanation, a solution to the problem under study that does not violate the "facts" that he and others have collected. This involves the formulation of one or more hypotheses, or models, and the testing of the models to identify the more accurate one.

A hypothesis is formulated through inductive reasoning.* In other words, from a number of isolated, detailed "facts" the scientist proposes an explana-

* The exact nature of hypothesis formulation is controversial. See Popper and Ghiselin for discussion of these problems.

tion that is of general validity, i.e., one that transcends the facts and will be valid in situations, places, and times not yet observed or tested. Philosophers have argued vehemently about whether induction can be justified logically and what its rational foundation is. Nevertheless, the formulation of new hypotheses through inductive reasoning appears to be the way by which new advances in science are made.

According to the classical idea of how the scientific method works, formulation of hypotheses follows data gathering. In reality both processes usually go hand-in-hand.

A hypothesis not only has to explain natural phenomena but must make certain predictions, predictions that must be testable. For example, to say that the offspring resembles its parents because they are related is not a true explanation, even though the statement is a truism. To say, however, that the offspring resembles its parents because they have transmitted to it a certain number and kind of genes according to certain rules is a valid scientific explanation, because the existence of the genes, the fact of their transmission, and the rules of their transmission can all be verified through experiments and observations.

The test of a hypothesis often takes the form of a planned laboratory experiment. Although this is the usual way of verifying hypotheses in physics, chemistry, and much of biology, it is not a requirement, and in certain fields, such as geology, laboratory verification is impossible. Observations in nature often replace laboratory experiments in verifying hypotheses in the field of evolution and population genetics. Other ways are through field experiments (i.e., agriculture); through modification of natural environments, so-called perturbation experiments (as in ecology, for example); through actual physical models that simulate a natural situation; or through numerical examples on the computer. Another very important way to test hypotheses is through comparison of equivalent or near-equivalent situations. Structures, organs, or processes are compared in different situations, different organisms, and/or different places when it is suspected that they have been subjected to similar forces, such as similar selective pressures. From the observed differences and similarities, insights are obtained regarding the operation of universal processes. This "comparative method" is very important in evolutionary studies.

The exact form of the verification is not crucial. What is essential is that the explanation of the problem always be verifiable through its predictions. These predictions are general *deductions* that follow logically from the hypothesis, and any hypothesis must be presented in such a way that predictions, or general statements of a testable or verifiable nature, follow. For example, at one time in the history of biology it was believed that something called "élan vital," vital essence, explained many unique properties of living organisms. Since this vital essence was supposed to be immaterial, it could not be tested in any way. Therefore "élan vital" was a pseudoexplanation, and it was eventually dropped in favor of a wholly materialistic view.

If the predictions deduced from the hypothesis are found to be true, that result is taken as temporary proof of the validity of the hypothesis. The question has been answered, and the investigator can repeat the process with a new question.

What we have just presented is the classical view of how the scientific method works. It assumes (a) that the scientist is completely impartial, (b) that in every case all the relevant facts are known, and (c) that experiments or observation designed to test a hypothesis always yield clear-cut results. These assumptions are seldom, if ever, found. In real life, hypotheses are formulated by more or less biased human beings, they are usually based on an incomplete knowledge of nature, and they are backed with observations and experiments that are less than conclusive. These conditions set the stage for the scientist-author to defend his or her hypothesis by some of the means explained before. Particularly if well known, the scientist can appeal to his or her authority to defend the theory. A good example comes from the early history of genetics. After Mendel published his paper on the genetics of garden peas, he sent a copy of it to the German botanist Carl Nägeli, considered by biologists of the time the greatest authority on heredity. He was Professor of Botany at the prestigious University of Berlin. Nägeli was skeptical of Mendel's work, not grasping the simplicity and elegance of the experiments. He consequently wrote to Mendel saying that he would believe the findings only if Mendel was able to repeat the work and obtain similar results with hawkweeds (*Hieracium* spp.), the plant with which Nägeli was experimenting at the time. Mendel tried but obtained negative results. Unknown to both investigators (and to everybody else at the time), hawkweeds form seeds asexually. Thus, Nägeli was able to use his prestige to impose conditions that effectively suppressed Mendel's ideas and kept them from being accepted. Although Nägeli's request for verification of Mendel's theories was perfectly valid, his request that Mendel work with hawkweeds is more questionable.

Another problem is that theories proposed to solve one problem are not independent of other theories proposed to solve different problems; rather, they are all part of a general view of the world. Theories are therefore really groups of theories, and the tendency when one faces a new problem is to propose an explanation consistent with other accepted theories. In other words, rather than a brand new theory, what is usually proposed is a subtheory of an already accepted theory. For example, a nonevolutionary theory to explain the structure of an organism would be considered unacceptable today, even if it agreed with the facts and the test of the hypothesis was positive. It is only when the whole construction of theories within theories becomes too unwieldy and complicated that the basic theory is challenged. A challenge usually creates a split in the scientific community. Typically, many of the older, established scientists refuse to believe the new general hypotheses, whereas others, particularly the younger scientists, enthusiastically endorse the new concepts. This difference of opinion results in clashes and controversies. The change from the Ptolemaic to the Copernican view of the world in the sixteenth century was such a clash. The change from a Newtonian to a relativistic cosmology at the beginning of this century was another such confrontation. In biology we can cite the change from a creationist to an evolutionary view of the world in the 1860s. In our days, the big change has been in geology. From a view holding that continents have always been in the same position relative to each other, we have changed to one proposing that continents "drift" over time.

Ideally the scientific method is a very effective procedure for obtaining information about nature. However, it is not necessarily free of errors, and it can actually be used to support false concepts about nature. Most scientists make an effort to apply the scientific method in its purest or ideal form, but most, if not all, fail to a lesser or greater extent for the reasons given above. More fundamentally, they fail because scientists are part of a cultural environment, and they cannot become entirely detached from the values of the society to which they belong. These values influence their thinking and the approach to their work.

NECESSARY COMPONENTS OF A HYPOTHESIS

The scientist is as much a creator as a discoverer, and imagination is as important as dedication and hard work. However, scientific work has more constraints than artistic endeavor, and hypotheses or models have to follow certain rules.

First, a hypothesis has to be formulated in such a way that verifiable deductions can be made, so that it can be decided whether the hypothesis does or does not explain the facts. Scientists and philosophers of science have argued vehemently about what the origin of a concept or hypothesis must be. Some philosophers of science, the so-called empiricists, argue that all concepts or hypotheses in science must originate from the scientist's past experience with specific objects or events and their relations. However, extreme empiricism could not explain the development of concepts such as the gene or natural selection, since Mendel and Darwin had no previous experience with either concept. Philosophers and logicians of science agree today that a hypothesis can have any of many alternative sources: intuitions, trial and error, past experience, accident, or imagination. What is crucial is that it be fruitful and confirmable. Often the hypothesis itself cannot be verified, but its consequences can (e.g., mutation theory, p. 119). One must always remember that a hypothesis that cannot be put to a test is meaningless.

Another important aspect of a hypothesis is its *logical form*. A hypothesis consists basically of three kinds of statements. First, there is a set of statements that specify the situation to be explained. These statements are called the *antecedent* of the hypothesis. For example, in Mendel's classical experiments (Chapters 1, 3), the *antecedent* would have been statements concerning the existence of lines of tall and short peas, with smooth and wrinkled seeds, white and red flowers, etc., that breed true to each other; the fact that any two lines can be crossed to each other with no loss of fertility; and the fact that all belong to the same species. The antecedent consists of statements indicating certain conditions that are realized prior to or at the same time as the phenomenon to be explained. The antecedent is the formalized "problem" of the scientist. Second, there is a set of statements of general and universal validity. They are natural laws that apply to the phenomenon. In our example they are the Mendelian laws. The antecedent and the appropriate general law explain (or attempt to explain) the phenomenon in question and consequently lead to a third kind of statement, or *prediction*, which can be verified by observation or experiment. Of the three kinds of statements, it is the second, the formulation of general statements, that is considered to be the hypothesis proper. But without an antecedent and without testable predictions, a hypothesis becomes a dogmatic statement of little utility.

In formulating and testing hypotheses, one must take competing hypotheses into consideration. The process of elimination consists of determining whether all competing hypotheses explain *all* the facts. Eventually one hypothesis is left standing because it alone accounts for all the facts. However, there is a qualification. The scientist testing alternative hypotheses can decide only among known hypotheses. Perhaps there is still another, not yet formulated, that accounts for all the facts just as well but more simply and economically, and that is of more general validity. Consequently, since we cannot be certain that all possible alternatives have been considered, a hypothesis can be proved only to be *probably* but not *necessarily true*. This restriction actually applies to all scientific knowledge: It is only probably, not necessarily, true.

There are further complications. Sometimes no single hypothesis explains the entire set of relevant facts, but two or more hypotheses will come close, explaining many but not all of the facts. Neither hypothesis is therefore entirely acceptable. If they cannot be combined into one because they make contradictory statements, a problem arises. Or we may have the opposite problem: Two or more existing hypotheses explain all the facts, and there is no way of discriminating between them. In these cases, the general rule is to decide in favor of the hypothesis that is *logically the simplest*. We choose the hypothesis that explains the facts with the least difficulty, i.e., with the fewest number of undefined terms and unproven assumptions. This rule is called "Occam's razor" in honor of William of Occam (or Ockham), an English Franciscan scholar of the early fourteenth century. As enunciated by Occam, the maxim states: "It is vain to do with more what can be done with fewer." Stated in modern language: "Entities are not to be multiplied without necessity."

Another important component of a model is its deductive development. By this we mean that a hypothesis is elaborated further so as to uncover *all* its implications. Many hypotheses are developed verbally, but more and more, mathematics is being used to develop hypotheses. A good example in population biology is the theory of population growth and regulation, which will be discussed in Chapter 13. Since mathematics is a deductive science, it is a perfect tool for the deductive elaboration of a new hypothesis, provided, of course, that the concepts of the hypothesis are quantifiable. But mathematics is only a tool, not an end. A hypothesis that is elaborated in mathematical terms is not intrinsically more correct than one that is verbally elaborated. But it is often easier to see inconsistencies as well as previously unthought-of predictions when a hypothesis can be stated and elaborated mathematically. Therein lies the strength of mathematics. However, in population biology we often deal with complex and incompletely understood facts (antecedents). To elaborate hypotheses mathematically regarding these facts often requires one to make simplifying assumptions, and that may be unrealistic. In short, although the scientist can use mathematics to great advantage in formulating and developing hypotheses, mathematics is no substitute for imagination and biological insight.

The last step in the formulation and development of a model or hypothesis is the observational and experimental test of the hypothesis, to verify whether or not predicted and actual consequences coincide. These tests have to be universal, repeatable, and public. Tests that cannot be repeated or can be repeated only in one laboratory or country or by one person or group of persons are not acceptable. Tests must have controls; that is, the relevant variables have to be accounted for. Since variables are easier to control in an artificial environment created by the experimenter, laboratory experiments are superior to observations in nature. However, since it is not possible to artificially recreate nature, laboratory experiments may result in a controlled but highly modified environment, invalidating the test of the hypothesis. This is particularly so with evolutionary hypotheses.

In the next chapter we will analyze the theory of evolution by natural selection in the light of the rules provided by the scientific method.

SUGGESTED FURTHER READING

Bunge, M. 1967. *Scientific Research.* 2 vol. Berlin: Springer-Verlag.

Cohen, M.R., and E. Nagel. 1934. *An Introduction to Logic and the Scientific Method.* New York: Harcourt, Brace.

Ghiselin, M.T. 1974. *The Economy of Nature and the Evolution of Sex.* Berkeley: University of California Press.

Kuhn, T.S. 1957. *The Copernican Revolution.* Cambridge, Mass.: Harvard University Press.

Kuhn, T.S. 1962. *The Structure of Scientific Revolutions.* Chicago: University of Chicago Press.

Madden, E.H. 1960. *The Structure of Scientific Thought.* Boston: Houghton Mifflin.

Popper, K.R. 1959. *The Logic of Scientific Discovery.* New York: Harper Torchbooks.

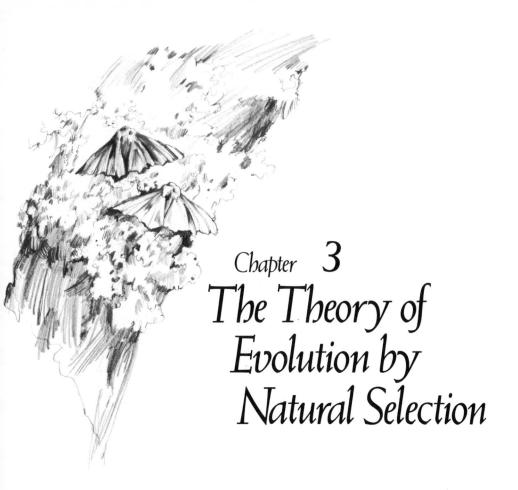

Chapter **3**

The Theory of Evolution by Natural Selection

Ideally the theory of evolution by natural selection should predict the state of the biota of the earth at some future time. For that to be possible, however, a complete description of the present state of the biota is needed, as well as a complete knowledge of the rules or laws by which organisms change in time. No such understanding exists, nor is it likely that it ever will exist. The number of individual organisms is so large and each is so complex that it is virtually impossible to describe them all. At present we possess only an incomplete and cursory description of the principal characteristics of the majority of the species of plants and animals (Chapter 18).

Not only do we not have a detailed description of the organisms that live in this world, but we do not understand yet the exact nature of the evolutionary changes that take place in a single lineage. Furthermore, another condition necessary to predict future evolutionary states is to know the exact changes that will take place in the environment. Clearly, a totally predictive theory of evolution is not possible.

But even if we had all the necessary requirements, such as a complete knowledge of the present state of the biota and an adequate knowledge of the rules that determine evolutionary change, as well as a means of predicting the changes in the environment, it still might not be possible to totally predict evolution. The reason that future evolutionary changes may be unpredictable in an ultimate sense has to do with the stochastic nature of certain phenomena. By this we refer to the fact that certain evolutionary events depend on chance, and consequently they are predictable only in a statistical sense (Chapter 7).

In this chapter we will review the theory of natural selection, its domain, and its shortcomings, thus setting the stage for the more detailed description that follows in subsequent chapters.

Darwin's goal in writing *Origin of Species* was to demonstrate that the variability present in natural populations was sufficient to give rise to new species under the right circumstances. His goal, therefore, was not to predict the future outcome of evolution but to develop a purely materialistic hypothesis to explain the origin of the diversity of organisms. To replace the idealistic concepts prevalent at the time, Darwin suggested a testable mechanism, natural selection, to explain the origin of species. Although it is not a requirement of the theory that all evolution proceed by natural selection, Darwin implied that evolution did proceed only in that way. In this sense his theory predicts that future species will originate only as a result of natural selection.

We will now use the concepts developed in the preceding chapter to restate the theory of evolution by natural selection in a more rigorous and precise way than we stated it in Chapter 1. We will divide the discussion into three parts. First we will discuss the fundamental observations on which the theory is based; in other words we will explore the "problem" and the "facts" that the theory of evolution by natural selection attempts to explain. Then we will explore the theory itself under the heading "Minimum Conditions Necessary for Evolution by Natural Selection." We close the chapter with a discussion of whether evolutionary theory is a general theory, a hypothesis, or a world view, and whether it can be rigorously disproved by experiment or observation.

EVOLUTION BY NATURAL SELECTION: THE PROBLEM AND THE FACTS

Wherever we look, whether up in the skies or down into the ocean depths, from the tropics to the polar regions, we find living organisms. Their multitude is enormous, their forms, shapes, and ways of life so varied that they often appear strange and incomprehensible to us. All this multiplicity is but an expression of an advanced level of chemical organization that we call life. As far back as there are records of human thought, there is evidence that humans tried to explain this diversity of life and for that purpose developed elaborate mythologies, tales, philosophies, and religious doctrines.

Starting in the sixteenth century, a number of European thinkers, most notably Bacon, Descartes, and Galileo, became dissatisfied with the way nature was explained. Aided by new instruments, especially the telescope and later the microscope, and by new mathematical tools, such as the differential calculus developed by Newton, they increasingly realized that nature often behaved in a counterintuitive manner; that is, things were not always as they appeared at first sight. Consequently these thinkers demanded proof of intuitive statements about nature. This was the origin of what we now call the scientific method, discussed in Chapter 2.

When natural historians applied the new methodology to the organic world, they made four important observations:

1. There is a certain similarity of structure, morphology, and behavior among organisms.
2. Plants and animals can be arranged into groups on the basis of such similarities, and these groups can in turn be ordered into series, going from the simplest to the most complex.
3. "Petrified" (fossilized) bones of extinct animals can also be distinctively grouped, and these groups can be fitted into the ordered series of living organisms. (Until the sixteenth century, petrified bones had been interpreted as belonging to animals that had perished at the time of the biblical Flood.)
4. Like begets like, but occasionally the offspring differs from its parents in small or occasionally in major ways. (These changes were called "mutations," but they should not be confused with what we call mutations today.)

These observations led logically to the idea that the diversity of living organisms could arise by a process of "evolution." Starting with a simple, primitive organism, by occasional "mutations" a slightly more evolved or "advanced" form could originate, which in turn could give rise to an even more "advanced" form, and so on, in a procession that would lead eventually to humans. And so the idea of evolution entered the field of natural history in the seventeenth and eighteenth centuries through the writings of Erasmus Darwin (Charles Darwin's grandfather), the German poet Johann Wolfgang Goethe, and especially the great French naturalist Jean Baptiste Pierre Antoine de Monet, better known as the Chevalier de Lamarck (1744–1829). Lamarck was the author of one of the first treatises in which species were not considered immutable but were supposed to change with time.

However, neither Lamarck nor any of the pre-Darwinian evolutionists provided a plausible reason for evolution or a credible, testable mechanism to account for evolutionary change. Theirs was an explanation for the existence of similarities between organisms, but since it did not make any testable prediction, that explanation did not constitute a rigorous scientific theory. The idea of evolution did not receive much favorable attention at the time.

Darwin provided the missing ingredient. He had noted that many more off-spring are produced than can possibly survive. Consequently there is competition or, to use Darwin's phrase, a "struggle for existence" among organisms. Any characteristic, he reasoned, that will increase the probability of survival of an individual and its offspring will become increasingly more abundant until eventually all individuals in the group will possess the trait (Fig. 1.3). This process Darwin called *natural selection,* because it is akin to the process of selecting the best seeds for sowing a crop or the best sires for breeding in animal husbandry. Selection exercised by humans Darwin called "artificial selection."

When the concept of natural selection was added to the observations of the pre-Darwinian naturalists as the driving force of evolution, a scientific hypothesis to explain the diversity of living organisms was produced. That hypothesis states specific necessary conditions, each of which is subject to verification (or negation) by observation and/or experimentation. These conditions have to do with reproduction, inheritance, the correlation between survival and the individual's characteristics, and the origin of variation. Let us now explore these conditions further.

MINIMUM CONDITIONS NECESSARY FOR EVOLUTION BY NATURAL SELECTION

Reproduction

For natural selection to occur, organisms must produce more offspring than the number necessary to replace them. The necessity of this condition should be clear. If every individual or every pair of individuals in a population produces the exact number of offspring needed to replace it—that is, exactly one per individual—and if each offspring survives to adulthood and has exactly one offspring, and so on, a perfect demographic equilibrium will ensue. Under such conditions evolution by natural selection in the way it is here defined could not occur, since there would be no differential reproduction or mortality. On the other hand, if individuals produce fewer than one offspring apiece, the population will inevitably become extinct.

Note that the emphasis is on producing exactly one offspring per individual that always survives to adulthood. Such a situation is unrealistic, however. Accidents of one sort or another will always occur, not to mention the fact that some species exist by consuming others, often concentrating on the more vulnerable young. If any deaths occur before the individuals reproduce, a species in which every adult produces exactly one offspring is on its way to extinction. Since by and large (over a short time span) species are not decreasing in numbers, it follows that most species produce more offspring than are strictly needed to replace them.

That statement can be verified by observation. Tables 3.1 and 3.2 show the number of offspring produced by females of various species. Two points are worth mentioning. First, we can see that this condition is met in all species. No known organism produces exactly one offspring per individual. All species produce more. The second point is that species vary tremendously in the number of offspring they produce. Some, such as the human species, have comparatively few offspring, but others, such as the proverbial oyster and many plants, produce thousands of offspring. Furthermore, the energetic investment in the offspring varies tremendously: Some species of plants invest more than 30 percent of their total body energy in the production of offspring, but others invest less than one percent.

Correlation between probability of survival and phenotype

Given that every species produces an abundance of offspring, two alternatives are possible regarding the probability of survival of individual offspring. (1) Survival is entirely due to chance. (2) Survival is at least in part related to the characteristics (morphological, physiological, biochemical) of the offspring. In both cases we would observe changes in the population (certain characteristics can be lost or fixed by chance alone), but in the first case, the observed changes should be random, and they should oscillate over time. Since we observe a certain degree of undeniable adaptation (such as the development of eyes to see) in lineages in the paleontological record, the hypothesis that evolution is solely

Table 3.1

Number of offspring per reproductive season for various animal species.

Homo sapiens	man	1/litter
Canis familiaris	dog	7 (1–22)/litter
Sturmus vulgaris	starling	4–6 eggs/clutch
Terrapene carolina	box turtle	2–7 eggs/clutch
Rana catesbeiana	bullfrog	6000–20,000 eggs/brood
Salmo trutta	trout	200–6000 eggs/spawning
S. aguabonita	trout	1380–2280 eggs/spawning
Homarus americanus	American lobster	8500 eggs/spawning
Argopecten irradians (*Pecten irradians*)	common scallop	2,000,000 eggs/spawning
Crassostrea virginica	eastern oyster	500,000–1,000,000 eggs/ spawning
Drosophila melanogaster	fruit fly	100 eggs/female
Vespula maculata	bald-faced hornet	25,000–35,000 eggs/queen
Musca domestica	housefly	75–200 eggs/female

Data from P.L. Altman and D.S. Dittmer (eds.), 1972. *Biology Data Book*, vol. 1, 2nd ed. Federation of American Societies for Experimental Biology.

Table 3.2

Number of seeds per plant per year for herbaceous species growing in a recently plowed field (1 yr), a weedy field (10 yr), and a forest.

	Number of Plants	Mean Number of Seeds/Plant	Range in Number of Seeds/Plant
1-Yr Field			
Ambrosia artemisiifolia	12	1190	110–2690
Chenopodium album	57	4820	90–50,100
10-Yr Field			
Solidago altissima and S. canadensis	11	3070	960–5330
Forest			
Dentaria laciniata	9	24	9–42
Cardamine bulbosa	10	33	5–90
Sanicula gregaria	27	26	3–50
Prenanthes alba	7	118	95–170

From Newell and Tramer, 1978. Reproductive strategies in herbaceous plant communities during succession. *Ecology* 59:228–234.

a chance phenomenon must be rejected. However, exactly what proportion of the changes in evolution are adaptive is an open question.

Inheritance

If the characteristics that promote the survival of some individuals are not transmitted to that individual's offspring, the descendants of a successful individual may all perish for lack of the right adaptive character. Observation and experimentation show that most characters of organisms are inherited, although they may be expressed by the offspring in a modified form because of environmental factors. We also know that this includes characters that are directly involved in increasing offspring survival.

A case in point is the moth *Biston betularia* in England. This animal, which is active at night as most moths are, rests during the day on the bark of trees with its wings open. Normally the trunks of the birches are covered with white lichens with gray specks. The color of the moth's wings is also white with some gray specks, so that it blends perfectly with the lichens in the background (Fig. 3.1). During the day moths are hunted by birds. The better the match of the moth with its background, the lower the probability that it will be seen by a bird and eaten (Table 3.3). About one hundred years ago, butterfly collectors around Birmingham, England, started to encounter a different form of the moth, which was given the name *carbonaria* because it had gray-black wings.

(a)   (b)

Figure 3.1

The peppered moth and its black form *carbonaria* (a) at rest on lichened tree trunk in unpolluted countryside, Dorset, England, and (b) at rest on oak trunk bare of lichens near Birmingham, England. (Photo courtesy of Dr. H.B.D. Kettlewell, Oxford University)

Table 3.3

A. Number of *Biston betularia* taken by birds over two days in a polluted forest after equal numbers of each were released.

f. *typica*	43
f. *carbonaria*	15
Total	58

B. Percentage of moths released which were recaptured at the end of the day over a seven-day period in a polluted forest.

f. *typica*	45.79
f. *carbonaria*	62.57

Data from B. Kettlewell, 1973. *The Evolution of Melanism.* Oxford: Oxford University Press.

An intermediate form, *insularia,* which at first was grouped with *carbonaria,* was discovered later. As years went by, f. *carbonaria* became more numerous in and around the industrialized areas of Britain, and it is today the prevalent form. What happened? With the advent of the industrial revolution, lichens were killed by air pollutants, so that the tree trunks and branches were dark. As a consequence, the moths lost the protection of their camouflage and became more vulnerable to bird predation. This was not true for f. *carbonaria,* since its dark color blended with the dark backgrounds that were becoming increasingly more numerous. Studies by Kettlewell (Chapter 8) showed that changes had also occurred in the genetic structure of the population.

Evolution occurs, then, because in populations of different organisms, each producing an excess of offspring and transmitting its characteristics to its descendants, some individuals and their offspring have a higher probability of surviving because they produce more offspring, or because they possess certain phenotypic characteristics that increase their survival, or both. These characteristics will be called *adaptive,* the processes *differential reproduction* and *differential survival.*

It is important to remember that natural selection will operate only on those characteristics that are transmitted to the offspring. If a dandelion seed chances upon a fertile patch of soil, it will produce more seed than will a plant growing in a more barren soil. However, since the reason for the excess seed production—the fertile soil—is not an inherited characteristic, no evolutionary change will result.

Environmental uniformity

If the process of differential reproduction and/or differential survival is to lead to systematic changes in the population, the same phenotypes must be favored for more than one generation. That is, natural selection requires a certain degree of environmental uniformity over time. This requirement is sometimes overlooked. Since the environment is *never* exactly uniform from generation to generation or from place to place, it is very difficult to assess the degree of required uniformity in the environment for natural selection to become operative. Some environmental aspects, such as weather, are very variable, but other characteristics of the environment, such as length of day, light intensity, or seasonality, are very uniform. Consequently, there is a greater probability that organisms will have structures directly related to those characteristics that are uniform over time (such as eyes in animals or leaves in plants whose functions are connected to light) than to unpredictable environmental characteristics, such as protection against hail. In an environment uniform over time, the process of differential reproduction and differential survival should produce a population of like individuals, as those having less adaptive characteristics and/or producing fewer offspring become eliminated. After that, survival should be entirely due to chance. The diversity that we observe in nature should there-fore not be encountered. Does this discrepancy deny evolution? To understand why there is so much diversity, we need to look at the sources of variability in populations.

SOURCES OF VARIABILITY

There are two sources of variability. One is internal and creates new charac-teristics. The other source is external, and although it does not "create" new characteristics, it contributes to maintaining the variation that is observed. As we will explain in detail in Chapters 4, 5, and 6, parents transmit to their off-spring chemical substances that will direct the development and growth of the offspring along the parental lines or very similar lines. However, changes in the chemical character of the genetic substance or in the quantity of the sub-stance can lead to novel forms in the offspring's characteristics. Such changes, which can be encompassed under the general heading of mutation and recom-bination (more detail is presented in Chapter 6), are the sources of new herita-ble characteristics. These processes result in the production of individuals with properties that are different from those of their ancestors. Some of the new characters will be adaptive and will spread in the population, replacing the characters that were present before. That is, if as a result of an internal change in the genes a phenotype results that can produce a larger number of surviving offspring by differential reproduction and/or differential survival, that pheno-type will spread. In an ultimate sense, this is the real source of new variation. However, much of the variation that we observe is maintained in the popula-tion because of *environmental* changes.

In effect, a certain character is advantageous to an organism only in a par-ticular environmental context. *Biston betularia* with light wings is superior in the context of an unpolluted environment; f. *carbonaria* is superior in a pol-luted environment. Consequently, both types exist because both kinds of en-vironments exist. The disappearance of one of the environments would prob-ably lead to the extinction of the form associated with that environment.

The environment is the sum of the surroundings of an organism. It in-cludes such things as soil, topography, weather, and other organisms. Of these factors, weather is very variable for a number of independent reasons, such as the rotation of the earth around its axis, the earth's orbit around the sun, atmospheric turbulence, and so on. In turn, weather, especially rainfall, affects soil formation. Consequently, different rainfall regimes result over time in dif-ferent soil types. Topography is in great part due to patterns of mountain formation and erosion that result from the drifting of continents and from weather. The plants and animals that live in a particular place, in turn, depend on all these factors. Given that "success"—that is, survival of an individual and its offspring—is correlated with a particular environment, and given that environments are themselves changing, "success" of any given character is a temporary thing. Furthermore, the environment changes in both time and space, so we expect organisms to be different from place to place and from one point in time to another.

No natural selection will take place in a population where all individuals are alike. But because of changes in the genetic substance (to be discussed in greater detail in Chapters 4, 5, and 6) and because of responses to temporal and spatial variation in the environment (Chapters 13, 14, and 15), all known populations of plants and animals are variable in their characters, and consequently, natural selection can take place.

In summary, the necessary conditions for evolution by natural selection are:

1. An excess of offspring in relation to the number that can normally survive.

2. Parents that transmit their characteristics to the offspring.

3. More than one kind of individual in the population in relation to one or more characteristics.

4. Increased probability of survival of offspring possessing a particular phenotypic characteristic.

5. Sufficient environmental uniformity over time.

Given these conditions, it will follow that individuals with the favorable characteristics will increase over time, and that is the essence of evolution by natural selection.

Natural selection is often defined as a process in which the organisms best fitted to the environment leave the largest number of offspring. If, then, the way to determine which organisms are the best fitted is to look at the survival of the offspring, the definition is a tautology. In essence, it is saying the organisms that produce the most offspring will produce the most offspring. What we want instead, given a certain environment, is to predict the characteristics that will enable the organism having them to leave the largest number of surviving offspring. Each of the conditions just enumerated can be verified or denied by observation and experimentation, as should be possible with any scientific theory.

The theory of evolution by natural selection, then, is the general framework to which more detailed and specific explanations of more restricted problems have to conform. Since it will never be possible to test the general theory in the broad context of the total biota, evolutionary biologists address themselves to the more limited problems of understanding specific systems in greater detail, in the hope that the solutions they encounter may be applicable to other organisms as well.

ON THE NATURE OF THE THEORY
OF NATURAL SELECTION

More than one hundred years have elapsed since Darwin made his original formulation. They have been years of careful observation, study, and experimentation in the natural sciences. A great mass of detailed knowledge has accumulated, the principal relevant aspects of which will be reviewed in the coming pages. These studies have shown that there are many dimensions and aspects that relate to the five premises of the theory of natural selection discussed in the previous section. These studies have revealed that occasionally there are severe restrictions in how adults interact with other adults, how they give rise to new gametes, and how they reproduce. We understand relatively well the rules by which gametes unite to produce zygotes (to be discussed in Chapters 4–6). We have started to develop some feeling for the interactions between adults (Chapters 12–15) and the rules relating to mating and gamete formation (Chapter 8), but our knowledge of the rules regarding the development of zygotes into adults is close to nil. However, in order to make realistic predictions about evolutionary events, one must understand the totality of the life cycle (Fig. 3.2). Consequently, the theory of natural selection serves at

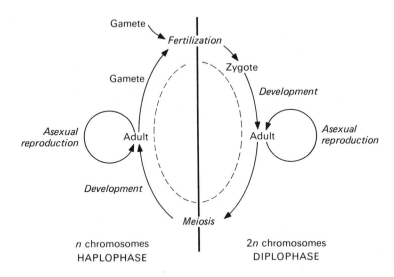

Figure 3.2

Generalized life cycle diagram. Not all stages of this general diagram apply to every organism. For instance, animals do not develop an adult in the haploid phase; they go directly from meiosis to fertilization. What we want to emphasize is that natural selection works during all parts of an organism's life cycle, not just in the adult stage.

present as the guideline for the development of more specific hypotheses. Only as the various aspects of the behavior of individuals and populations are understood and generalized will the theory of natural selection become a truly predictive theory. That moment is still far away, and given the complexity of the world of organisms, it may never come. To open new avenues, to tread uncharted paths—therein lies the excitement and the challenge of population biology.

SUGGESTED FURTHER READING

Darwin, C. 1964. *On the Origin of Species by Means of Natural Selection.* Facsimile of the first edition (1859) with introduction by E. Mayr. Cambridge, Mass.: Harvard University Press.

Darwin, C. 1975. *Charles Darwin's Natural Selection.* Edited from manuscript by R.C. Stauffer. Cambridge: Cambridge University Press, pp. 692f.

Ghiselin, M.T. 1969. *The Triumph of the Darwinian Method.* Berkeley: University of California Press, pp. 287f.

Ghiselin, M.T. 1974. *The Economy of Nature and the Evolution of Sex.* Chapters 1 and 2, pp. 1–48. Berkeley: University of California Press.

Horn, H. 1979. On the concept of optimality. In O. T. Solbrig *et al.* (eds.), *Topics in Plant Population Biology.* New York. Columbia University Press.

Kettlewell, B. 1973. *The Evolution of Melanism.* Oxford: Clarendon Press, pp. 423f.

Mayr, E. 1976. *Evolution and the Diversity of Life.* Chapter 1, Basic concepts of evolutionary biology, pp. 9–16. Cambridge, Mass.: Belknap Press of Harvard University Press.

Solbrig, O.T. 1976. Plant population biology: an overview. *Systematic Botany* 1: 202–208.

Part II

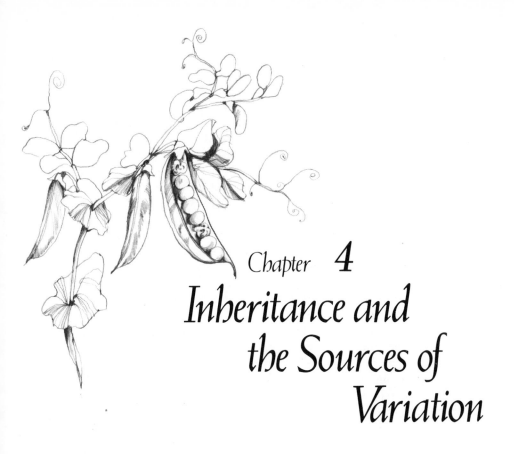

Chapter *4*

Inheritance and the Sources of Variation

Evolution depends on the transmission to the offspring of the characteristics of the parents. Only if this occurs can traits spread in the population by differential survival and/or reproduction. Darwin realized that inheritance played an important role in evolution. He was very interested in the mechanisms underlying inheritance and reproduction, and he conducted many experiments to learn about them, but he never discerned the fundamental genetic rules. Darwin's discussion of inheritance is the weak aspect of his original enunciation of the theory of evolution.

Although Mendel proposed his principles of inheritance in 1865, they had no impact on biology until their rediscovery at the turn of the century. Since then there has been a real explosion in our understanding of inheritance. In this and the following three chapters the main concepts of genetics and the mechanisms of inheritance, as well as the way they affect and constrain evolution by natural selection, will be explored.

PARTICULATE AND BLENDING INHERITANCE

Very often an individual is not an exact copy of its parents, sometimes resembling one parent more than the other or occasionally resembling a more
remote ancestor. Some offspring are found to have entirely new attributes.
Over the years elaborate theories have been brought forth to explain these
observations. For example, it has been proposed that the characteristics of
certain parts of the plant or animal are inherited exclusively from the maternal
line, other attributes from the paternal line. The seasons of the year have been
suspected of causing certain characteristics in the offspring. Foods ingested by
the mother have been proposed as the basis for certain effects on the offspring,
and so on. At the base of all these ideas lay the concept—called the *theory of
blending inheritance*—that heredity somehow consists of a "mixing of the
bloods" of the ancestors. This notion still persists in popular thought. It is
reflected in such expressions as a "pure blood" for a purebred animal or in
references to having one's father's or mother's "blood." According to the
theory of blending inheritance, the totality of the characteristics of each parent
is transmitted as a unit to the offspring, where the two sets mix and lose their
individuality. The fact that the offspring differs from either of its parents is
supposed to be due to this mixing effect. Differences among siblings are explained by the assignment of variable strength to the "humors" of the parents.
An often cited further justification of blending inheritance is that the charac-

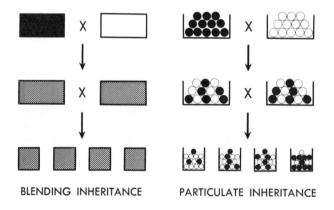

BLENDING INHERITANCE   PARTICULATE INHERITANCE

Figure 4.1

Effect of blending and particulate inheritance on variability. When two extremes are
crossed, under blending inheritance all the offspring of coming generations are
intermediate; under particulate inheritance there will be segregation after the first
generation. What we observe in nature is best explained by particulate inheritance.
(From O.T. Solbrig, 1966. *Evolution and Systematics.* New York: Macmillan)

teristics of the offspring in general tend to be in between those of the parents.

The last corollary furnishes a test for the theory. Natural populations of plants and animals are formed by individuals that are not totally alike. This variability furnishes the raw material for natural selection. But under the theory of blending inheritance, this necessary variability would be reduced by one half in each generation, and in a very few generations the members of a population would be completely uniform in all their characteristics (Fig. 4.1). No more evolution would be possible unless a new source of variability was introduced. The proponents of the blending theory of heredity postulated that new variability was introduced at a rate equivalent to the loss of variability due to blending. The mechanism by which variability was supposedly introduced was believed to be the inheritance of acquired characters.

THE THEORY OF PANGENESIS

Charles Darwin was aware that a store of variability in a population was necessary for natural selection to act on and thus bring about evolution. In a population in which all individuals are exactly alike, nothing is changed from generation to generation, even when there is differential reproduction and/or mortality. In his travels around the world and in his observations on the flora and fauna of his native England, Darwin had realized that animal and plant populations are not uniform but quite variable. Furthermore, his extensive breeding experiments with plants had shown him that this variability was inherited and not due to accidents of the environment. Evolution through natural selection required the presence of a large store of variability, but under the theory of blending inheritance, almost universally accepted at the time, variability should be practically nonexistent within a breeding population. Darwin tried to reconcile this conflict by proposing his theory of pangenesis. It stated that representative particles, pangenes, coming from all parts of the body and carrying information on inherited and acquired characteristics, are incorporated into the gametes and transmitted to the offspring (Fig. 4.2). This complicated theory could never be verified experimentally, and furthermore it did not entirely explain all the observed facts. It was never quite accepted.

GERM LINE AND SOMATIC LINE

August Weismann (1834–1914), Professor of Zoology at Freiburg, Germany, proposed in 1883 his theory of the germ plasm to reconcile the theoretical need for a great deal of variation in populations and his observation that acquired characters are not inherited. Weismann had performed an experiment in which for 22 generations he cut off the tails of mice and observed the effect of that treatment on the tail length of their offspring. He observed that amputating the tail of the parents had no effect whatsoever on the length of the tail of the descendants. From this finding he deduced that acquired characters are

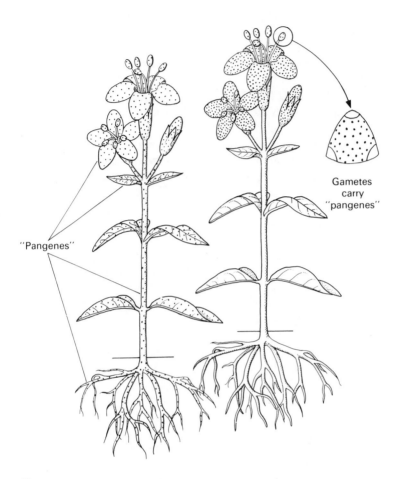

Figure 4.2

The theory of pangenesis. According to this theory, particles called pangenes, which carry information on all of an organism's characteristics, are produced in each part of the body. They move to the reproductive organs, where they are incorporated into the gametes and passed to the offspring. Since pangenes carry information on all characteristics, the theory of pangenesis implies that characteristics an organism acquired during its lifetime can be transmitted to its offspring.

usually not inherited. Weismann also observed that the most primitive organisms were unicellular and that they reproduced by fission, transmitting to their offspring all the content of their bodies. On the other hand, the more advanced multicellular organisms transmitted to their offspring only the content of their sex cells, or *gametes*, and the rest of their bodies died and decayed.

He concluded that there is a part of any organism, which he labeled the *germ line*, that is eternal and one other part, derived from the former, that is mortal. He labeled the mortal part the *somatic line* (Fig. 4.3). With his theory of the germ line Weismann was able to reconcile his mutilation experiments with the existence of variation in nature. Only those changes affecting the germ line are transmitted to the offspring; those that affect the somatic line (such as mutilations) do not. Thus, although populations cannot maintain variability through the inheritance of acquired characteristics as that process is usually thought of, Weismann's theory leaves open the possibility of changes in the germ line that could yield variability in the population.

Weismann's theory had a great deal of influence in his time, and it was an important step in the history of biology. Today we still accept a modified concept of the germ line and somatic line. An important contribution of his theory was the idea that not all the acquired characteristics of the parents are transmitted to the offspring. However, although Weismann could explain why many acquired characteristics were not transmitted to the offspring, his theory failed to explain the origin and maintenance of the variability necessary for evolution.

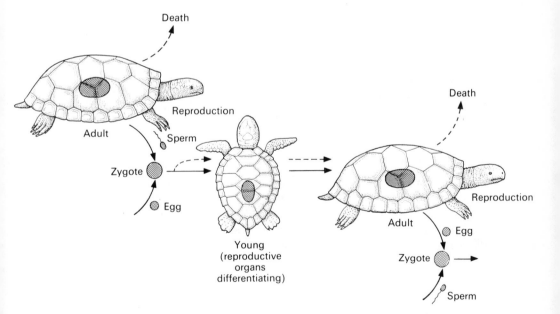

Figure 4.3

The germ line and the somatic line. The gametes and gamete-producing structures, the germ line of an organism (solid arrows), are continuous through the generations. Bodies, which are the somatic line (dashed arrows), are temporary supporting structures.

THE PARTICULATE THEORY OF INHERITANCE

The Austrian monk Gregor Mendel in 1865 proposed the *particulate theory of inheritance*. According to this theory, parents transmit to their offspring particles, now called *genes*, that carry the information necessary for reproducing the characteristics of the parents. The genes are blueprints for the development of characters.

We know today that genes are located on subcellular organelles called *chromosomes*. A given gene is normally in the same precise position on a chromosome, called the *locus* (Latin for "place"; plural, "*loci*") of the gene. Most plants and animals are *diploid*, which means that they have two sets of chromosomes, one set received from each parent. Diploid organisms, then, have two of each kind of gene, occupying the same locus on their respective chromosomes. If two genes at the same locus influence the organism's characters differently, they are called different *alleles* of the same gene.

The number of alleles can vary for a certain gene from one to many, but it is normally less than ten. Each diploid individual can have not more than two at a time, one received from the father and one from the mother. When the alleles received from both parents are the same, the individual is called *homozygous;* when they are different, the individual is *heterozygous*. A heterozygous individual may resemble an individual homozygous for one of its two alleles; that allele is then said to be *dominant*. Alternatively, a heterozygous individual can be intermediate between individuals homozygous for the two alleles it carries. The combination of dominance at some loci and intermediate situations at others accounts in part for the similarities and differences between the parents and their offspring. If in a population there is only one allele for each of the genes affecting a character, all members of the population will have the same form of that character.

Particulate inheritance, or Mendelian inheritance, as it is often called, incorporates three main concepts:

1. Inheritance is particulate and discrete, and each parent contributes to its offspring an equal number of genes, with the exception of those animals and plants that have sex chromosomes.* In those organisms the contribution of the sex chromosomes only is unequal. The existence of genes was deduced theoretically, but more recently they have been identified biochemically.

2. Although some genes may suppress, mask, or alter the effects of other genes, the inherited factors do not contaminate one another. That is, the genes are not changed or altered in their fundamental structure by association with other genes, and they will be transmitted to the next generation in the same form in which they were received.

* *Sex chromosomes* determine the sex of an individual. There may be two sex chromosomes in one sex (XX) and one in the other (XO), as in the squash bug *Anasa tristis*. In many species, including humans and *Drosophila*, one sex has a chromosome (Y) not present in the other sex, which pairs (XY) with a chromosome present in a double dose in the second sex (XX).

3. The genes are exceptionally stable in their composition and function. Nevertheless they can change, or *mutate,* to a different form called a *mutant,* in which case their function may also be altered. Genes can mutate because of ever-present natural causes; the process is then called *spontaneous mutation.* They may also mutate as a result of artificially applied outside agents, such as certain chemicals or certain radiations, particularly X rays (and the atmospheric radiation that results from atomic fallout!). Substances capable of inducing mutations are called mutagens, and the type of mutation they cause is called *induced mutation.*

Mendel proposed two main corollaries to his theory that are commonly called "Mendel's laws of inheritance." The first is the *law of segregation* (Figs. 4.4, 4.5). Stated in modern terminology, it says that if two individuals, called the *parental generation* (P), each homozygous for a different allele of the same gene, are crossed to each other, and if the progeny, or *first filial generation* (F_1), are mated to each other, their progeny, or the *second filial generation* (F_2), will consist of three types of individuals in the following proportion: One-fourth will resemble one of the grandparents (P), one-half will resemble the parent (F_1), and one-fourth will resemble the other grandparent (P) (Fig. 4.4). In those cases where complete dominance is present in the first filial generation (F_1)— that is, when only one of the two alleles is expressed—the class in the F_2 that

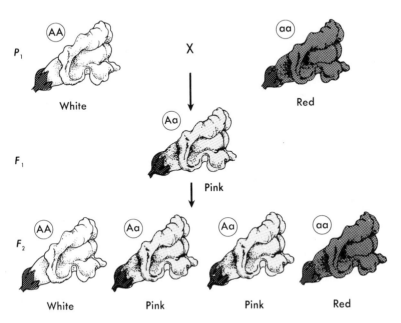

Figure 4.4

Mendel's law of segregation. In flower color in snapdragons, there is no dominance. The F_1 is intermediate between the parents—that is, pink. (From O.T. Solbrig, 1966. *Evolution and Systematics.* New York: Macmillan)

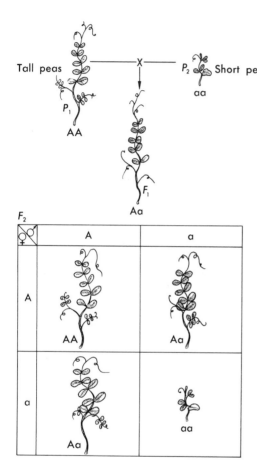

Figure 4.5

Mendel's law of segregation. In plant size in peas there is dominance. The F_1 resembles one of the parents (the tall one). (From O.T. Solbrig, 1970. *Principles and Methods of Plant Biosystematics*. New York: Macmillan)

resembles the F_1 will not be distinguishable morphologically from the class representing the dominant grandparent (Fig. 4.5). In such cases there is a 3:1 ratio in the F_2, with three-fourths of the offspring resembling the dominant grandparent and one-fourth resembling the other grandparent, whose allele is called *recessive*. The law of segregation, which has been demonstrated many times, is a logical corollary of the stability of genes, as discussed above. The law of segregation also helps to explain the fact that often a character in some of the offspring does not resemble either parent but is more like one of the grandparents.

The second Mendelian law, the *law of independent assortment*, deals with inheritance at two loci. It says that the alleles of one gene will segregate according to the first law, regardless of which allele of the second gene it was associated with in the parental generation. For example, consider two individuals, each homozygous at two loci, **A** and **B**, for different alleles of the

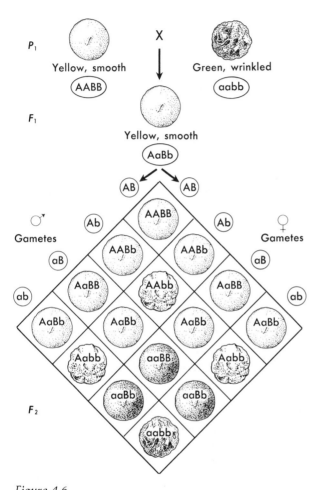

Figure 4.6
Mendel's law of independent assortment. Note that the segregation of yellow versus green color in peas is independent from smooth versus wrinkled surface. In this case there is dominance, the F_1 being like one parent—that is, yellow and smooth. (From O.T. Solbrig, 1966. *Evolution and Systematics.* New York: Macmillan)

genes, 1 and 2. Call their genotypes $A_1A_1B_1B_1$ and $A_2A_2B_2B_2$. The F_1 of a cross between them will be $A_1A_2B_1B_2$. If individuals of the F_1 are crossed, the F_2 will include some individuals with the parental genotypes; some with both alleles of one gene from one original parent (P) and both alleles of the other gene from the other original parent (e.g., $A_1A_1B_2B_2$); and some with other combinations (Fig. 4.6). For each combination of alleles of one gene, the alleles of the other gene segregate according to the first law. For instance, of those individuals with A_1A_2, one-fourth are $A_1A_2B_1B_1$, one-half are $A_1A_2B_1B_2$, and one-fourth are $A_1A_2B_2B_2$.

In many cases there is no independent assortment. The reason is that genes are not physically independent bodies but are part of the chromosomes. Whenever two genes are on the same chromosome, they are transmitted together and can no longer assort independently. But as we will see in the next chapter, certain special mechanisms exist that allow the exchange of alleles between homologous chromosomes during gamete formation (see Linkage and Crossing Over).

SINGLE GENES AND MULTIPLE GENE SYSTEMS

According to the original formulation of the particulate theory of inheritance by Mendel, each character was determined by a pair of alleles of the same gene. Although there are instances in which a character is controlled by a single gene, the rule is that characters are affected by many genes. Mendel also implied that each gene would affect only one character. Today we know that most genes affect many characters, although their major action may be evident only in a particular structure or at only one time of development. This statement may appear to contradict what has been said about particulate inheritance. How can a gene that is supposed to store some precise piece of information affect so many aspects of the life of an organism? The answer lies in the biochemistry of the gene and in the development of a character.

Genes are located on chromosomes, which are found in every nucleus of the cells of an organism. Chemically genes are *deoxyribose nucleic acid*, or DNA (RNA in some viruses). This complex molecule (Fig. 4.7) consists of six simpler molecules: deoxyribose (a sugar), phosphoric acid, and four nitrogenous bases (two purines, adenine and guanine, and two pyrimidines, cytosine and thymine). These six molecules polymerize to form DNA. Each sugar molecule bonds to a phosphate molecule and to a nitrogenous base (any of the four), forming a *nucleotide*. Nucleotides in turn combine through bonding of the deoxyribose with the phosphate of another nucleotide to form long fibers in the shape of helixes, or corkscrews, with the bases projecting inward. Two such strands in turn can combine by hydrogen bonding of nitrogenous bases in opposite strands to form a double-stranded helix (the normal or stable form of DNA found in chromosomes), provided that the following condition is met. Opposite each thymine in the first strand there must be an adenine in the second, and opposite each cytosine a guanine, and vice versa. Thus the order of the nitrogenous bases in one chain uniquely determines the order of bases in the other. When the DNA molecule replicates (Fig. 4.8), the hydrogen bonds between bases are broken, and the two strands unwind and separate. Each strand then serves as a template in the synthesis of its complementary chain, resulting in two new double strands with the exact same order of nitrogenous base pairs.

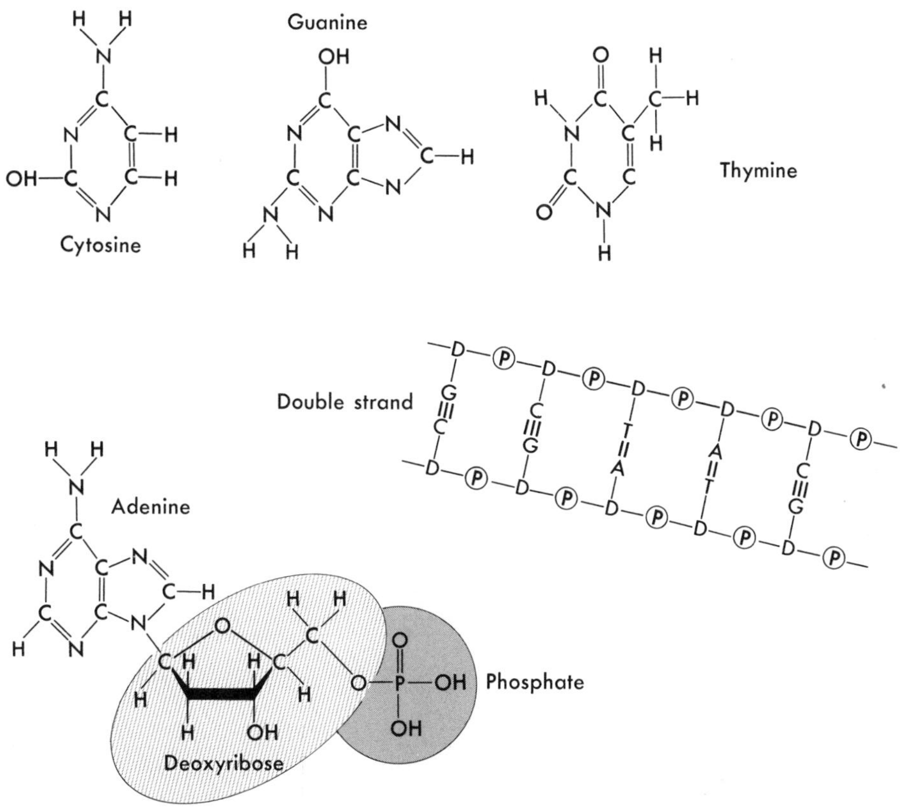

Figure 4.7

The components of the DNA molecule. Each backbone of the double-stranded DNA molecule consists of alternating phosphate and deoxyribose molecules. In turn, the two strands are connected by links consisting of guanine-cytosine or thymine-adenine molecules anchored on deoxyribose molecules on each strand. No other combination of the purine bases has the proper configuration. (From O.T. Solbrig. 1970. *Principles and Methods of Plant Biosystematics.* New York: Macmillan)

The order of the base pairs along the DNA molecule acts as a code. Three adjacent nucleotide pairs, known as a "triplet," determine uniquely (through some intermediate reactions) the assembly of an amino acid in the cytoplasm. Amino acids are the building blocks of proteins. Enzymes are an important class of proteins that act as catalysts; that is, they control specific chemical reactions within the cell. To produce a structure such as a flower or a limb, or even a hair or the color in a petal, a whole series of complex reactions is required. The timing and the rate of each reaction are controlled by an enzyme, which is produced by a particular gene or set of genes. But most of the chemi-

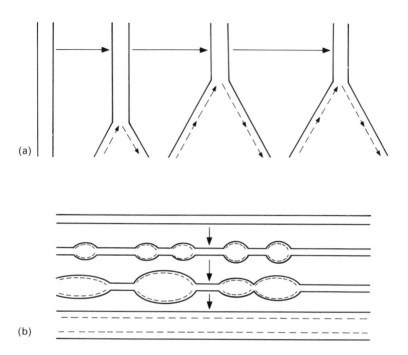

(a)

(b)

Figure 4.8

DNA replication. (a) In replicating DNA the strands of the double helix pull apart, and daughter strands are synthesized in segments (which are later joined), using the parent strand as a template. (b) In eukaryotic DNA, replication is initiated at many points along the molecule.

cal reactions taking place in the cells during development of an insect wing, for example, are not unique to that structure. Only their timing and relative importance and rates are probably unique. Consequently we can easily see that more than one gene may be involved in the development of a character (Fig. 4.9), and that a gene can affect more than one character.*

The timing and the rates of production of an enzyme are usually under the control of at least two genes. One, called the *structural gene*, carries the code for the amino acid sequence of that enzyme. The other, called a *regulatory gene*, produces an enzyme (or some other substance) that affects the enzyme produced by the structural gene (or the structural gene itself), thereby regulating (i.e., determining the action of) the structural gene. How this regu-

---

* The nature and action of the units of heredity (cistrons, regulator genes, operons, and so on) at the subcellular and cellular level are now fairly well understood. The use of the general term "gene" is nevertheless justified in explaining genetic phenomena at the populational level, where our knowledge of gene action is still incomplete.

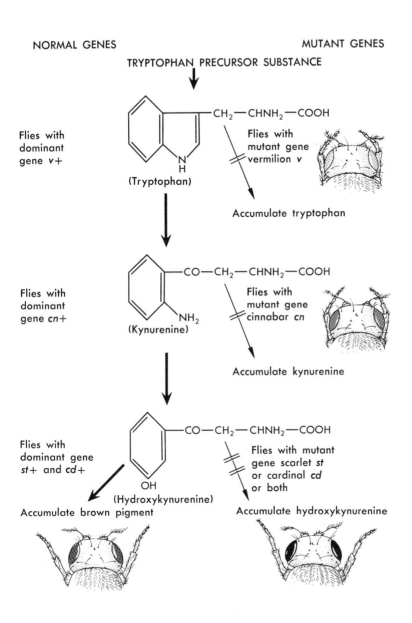

NORMAL GENES

MUTANT GENES

TRYPTOPHAN PRECURSOR SUBSTANCE

Flies with
dominant
gene v+

(Tryptophan)

Flies with
mutant gene
vermilion *v*

Accumulate tryptophan

Flies with
dominant
gene cn+

(Kynurenine)

Flies with
mutant gene
cinnabar *cn*

Accumulate kynurenine

Flies with
dominant gene
st+ and cd+

(Hydroxykynurenine)

Accumulate brown pigment

Flies with mutant
gene scarlet *st*
or cardinal *cd*
or both

Accumulate hydroxykynurenine

Figure 4.9

Some of the steps involved in the formation of the brown eye pigment in *Drosophila*, and the genes that control them. At left, the normal situation; at right, the action of the mutant genes. Although the obvious effect is a morphological one, the genes act by affecting specific chemical reactions in the cell. (From O.T. Solbrig, 1966. *Evolution and Systematics.* New York: Macmillan)

lation operates is known with some precision only in bacteria, where regulatory genes can inhibit the action of structural genes. Apparently such a mechanism of regulation does not operate in higher plants and animals. Although the exact mechanism of regulation is obscure, some evolutionists have argued that changes in regulatory genes may be more important than changes in structural genes. We will return to the subject in Chapter 9.

If the effect of an allele on a character is minor, its presence or absence will produce only a minor effect on the chemical reactions leading to the formation of that character. If we measure the character, the variation between an organism not having the particular allele and one having it will be minor. The same applies when a series of different, nonallelic genes affect a character. In such a case, the variation within a population appears to be continuous between the two extremes. Systems of genes that individually have small cumulative effects and together control continuous variation are called *multiple gene systems*. Genes that individually have drastic effects and produce discontinuous effects are called *major* or, sometimes, *Mendelian genes*.

There is no absolute distinction between multiple gene systems and major genes, since groups of genes with all degrees of intermediate effects are known. Moreover, a gene may drastically affect one character and be classified as a major gene, but it may only slightly affect another and consequently be also classed as part of a multiple gene system. Nevertheless, despite the absence of a clear distinction between major genes and systems of multiple genes, the latter concept is useful and quite adequate for most purposes.

One might think that a character such as height, which varies continuously in a population, or color intensity, which can have all kinds of gradations between the lightest and the darkest, could not be controlled by particulate genes having defined and distinct effects. Alleles of major genes certainly cannot produce such an effect, but a large number of genes, each with a small cumulative effect, can. This was first demonstrated in 1911 by the Swedish geneticist Nilsson-Ehle. Working with wheat, he discovered three nonallelic genes that affected the color of the kernels. Plants that were double recessive for all three genes had white kernels; plants with dominant alleles in all three genes had dark red kernels; kernels of plants with some, but not all, dominant alleles were also red but not so dark. The action of each gene was similar to that of the others. When each plant was studied independently, plants homozygous recessive for two loci and heterozygous for one produced a ratio of 3 red kernels: 1 white (Fig. 4.10a). Plants heterozygous for two genes produced a 15 red: 1 white ratio (Fig. 4.10b). Plants heterozygous for all three gave a 63 red: 1 white ratio (Fig. 4.10c). When the red kernels of the F_2 plants from a cross that combined all three genes in heterozygous conditions were grown, some plants segregated in a 3:1 ratio, others in a 15:1 ratio, and still others in a 63:1 ratio, as expected. It is clear that each gene behaves as a true Mendelian factor and that all three follow the first and second Mendelian laws.

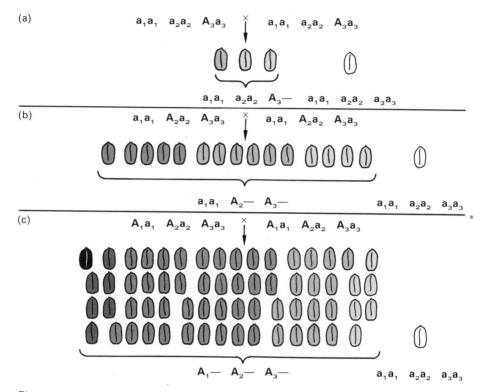

Figure 4.10

Genetics of wheat kernel color. (a) The offspring of plants homozygous recessive at two loci and heterozygous at the (same) third locus are 3 red : 1 white. (b) For plants recessive at one locus and heterozygous at the other two, the offspring are 15 red : 1 white. (c) If all loci are heterozygous, the offspring are 63 red : 1 white. Red kernels differ in the intensity of the red color in proportion to the number of dominant alleles they possess.

That the effect of the three genes is additive is further borne out by variations in the intensity of the red kernels. Series of such genes, each with a small effect, account for the inheritance of quantitative characters with a seemingly continuous variation. The existence of multiple genes was conclusively demonstrated by the Harvard geneticist East in 1916 working with tobacco (Fig. 4.11).

Nevertheless, even when more than one gene is involved, there still ought to be some discontinuity between genetically heterogenous groups, although there may be some problem in distinguishing the genotypic classes. It is not always possible to make the distinctions because of the noninheritable component of variation—that is, the effects of such environmental factors as food, tempera-

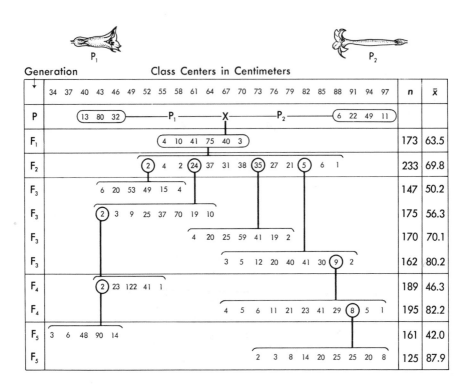

Generation	34	37	40	43	46	49	52	55	58	61	64	67	70	73	76	79	82	85	88	91	94	97	n	x̄
P				13	80	32		—P₁—			—X—		—P₂—					6	22	49	11			
F₁									4	10	41	75	40	3									173	63.5
F₂								(2)	4	2	(24)	37	31	38	(35)	27	21	(5)	6	1			233	69.8
F₃				6	20	53	49	15	4														147	50.2
F₃			(2)	3	9	25	37	70	19	10													175	56.3
F₃								4	20	25	59	41	19	2									170	70.1
F₃												3	5	12	20	40	41	30	(9)	2			162	80.2
F₄			(2)	23	122	41	1																189	46.3
F₄											4	5	6	11	21	23	41	29	(8)	5	1		195	82.2
F₅	3	6	48	90	14																		161	42.0
F₅														2	3	8	14	20	25	25	20	8	125	87.9

Figure 4.11

Results of the experiments of East with *Nicotiana longiflora* that demonstrated the multiple-factor hypothesis. Plants with short (40–46 cm) and long (88–97 cm) flowers were crossed. The F₁ plants were intermediate (55–70 cm), and the F₂ segregated for this character (52–88 cm). Plants of the F₂ with different values (52, 61, 73, and 82 cm) were self-pollinated. As expected, their progeny segregated, but the F₃ petal length values revolved around those of the parent. The same results were obtained in the F₄ and F₅. If the variation had been environmental rather than genetic, all generations would have had similar values; if due to only one pair of segregating alleles, no segregation would have been expected in some lines after the second filial generation. (Numbers in the table are numbers of plants of a certain size in a given generation. Note the distribution within each generation.) (From O.T. Solbrig, 1970. *Principles and Methods of Plant Biosystematics*. New York: Macmillan)

ture, light, shelter. These forces produce variation among genetically identical individuals. For example, a well-fed animal will be heavier and possibly bigger than an undernourished one of the same genetic composition. Since the off-spring of the well-fed animal will not be any heavier than those of the under-nourished one when both kinds of offspring are given the same amount of food, this component does not play any major evolutionary role. However, it has the effect of creating continuous phenotypic variation on discontinuous underlying genotypes. In order to avoid this environmental effect, the geneticist tries to grow his plants and animals in uniform environments. But it is almost never possible in a field to avoid small variations in such things as rainfall, drainage, and minerals in the soil or to feed animals exactly the same amount. Many discrete genes that affect a character, each in a small degree, combine with environmental noninheritable changes to produce truly continuous variation (Fig. 4.12).

EPISTASIS AND PLEIOTROPY

Occasionally the allele of one gene masks or modifies the phenotypic effect of an allele at another locus. Such interaction is termed *epistasis*. This multiple effect of an allele is due to the fact that the primary gene products, the enzymes, catalyze a specific chemical reaction in the metabolism of the cell. When two different enzymes act on the same metabolic pathway, epistatic interactions follow.

Two independent genes may affect the same metabolic pathway (see Fig. 4.9). When that is so, the rate and timing of gene action in one gene has to be in phase with the rate and timing of gene action of the second gene. Strictly speaking, epistasis exists between alleles of different genes when one masks the effect of the other so that the phenotype is determined by the former gene and not by the latter. The gene that thus masks or prevents the expression of another is said to be *epistatic* to it, and the gene that is hidden is said to be *hypostatic*.

A case of epistasis is found in a wild species of violet, *Viola tricolor*, studied by the Danish geneticist Jens Clausen in 1926. The normal color of the flowers in this species is violet. There are five genes that affect flower color, known as M_1, M_2, M_3, M_4, and M_5. The first gene (M_1), when present in a dominant form, produces violet flowers regardless of whether the other four genes are in a dominant or recessive form. On the other hand, if the first gene is present in a double recessive form (m_1m_1), flower color will be light purple whenever the second gene is present in a dominant form (m_1m_1, M_2M_2, ... or m_1m_1, M_2m_2, ...). Color will be purple if the first two genes are recessive and the third gene dominant (m_1m_1, m_2m_2, M_3 ...), deep purple if the first three are recessive and the fourth dominant (m_1m_1, m_2m_2, m_3m_3, M_4 ...), velvety black if the first four are recessive and the fifth dominant (m_1m_1, m_2m_2, m_3m_3, m_4m_4, M_5 ...), and jet black if all five genes are recessive (m_1m_1, m_2m_2, m_3m_3, m_4m_4, m_5m_5). Another well-known case is the inheritance of color in squash (Fig. 4.13).

The term *pleiotropy* denotes the occurrence of diverse phenotypic effects by an allele of a single gene (Fig. 4.14). An example of pleiotropy is the **S** gene in the tobacco plant. When the dominant allele, **S**, is present, the plant produces leaves with long petioles and pointed tips, calyces with long slender

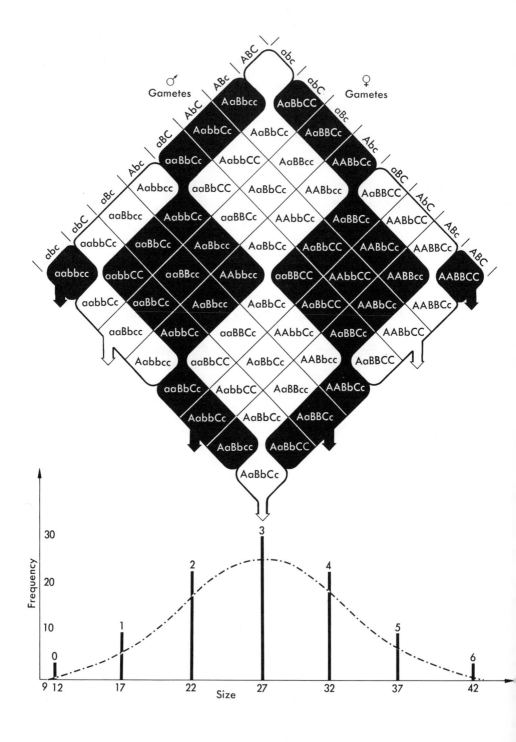

◀ *Figure 4.12*

Multiple genes and plasticity combine to produce a continuous distribution (dashed curve) rather than a discontinuous distribution (bar graph). In effect, if we assume that each dominant allele in the checkerboard has an effect on the size of a hypothetical character, x, of 7 units, and that each recessive allele in the checkerboard has an effect of 2, we obtain seven discrete size categories measuring 12, 17, 22, 27, 32, 37, and 42 units, respectively, with a frequency of 1.6%, 9.4%, 23.4%, 31.3%, 23.4%, 9.4%, and 1.6%, respectively, and corresponding to individuals with 0, 1, 2, 3, 4, 5, and 6 dominant alleles. However, if because of the effects of the environment, an individual may deviate by up to 3 units around the value expected from its genotype, a series of phenotypes with values ranging continuously from 9 to 45 units is produced, with expected frequencies as shown by the curve. The larger the number of genes affecting the character, the more the distribution will tend to be continuous. (From O.T. Solbrig, 1970. *Principles and Methods of Plant Biosystematics*. New York: Macmillan)

Figure 4.13

Inheritance of skin color in summer squash. In summer squashes there are three common fruit colors, white, yellow, and green. In crosses between white and yellow and between white and green, white is found to be always dominant. In crosses between yellow and green, yellow is found to be always dominant. Yellow thus acts as a recessive in relation to white but as a dominant in relation to green. Therefore there is a gene, **W**, which is epistatic to the gene **Y**. Whenever the gene **W** is present in its dominant form, the fruit will be white. When **W** is lacking (in a ww genotype), the fruit color will be yellow if **Y** is present and green if it is absent. The **Y** blocks the formation of green pigment but not of yellow carotenoids, and **W** blocks the formation of both pigments.

▼

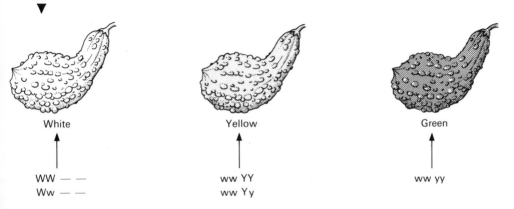

White	Yellow	Green
WW — —	ww YY	ww yy
Ww — —	ww Yy	

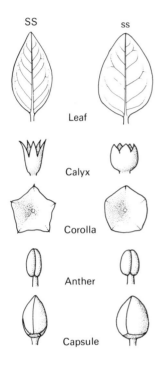

Figure 4.14

The "petioled" gene (**S**) in tobacco (*Nicotiana taba-cum*) has a number of pleiotropic effects, such as the formation of leaves with long petioles and pointed tips, calyces with long slender teeth, petal lobes with slight tips, long anthers, and elongated fruits.

teeth, petal lobes with slight tips, long anthers, and elongated fruits. On the other hand, plants homozygous for the recessive allele have sessile leaves without sharply pointed tips, calyces with short teeth, petals without points, short anthers, and roundish fruits. Pleiotropy is the result of gene action at an early developmental stage, so that many structures are affected. Most, if not all, genes are pleiotropic, but only in few cases are the effects as drastic as in this example. Consequently the term *pleiotropy* is reserved for the more extreme cases.

GENES, GENOTYPES, AND PHENOTYPES

So far we have described the inheritance of characters in organisms as if there was a simple relationship between the hereditary character and the genes that ultimately determine it. In some instances, especially in experimental situations when the environment and alleles at other genes are kept invariant, substitution of one allele of a gene for another in an organism produces simple changes in characters, as when Mendel crossed short and tall pea plants. There are thousands of examples known where substitution of one allele by another changes characters. In plants the changed characters may be plant size, shape of leaves, color of flowers, amount of starch in seeds. In animals they may be color of pellage, stature, type of hemoglobin in blood, susceptibility to disease, ability to taste certain substances, etc. However, the relationship between genes and characters is not a simple one, since the expression of the gene depends on a number of other factors, primarily the environment in which the organism grows and the other genes in the organism, or what is usually called the *genetic background.*

An example of the effect of environmental influence on the expression of a gene is provided by the gene *Curly* (**Cy**) of the fruit fly *Drosophila melanogaster* (Fig. 4.15). This dominant gene gives rise to flies with curled-up wings when the pupae are maintained at 25°C. However, when pupae of exactly the same genetic constitution are maintained at 19°C, many flies have normal wings, and the rest have a variety of abnormalities ranging from slightly to very curled wings. It appears that the *Curly* gene produces wings that are unusually sensitive to heat at the time the fly emerges from the moist pupal case. At the high temperature the upper and lower portions of the wings dry at different rates resulting in a *Curly* wing; at the lower temperature the rate of drying is slower, and fewer wings curl.

The effect of genetic background is harder to study. Figure 4.16 illustrates the partial pedigree of five generations of a human family that carried a dominant gene for polydactyly (meaning an excess of fingers and/or toes). As the pedigree shows, the expression of the gene varied widely. Some individuals have the normal number of fingers but six-toed feet; others have six-fingered hands and normal feet; one has six digits on all four extremities; one has more fingers and toes on his right than on his left extremities; and two appear to have normal hands and toes. In this case it is not possible to state whether the observed differences are due to environment or genetic background, but probably both are influential.

Figure 4.15
Curly wing in *Drosophila melanogaster.* This character affects the shape of the wing. (a) Flies with normal wings, when resting, fold their wings over the abdomen. (b) Flies that carry the dominant gene Curly (**Cy**) cannot do so because their wings curl upward.

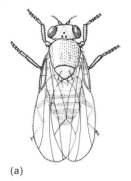
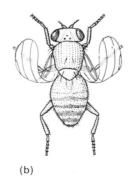

(a)   (b)

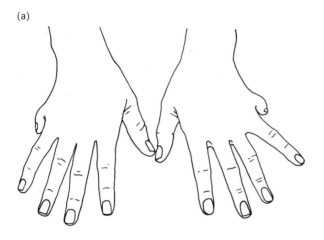

(a)

(b)

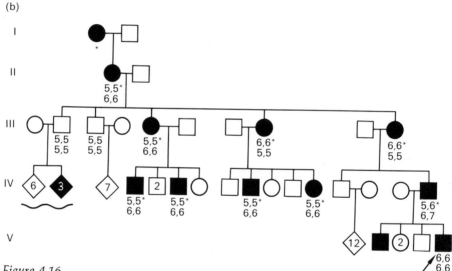

Figure 4.16

(a) The hands of a polydactylous girl. This individual does not belong to the poly-dactylous kindred presented in (b). (b) Selected individuals in a pedigree of polydac-tyly. In each group of four numbers, the upper two represent the numbers of fingers on the left and right hand, respectively, and the lower two the numbers of toes. An asterisk indicates that the type of polydactyly was not stated unequivocally in the original report. (After Lucas, *Guy's Hosp. Rpts.* 3rd ser. **25, 1881**)

Natural selection is the process of differential survival and/or reproduc-tion of individuals. If an organism possesses a certain character that enhances its survival and/or reproduction, the alleles that ultimately determine that character are also selected. However, if they are expressed in only some in-dividuals and under only some environmental circumstances, as in polydactyly in humans or *Curly* wing in *Drosophila,* then they will be selected only when they express themselves. Let us consider the situation of a character that decreases the survival and/or reproduction of individuals possessing it. Individuals with such a character will leave fewer offspring than the average. The allele or alleles responsible will become less frequent, and eventually they will be eliminated. However, the rate at which they will disappear depends on how often they express themselves. If certain special environmental conditions are required before individuals with that allele produce that character, the allele in question, though harmful, could persist in the population as a result of its lack of expression.

Now let us imagine still another situation. Under one set of environmental conditions a character is favorable to the individual, and under a second set of conditions that character is detrimental. An example is leaf size in plants. In a moist environment large leaves are favored, but under drought conditions smaller leaves are usually better for the plant. A plant that could produce large leaves when it rained, and small leaves when it was dry, would consequently have an advantage. One such plant is the desert shrub *Encelia farinosa,* which grows in the southwestern United States (Fig. 4.17). When there is plenty of water in the soil, the cells in the growing leaf expand, giving rise to a large leaf. When water is scarce, cells expand much less, resulting in a small leaf. The exact genetic mechanism responsible for this behavior is not known, but here is a case in which expressing the character in one environment and not in another produces two favorable characters. Obviously alleles that control this variable expression are favored by natural selection.

The sum total of the genes of an organism is called the *genotype;* the sum of its characteristics is its *phenotype.* Natural selection acts on phenotypes and only indirectly on genotypes. When the expression of a character, such as leaf size in *Encelia farinosa,* depends on environmental factors, it is said to be *plastic.* When a gene produces a character only in certain situations and not in others, it is said to have *incomplete penetrance;* when the resulting character is variable, as with the *Curly* wing, the gene is said to have *variable expressivity.* A plastic character is consequently the phenotypic manifestation of genes with incomplete penetrance and/or variable expressivity.

Because of dominance, epistasis, incomplete penetrance, and variable expressivity, dissimilar genotypes can produce similar phenotypes. These phenomena also account for the fact that organisms that are phenotypically similar (but genotypically dissimilar) in one environment can be phenotypically dissimilar in another environment. Consequently, understanding how characters are inherited under all situations requires much more knowledge than determining which alleles affect it. The intermediate steps between the DNA molecule and the phenotypic character—an area of study known as growth and development—and the rules determining those steps must also be understood. At present we do not understand them, and it will be some time before we acquire that knowledge. Consequently, at present the evolutionist and

Figure 4.17

The desert shrub *Encelia farinosa* produces small hairy leaves in times of drought and larger, less hairy leaves during rainy periods of the year. (Photo courtesy of Dr. James Ehrelinger)

population geneticist is tremendously constrained in his quest for knowledge about the evolutionary implications of the genetic structure of populations. In models of evolution a number of simplifying assumptions are made to solve this problem, such as that selection operates on genes, or that selection co-efficients are constants, or that characters are uniquely determined by geno-types. We will discuss these assumptions in this book, but the reader should be aware that present ignorance regarding the exact relation between genotype and phenotype has the effect of making much of present day evolutionary theory temporary and contingent. We therefore must make certain that we understand the natural situation—the facts—and satisfy ourselves that the theory is consistent with the facts.

THE HARDY-WEINBERG LAW
AND THE MAINTENANCE OF VARIATION

The importance of variability in populations can hardly be overemphasized. Unless members of populations are different, natural selection cannot operate. We have seen that the sources of variability are two: genetic and environmental, or nongenetic. In evolution the first source is the more important one. Continuous as well as discontinuous variation, small or large differences—all have exactly the same kind of genetic basis; i.e., they are determined by genes. According to the theory of blending inheritance, this genetic variability is lost as a result of hybridization. The fact that genetic variability is observed in nature is one of the reasons for rejecting that particular hypothesis. Particulate inheritance, however, offers an explanation for the transmission of individual characteristics from parents to offspring without loss of variability. G. H. Hardy, a British mathematician, and G. Weinberg, a German geneticist, pointed this out independently in 1908. They demonstrated that the original variability in a population will be maintained in the absence of forces that tend to decrease or increase this variability. Their demonstration is known as the *Hardy-Weinberg* theorem (or sometimes law).

Let us consider a population of a very large number of individuals (theoretically an infinitely large number), in which breeding is strictly random. Suppose that A_1 and A_2 are a pair of alleles of the same gene present in the population with a frequency of p and q, respectively, such that $p + q = 1$. *Gametes*, special sex cells that carry one of the organism's two sets of genes, with the allele A_1 occur with frequency p. If they combine at random with other gametes, a fraction p will combine with A_1 gametes and a fraction q will combine with A_2 gametes. The *zygotes* produced by the fusion of two gametes will have the genotypes A_1A_1 in the frequency p^2 and A_1A_2 in the frequency pq (Fig. 4.18). Similarly, gametes with A_2 will combine with A_2 or A_1 gametes to give rise to new individuals with genotypes A_2A_2 in the frequency q^2 and A_2A_1 in the frequency qp. Since genetically A_1A_2 and A_2A_1 are identical, their combined frequency is $2pq$. Consequently in such a population the frequency of genotypes will be as follows:

genotype	*frequency*
A_1A_1	p^2
A_1A_2	$2pq$
A_2A_2	q^2

In the next generation the homozygous genotypes will produce only one kind of gamete, whereas the heterozygous genotypes will produce both kinds of gametes in equal numbers. If each organism in the population on the average produces the same number of gametes, the proportion of each kind will depend on the frequency of the different genotypes in the population, so that

proportion of A_1 gametes: $p^2 + 1/2\ (2pq) = p(p + q) = p$

proportion of A_2 gametes: $q^2 + 1/2\ (2pq) = q(q + p) = q$

The frequency of A_1 and A_2 gametes produced by the population after one generation of random mating without selection or mutation is the same as that of the gametes that gave rise to it. It follows that the proportion of genotypes

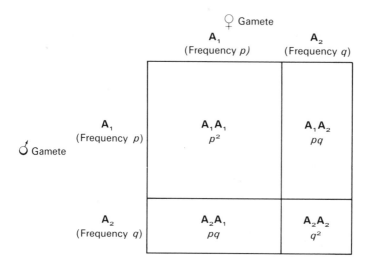

Figure 4.18

Zygote frequencies in a population. Where there are two kinds of gametes, A_1 and A_2, with frequencies p and q, respectively, there will be three kinds of zygotes: A_1A_1, in the frequency p^2; A_1A_2, in the frequency $2pq$; and A_2A_2, in the frequency q^2.

in subsequent generations is not going to change, either, and consequently the initial variability has been maintained.

The Hardy-Weinberg theorem demonstrates that the initial gene frequencies in a population will be maintained and that any changes in gene frequencies are brought about by outside forces. There is no need for another hypothesis, such as the inheritance of acquired characters, to explain the observed maintenance of variability. Essentially the Hardy-Weinberg theorem demonstrates the compatibility between Mendelian genetics and Darwinian evolution. Because in most instances natural selection decreases the variability in the population, a source of new variability is still a necessary condition for continued evolution, even granting particulate inheritance. However, the magnitude of the new variation is much less than it would be with blending inheritance.

PHENOTYPIC VARIATION AND HERITABILITY

For evolution by natural selection to operate there must be genetically determined variation of phenotypic characters in the population; that is, the individuals in the population cannot be genetically identical (Chapter 3). Since at present we normally cannot determine that variation by investigating the genotype directly, the variation of phenotypic values must be measured by means of statistical techniques. These techniques are used by plant and animal breeders, and they are very useful in that context. At times, however, they have been misused, as, for example, when applied to estimate intelligence differences between races or between poor and rich. It is therefore important to have some understanding of how these measures are obtained.

The problem is to estimate how much of the variation that is observed in a population living in a certain environment can be attributed to genetic factors and how much to environmental interactions.

The genetic properties of a population have often been expressed in terms of gene frequencies. However, what one observes when working with populations is quantitative differences in some character or characters between individuals. Any and all observations and measurements made on individuals in a population are made on their phenotypes. When one is dealing with a quantitative character, what one usually calculates is the mean and the variance of that character. The *mean* of a character $(\overline{X})$ in a population is equal to the sum of the individual values divided by the number of observations; that is,

$$\overline{X} = \frac{\sum\limits_{i=1}^{i=N} X_i}{N},$$

and the variance (s^2) is equal to the sum of the squares of the deviations from the mean, divided by the number of observations minus 1,

$$s^2 = \frac{\sum\limits_{i=1}^{i=N} (X_i - \overline{X})^2}{N - 1}.$$

(The *standard deviation* is the square root of the variance.)

The mean is a measure of the central value of the character, whereas the variance measures the dispersion around the mean in the population. The problem we face is to divide the variance of phenotypic values into proportions attributable to the different genetic and environmental factors. For example, how much of the stature of individuals in the population is due to environment and how much to genetics? Let us briefly review the fundamentals of this operation.

The genetics of a metric (continuous) character is centered on the study of its *variation*. The reason is that, when confronted with a population that has one or more characters that are varying continuously, the geneticist cannot count classes as when studying discontinuous characters. What *can* be determined are the *mean*, the *range*, and the *variance* of the population. The problem is to determine the underlying genetic mechanism when dealing with only those statistics. The basic idea in the study of variation is to divide this variation into components attributable to different causes. Once the relative magnitudes of the various components of the variation of a population are known, the geneticist can infer from them the genetic structure of the population.

The amount of variation is measured and expressed as the variance. When values are expressed as deviations from the population mean, the variance is simply the mean of the squared values. (Note: in this case the variance = $\Sigma(x_i - \bar{x})^2/N$.) The total variance is the phenotypic variance (what is measured), and it is equal to the sum of the genetic and environmental variances, since the observed value of a character is related to its underlying genes and to the environment where it lived.

To determine the genetic variance, the geneticist could grow plants or rear animals in perfectly uniform environments and, by making the further assumption that in that environment all nongenetic deviations would equal zero, could equate the obtained variance with the genetic variance. This is seldom done, however, because it is practically impossible to produce uniform environments.

One of the properties of metric characters determined by multiple gene systems is the resemblance between relatives, which is easily measured and provides a means for estimating the genetic variance. To make this measurement, the geneticist divides the population into groups of equally related organisms, called *genetic families*, such as full sibs, half sibs, offspring and one or both parents, etc. By a statistical technique known as *analysis of variance*, the variation observed in the population can be divided into two components: (1) *within*-family variance, which is the variance of individuals about the mean of their family; and (2) the *between*-families variance, which is the variance of the family means about the population mean. If all the individuals within a family are identical, then all the variation in the population will be the variation that exists between families. On the other hand, if the individuals in each family are variable and all families vary in the same way, the differences between families will be minimal, and the variation within families will be maximal. The degree of resemblance in the population can therefore be expressed as a ratio of the "between-family" variance to the sum of the "within- and between-" family variance, as follows:

$$t = \frac{B^2}{B^2 + W^2},$$

where B^2 = variance between families, W^2 = variance within families, and t = intraclass correlation coefficient. The larger t is, the more similar the members of a family are, that is, the more closely related. By a series of complicated statistical techniques, it is possible by the use of this general approach to estimate the genetic variance in the population.

Knowing the total phenotypic variance and the genetic variance, one can calculate another measure, called the heritability,

$$h^2 = \frac{V_G}{V_P},$$

where h^2 = heritability (note, *not* the square of heritability), V_G = genetic variance, and V_P = total phenotypic variance. Since V_P is always larger than V_G, h^2 will always have a value less than 1 (Table 4.1). The closer the value is to 1,

Table 4.1

Approximate values of the heritability of various characters in domestic and laboratory animals.

Cattle	
Amount of white spotting in Friesian cows	0.95
Percentage of butterfat in milk	0.60
Milk yield	0.30
Conception rate in first service	0.01
Pigs	
Thickness of back fat	0.55
Body length	0.50
Weight at 180 days	0.30
Litter size	0.15
Poultry	
Egg weight	0.60
Age at laying of first egg	0.50
Annual egg production	0.30
Body weight	0.20
Viability	0.10
Mice	
Tail length at six weeks	0.60
Body weight at six weeks	0.35
Litter size	0.15

From D. S. Falconer, 1960. *Introduction to Quantitative Genetics.* New York: Ronald Press.

the greater the genetic component determining it. Note, however, that since V_G is obtained from correlation of relatives *within a population*, comparisons of heritabilities *between populations* are invalid. Heritability is a measure of the degree to which genetics contributes to the variation of a character within a population. As such, it can tell the plant or animal breeder whether or not there is a reservoir of genetic variation amenable to selection and how great that reservoir is. By extension, it can be stated that natural selection will act only on phenotypic characters that are *heritable* (i.e., where $h^2 > 0$), regardless of how variable they may be in nature.

We have dealt with the topic of heritability in some detail because it is a poorly understood and much misused concept. Note that the heritability is the ratio between the genetic variance and the total phenotypic variance. Neither of these numbers is an absolute property of an individual or a character, but both are statistical properties valid only in the context of a given environment and a given genetic composition of a population. Therefore, we cannot speak of the *heritability* of a character without making reference to the environment and the population from which the data were obtained. Especially dangerous are statements regarding the heritability of human characters, particularly when comparisons between groups are made. Such statements are meaningless and should be avoided.

The concept of heritability is also important in the context of natural selection, as we will see in Chapter 6, when Fisher's fundamental theorem of natural selection is considered, since the greater the heritability of a character, the greater is the potential for change under natural selection.

SUGGESTED FURTHER READING

General references

Falconer, D.S. 1960. *Introduction to Quantitative Genetics.* New York: Ronald Press.

Goodenough, U., and R.P. Levine. 1978. *Genetics,* 2nd ed. New York: Holt, Rinehart and Winston.

Spiess, E.B. 1977. *Genes in Populations.* New York: Wiley.

Strickberger, M.W. 1976. *Genetics,* 2nd ed. New York: Macmillan.

Watson, J.D. 1976. *Molecular Biology of the Gene,* 3rd ed. Menlo Park, Cal.: Benjamin.

Specialized articles and reviews

Britten, R.J., and E.H. Davidson. 1969. Gene regulation for higher cells: a theory. *Science* 165:349–357. (gene regulation)

Calvo, J.M., and G.R. Fink. 1971. Regulation of biosynthetic pathways in bacteria and fungi. *Ann. Rev. Biochem.* 40:943–968. (gene regulation)

Clausen, J. 1926. Genetical and cytological investigations on *Viola tricolor* L. and *V. arvensis* Murr. *Hereditas* 8:1–156. (epistasis)

East, E.M. 1916. Studies on size inheritance in Nicotiana. *Genetics* 1:164–176. (quantitative inheritance)

Ehrelinger, J.R., O. Björkman, and H.A. Mooney. 1976. Leaf pubescence: effects on absorbance and photosynthesis in a desert species. *Science* 192:376–377. (plasticity)

Hardy, G.H. 1908. Mendelian proportions in a mixed population. *Science* 28:49–50 (Hardy-Weinberg law)

Nilsson-Ehle, H. 1909. Krenzungsuntersuchungen an Hafer und Weizen. *Lunds Univ. Arsskr. Afd.* 2, 7(2):1–84. (quantitative inheritance in wheat)

Nirenberg, M.W. 1963. The genetic code: II. *Scientific American* 190 (March): 80–94 (genetic code)

Nirenberg, M.W., and P. Leder. 1964. RNA codewords and protein synthesis: the effect of trinucleotides upon the binding of sRNA to ribosomes. *Science* 145:1399–1407 (genetic code)

Stebbins, G.L. 1959. Genes, chromosomes and evolution. In W. Turrill (ed.), *Vistas in Botany*. London: Pergamon. (pleiotropy)

Stern, C. 1973. *Principles of Human Genetics*, 3rd ed. San Francisco: Freeman. (polydactyly in humans)

Watson, J.D., and F.N.C. Crick. 1953. A structure for deoxyribose nucleic acids. *Nature* 171:737–738. (structure of DNA)

Historical works

Brink, R.A. (ed.). 1967. *Heritage from Mendel*. Madison: University of Wisconsin Press. (papers presented at a centennial celebration of Mendel's paper)

Carlson, E.A. 1966. *The Gene: A Critical History*. Philadelphia: Saunders.

Sturtevant, A.H. 1965. *A History of Genetics*. New York: Harper and Row. (a history of genetics by one of the best of the early geneticists)

Watson, J.D. 1968. *The Double Helix*. New York: Atheneum. Also 1969, New York: The New American Library. (the history of the discovery of the structure of DNA by one of the codiscoverers)

Chapter **5**

Genes and Chromosomes

The importance of the study of chromosomes is twofold: Chromosomes can give an insight into genetic phenomena and the evolutionary processes that have brought them about; and their behavior and morphological and chemical characteristics can be used in the classification of species. We will review in this chapter the chromosomal characters that are particularly important in these connections, and we will consider some examples in which cytology has been especially helpful as a tool to solve evolutionary problems.

Like all characteristics of an organism, chromosomes and the mechanisms of cell division and of gametic formation are subject to natural selection. Therefore they will vary from species to species as a result of different evolutionary histories. This variation is usually small in related forms, but it can be quite large between distantly related groups. Consequently, the study of cytological differences provides good clues as to the probable evolution of a species. But since chromosomes are the carriers of the genetic information, they also influence the evolutionary potential of a species.

CHROMOSOMAL CHARACTERS

Chromosomes are discrete bodies found in the nucleus of every cell (Fig. 5.1), with the possible exception of bacteria and blue-green algae (these groups of primitive organisms presumably have not evolved chromosomes with the complex structure of DNA and protein found in more advanced species). Chromosomes are elongate in shape, with a constriction called the *centromere* somewhere along their length. The centromere divides the chromosomes into two parts, called *chromosomal arms*. If the chromosomal arms are of the same length, we classify the chromosome as having a *median* centromere. If the centromere is not exactly in the middle, we call it a *submedian* centromere; if near one end, it is a *subterminal* centromere; and if almost at the end (apparently it is never on the very end), it is a *terminal* centromere. The position of the centromere, then, provides a basis for the classification and identification of chromosomes.

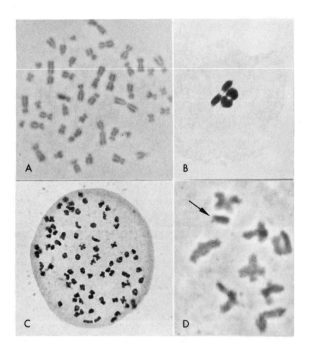

Figure 5.1

(A) Human somatic chromosomes ($2n = 46$). (B) Metaphase I division in *Haplopappus gracilis* ($n = 2$); this is the lowest chromosome number so far known in a plant. (Photo courtesy of Dr. Ray Jackson) (C) Diakinesis in the fern *Polystichum californicum* ($n = 82$); ferns are often characterized by very high numbers of chromosomes. (Photo courtesy of Dr. Warren H. Wagner) (D) Diakinesis in *Gutierrezia bracteata* ($n = 8 + 1$); arrow points to supernumerary chromosome.

A second way to identify chromosomes is by their length. The length of the chromosomes varies from about one micron to 30 microns, and most chromosomes are less than 10 microns in length during the stage of maximum contraction in mitosis. Absolute length and relative length of the two chromosomal arms are the main—sometimes the only—ways to recognize individual chromosomes in a cell.

Sometimes chromosomes can be identified by additional characteristics. A very frequent identifying characteristic is the possession at one end of a small, usually rounded body called a *satellite*, which is united to the main body of the chromosome by a thin, threadlike filament called a secondary constriction. Usually only one pair of homologous chromosomes has a satellite. An individual chromosome does not have more than one satellite. Another characteristic by which chromosomes can sometimes be distinguished is the presence of distinct areas of strongly staining material called *heterochromatin*. A number of different staining techniques reveal that the chromosome during division has a banded structure. One of these techniques, cold treatment prior to fixation (chemical killing) and staining has been used for many years, but most of the approaches have been developed within the last ten years. One of the new techniques, called *Q-banding*, relies on the fact that, in humans and some other organisms, certain fluorescent dyes have a specificity for only certain heterochromatic regions. Other dyes, designated *G-* or *C-banding*, also bond specifically with regions of the chromosome, often corresponding to the Q-bands. By applying these different staining techniques, one can divide the chromosome into distinct regions, each having a repeatable pattern of staining, allowing the identification not only of individual chromosomes but of chromosomal arms and chromosome segments (Fig. 5.2).

ULTRASTRUCTURE OF THE CHROMOSOME

In the cells of most organisms chromosomes are visible only during cell division. After mitosis they elongate until they become so thin that they can no longer be distinguished even with the most powerful light microscope. This fact has puzzled biologists for a long time. Only recently has the puzzle been partly solved as a result of a combination of patient chemical analysis and studies using the electron microscope.

Chemically the chromosome consists primarily of three substances:

1. Protein. More than one half of the mass of the chromosome is a special kind of protein called *histone*, characterized by being basic (positively charged at neutral pH) because of a high proportion of the amino acids arginine and lysine, and with molecular weights of 11,000 to 21,000. In addition, the chromosome contains some acidic (negatively charged at neutral pH) protein as well.

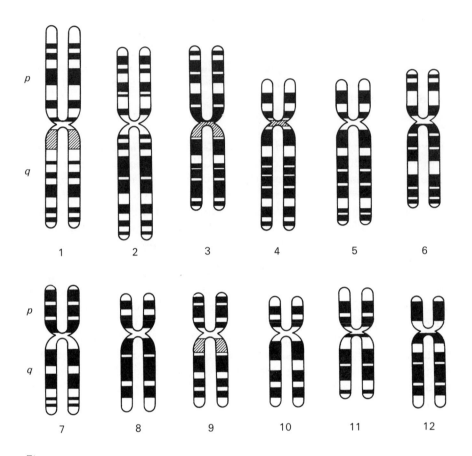

p

q

1   2   3   4   5   6

p

q

7   8   9   10   11   12

Figure 5.2

Differentially stained human chromosomes. Human chromosomes can be identified by patterns of bands, which appear with certain staining techniques. The dark areas indicate positive-staining Q- and G-bands (which correspond fairly closely), and the shaded areas indicate variable bands. (After Paris Conference (1971): *Standardization in Human Cytogenetics*. Birth Defects: Original Article Series, v.8: 7, 1972)

2. DNA. The structure of DNA has already been described (p. 56), and we have also discussed its function as the genetic code.

3. RNA. Ribose nucleic acid, a nucleic acid that transcribes the basic genetic message, is present in small but measurable amounts.

The problem so far has been to determine precisely how these substances are associated in the chromosome. Available evidence now favors the following model.

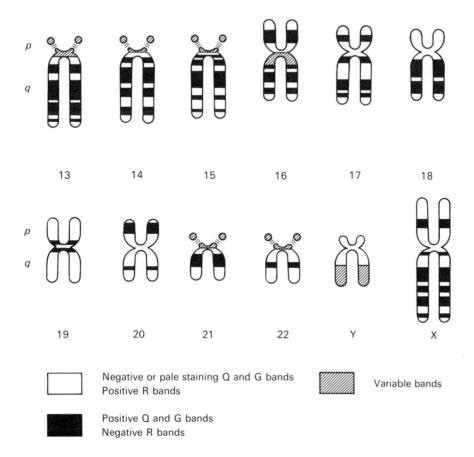

	Negative or pale staining Q and G bands Positive R bands		Variable bands
	Positive Q and G bands Negative R bands		

Histone and DNA associate together into a permanent structure called a *chromatin fiber*. A chromatin fiber contains a double helix of DNA surrounding a core of histone in repeating units (*nucleosomes*) of about 200 base pairs of DNA and two molecules each of four histones, known as H2A, H2B, H3, and H4 (Fig. 5.3). These eight histone molecules are believed to form a spherical unit. The DNA double helix passes around the histones in a manner not yet known. In many but not all cells, one molecule of another histone, H1, is also associated with each nucleosome. Figure 5.4 is an electron micrograph of a chromatin fiber showing the nucleosomes.

There is also a tendency for the chromatin fiber to form a coil about 250 A in diameter, which is barely within the resolving power of the best light microscope. These fine threads are called *euchromatin* to distinguish them from the darker, more compact, and denser staining *heterochromatin*. When these names were first proposed, it was believed that euchromatin and heterochromatin were different entities, rather than different configurations of the same material. However, the distinction is still valid as a descriptive characteristic

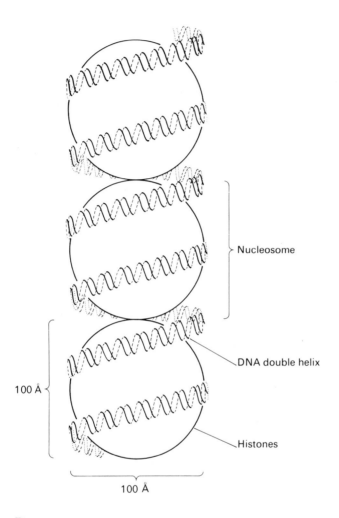

Figure 5.3

Chromatin structure (hypothetical). Chromatin consists of repeating units called nucleosomes. Each nucleosome contains eight histone molecules arranged in a sphere, surrounded by a double helix of DNA about 200 base pairs long. The configuration of the DNA is uncertain. (After R.D. Kornberg, 1977. Structure of chromatin. *Ann. Rev. Biochem.* 46:931-954)

of chromosomes, and there is ample evidence that heterochromatin has special genetic properties.

Prior to cell division, most chromatin condenses to form the chromosomes. When viewed under the electron microscope (Fig. 5.5) the chromosome can be seen to consist of tightly folded chromatin, where the larger 250 A units prevail. Although it is clear from the light microscope studies that the pattern of

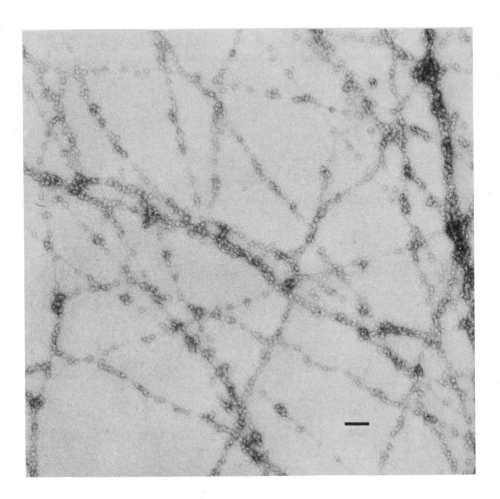

Figure 5.4

Chicken erythrocyte chromatin. The beads along the fiber are nucleosomes. The scale marker represents 500 A. (From C.L.F. Woodcock, J.P. Safer, and J.E. Stanchfield, 1976. Structural repeating units in chromatin. I. Evidence for their general occurrence. *Experimental Cell Research* 97:101–110)

aggregation is highly consistent, it is not clear how the fibers aggregate. Furthermore, the pattern clearly varies from species to species and sometimes from tissue to tissue (Fig. 5.1).

CHROMOSOME NUMBER

The number of chromosomes in the nucleus of the cells of the individuals of a species is constant and characteristic for that species.

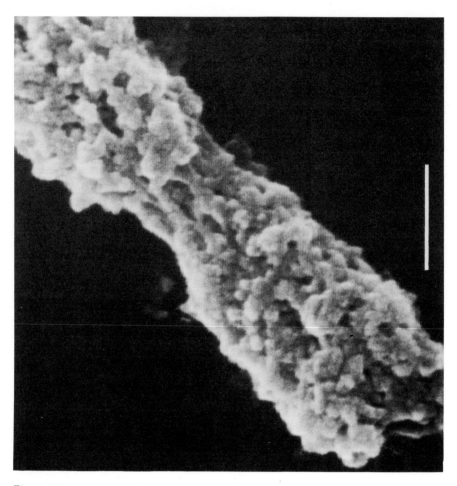

Figure 5.5

Scanning electron micrograph of a metaphase chromosome from a cell of a Chinese hamster. The middle constricted zone marks the location of the centromere. The white bar represents 1μ. (From D.I. Patt and G.R. Patt, 1975. *An Introduction to Modern Genetics.* Reading, Mass.: Addison-Wesley)

All the cells of an organism are derived from the product of the fusion of two gametes. That first cell is called a zygote. It has two sets of chromosomes. In each division of the zygote and its products, the chromosomes split longitudinally, and consequently all the cells of an organism have the same number and type of chromosomes. In certain tissues longitudinal division of the chromosomes takes place without the corresponding division of the cell. This produces cells with twice or even four times the number of chromosomes present in the zygote. In this process, called *endomitosis,* the number but not the morphology of the chromosomes is changed. Endomitosis is of physiological but not of genetic importance, since it does not occur regularly in the cells of tissues that lead to the formation of gametes. The number of chromosomes in somatic cells varies from a minimum of two pairs—found in one plant, *Haplopappus gracilis* of the sunflower family (Fig. 5.1), and in several insects and lower animals—to several hundred (Table 5.1). The majority of species of plants and animals have between five and 30 pairs of chromosomes. Human beings have 23 pairs (Fig. 5.1). One animal, *Ascaris megalocephala* var. *univalens*, has one chromosome in meiosis and in the germ line, but the chromosome fragments into several in the somatic line. Although the number of chromosomes is constant, occasional changes do occur in the gametes or in tissues that lead to the formation of gametes. These changes can occur in two main ways: by the division of the chromosomes in a cell that is going to give rise to the gametes without the division of the cell itself, yielding a cell with exactly double the original number of chromosomes; or by the fusion or breakage of individual chromosomes. The latter way gives rise to cells with one or two chromosomes in excess or in deficiency (Fig. 5.6). For a viable cell, in-

Table 5.1

Chromosome numbers of some common plants and animals.

Species	Common Name	Somatic Chromosome Number	Polyploid Level
Triticum aestivum	wheat	42	hexaploid
Zea mays	corn	20	diploid
Lactuca sativa	lettuce	18	diploid
Lycopersicum esculentum	tomato	12	diploid
Gossypium barbadense	New World cotton	52	tetraploid
Phaseolus vulgaris	kidney beans	22	diploid
Glycine max	soy beans	40	tetraploid
Pyrus communis	apple	34	diploid
Ipomea batatus	sweet potato	90	hexaploid
Ophioglossum vulgatum	fern	480	?
Apis mellifera	honey bee	16	diploid
Bombyx mori	silk worm	56	diploid
Cyprinus carpio	carp	100	diploid
Rana pipiens	green frog	26	diploid
Alligator mississippiensis	alligator	32	diploid
Anas platyrhynchos	mallard duck	78	diploid
Bos taurus	cattle	60	diploid
Equus caballus	horse	64	diploid
Felis felis	cat	38	diploid
Canis familiaris	dog	78	diploid

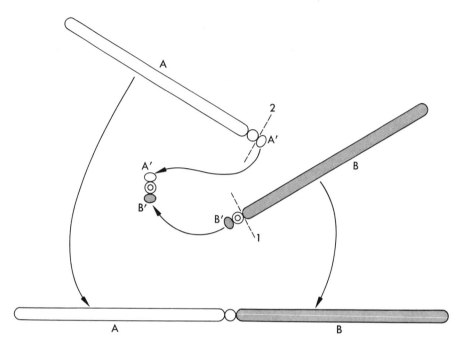

Figure 5.6

Chromosome number reduction. The two chromosomes, AA′ and BB′, each with terminal centromeres, fuse to form one chromosome, AB, with a median centromere, and what is essentially a single centromere with the two short arms A′B′. The latter can be lost with no genetic effect on the organism. This mechanism requires simultaneous breaks at (1) and (2). (From O.T. Solbrig, 1970. *Principles and Methods of Plant Biosystematics.* New York: Macmillan.)

creases and decreases in the number of chromosomes have to take place without any appreciable loss of genetic material. Loss of entire chromosomes is therefore deleterious unless the individual has already doubled its number of chromosomes. When two chromosomes fuse, one of the two centromeres has to be lost. Otherwise during division the chromosome would be attached to two spindle fibers (the spindle fibers are involved in the movement of the chromosomes during cell division), and this would have deleterious effects. Fusion occurs most commonly by the loss of the extremely short arms and portions of the centromere of two terminal chromosomes, which then unite by the remaining parts of the centromere to give rise to a median or submedian chromosome. The resultant chromosome is approximately as long as the sum of the lengths of the two original chromosomes. Species with a low number of chromosomes often have median or submedian chromosomes, indicating a reduction from ancestors with more chromosomes. The plant genus *Crepis* illustrates this very well. In this group of plants it has been possible to reconstruct the evolution of the *karyotype*, by which name the basic chromosome set of a species is designated. In this genus, species that from an evolutionary point of view are primitive have a basic chromosome number ($n = 6$) higher than related specialized forms, which have base numbers of 5, 4, and 3 (see Fig. 5.7 and Table 5.2).

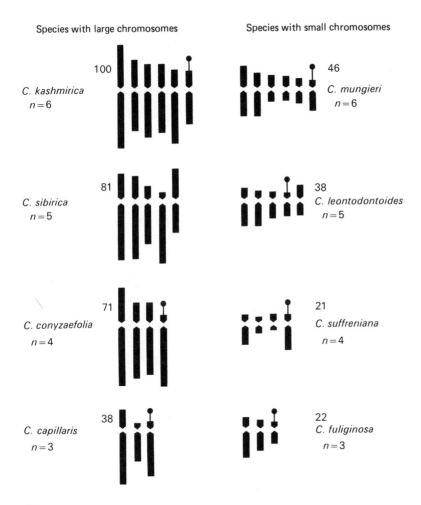

Figure 5.7

Evolution of the chromosomes in two lines of *Crepis*, one with large chromosomes and one with small chromosomes. The most primitive species in each line are at the top of the figure, the most advanced at the bottom. Numbers refer to the total length of all the chromosomes for each species, relative to *Crepis kashmirica* with 100. (From E.B. Babcock, 1947. *The Genus Crepis*, Part I. Berkeley: University of California Press, Botany series, Vol. 21)

Table 5.2

Chromosome numbers of selected species of *Crepis*.

Species	Somatic Chromosome Number	Polyploid Level	Reproduction Type	Habit
C. elegans	14	diploid	sexual	perennial
C. paludosa	12	diploid	sexual	perennial
C. sibirica	10	diploid	sexual	perennial
C. chrysantha	8	diploid	sexual	perennial
C. polytricha	16	tetraploid	sexual	perennial
C. ciliata	40	octoploid	sexual	biennial
C. alpina	10	diploid	sexual	annual
C. patula	8	diploid	sexual	annual
C. capillaris	6	diploid	sexual	annual
C. monticola	22–88	allopolyploid	asexual	perennial
C. intermedia	33–88	allopolyploid	asexual	perennial
C. barbigera	44–88	allopolyploid	asexual	perennial

Data from E.B. Babcock, 1947. *The Genus Crepis*, Part 1. Berkeley: University of California Press.

MITOSIS AND MEIOSIS

Cells are all of about the same minute size, and millions and billions of cells form the body of an animal the size of a horse or of a plant the size of an oak. Since many cells are relatively short-lived, they have to be replaced constantly. Nuclear division preceding cell division is a complex process known as *mitosis* (Fig. 5.8).

Cell division varies widely from species to species in a number of ways, and there are also some striking differences of detail between plants and animals. The essential process of mitosis is nevertheless basically similar in all organisms. During division of the nucleus a mechanism has to be provided for an exact distribution to the daughter cells of the particles of heredity, the genes. Although mitotic peculiarities can sometimes be used to identify groups of animals or plants, such as in species with diffuse centromeres, these instances are rare.

From a genetic point of view, the most important feature of mitosis is the exact copying of the chromosome and of its main constituents, so that both daughter nuclei receive the same amount of genetic information. This is necessary if genetic continuity is to be preserved in the species. The physicochemical properties of DNA provide a mechanism for the exact copying of the genetic content of the chromosomes.

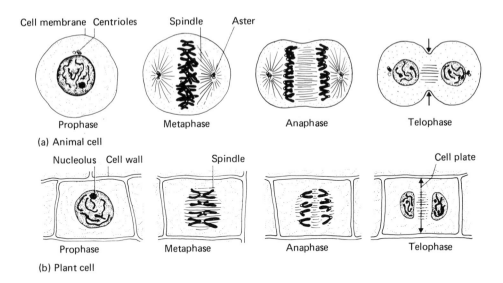

Cell membrane Centrioles   Spindle   Aster

Prophase   Metaphase   Anaphase   Telophase

(a) Animal cell

Nucleolus Cell wall   Spindle   Cell plate

Prophase   Metaphase   Anaphase   Telophase

(b) Plant cell

Figure 5.8

Mitosis, the process that provides for the distribution to daughter cells of the duplicated genetic material. In the early stages (prophase), chromatin condenses and becomes visible with the light microscope. By late prophase the chromosomes consist of two sister chromatids. In metaphase, the nuclear membrane disappears in most cells. A spindle appears, stretching between the two poles of the cell. The chromosomes line up at the equator of the cell. During anaphase, the centromeres divide and the sister chromatids separate and move to opposite poles. In telophase, the nuclear membrane reforms around each set of daughter chromosomes, the spindle disappears, and the chromosomes gradually uncoil. In most cells, mitosis is followed by cell division. (a) Mitosis in animal cells and (b) in plant cells. (From D.I. Patt and G.R. Patt, 1975. *An Introduction to Modern Genetics.* Reading, Mass.: Addison-Wesley)

Cell division provides for growth and maintenance of life in individual organisms. In unicellular organisms, cell division means increase in the number of individuals also. In multicellular organisms, special cells called gametes fuse to form a zygote. By mitosis and cell division the zygote gives rise to a new individual of the same species. Since during fertilization—the fusion of two gametes—the chromosomes of the two cells are added together, the number of chromosomes has to be halved before the next gametes are formed. If not, in a very few generations the number of chromosomes would be such that their volume alone would be greater than the volume of the nucleus and even greater than the volume of the cell itself. The process by which the number of chromosomes is restored to its original value is known as *meiosis*.

Meiosis consists of two consecutive divisions and results in the formation of four daughter nuclei (Fig. 5.9). However, it involves only one cycle of chro-

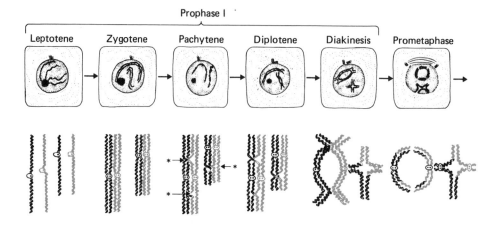

Figure 5.9

The stages of meiosis in a schematized animal cell. The physical states of the chromosomes and of their chromonemata in each stage of prophase I are diagrammed separately to show more clearly the phenomena of synapsis, recombination, and chiasma formation. Points of breakage and exchange are indicated by asterisks. (From D.I. Patt and G.R. Patt, 1975. *An Introduction to Modern Genetics*. Reading, Mass.: Addison-Wesley)

mosome duplication. Nuclear division is followed by cytoplasmic division in most organisms. Meiosis is a process that has far-reaching consequences for both heredity and evolution. From a genetic point of view the most important features of meiosis are the exact halving of the number of chromosomes, the independent assortment of the homologous chromosomes to the daughter nuclei, and the exchange of segments of homologous chromosomes through crossing over. *Homologous chromosomes* are those carrying alleles of the same genes. When two gametes fuse, for each chromosome from one gamete, there will be a homologue from the other gamete.

Meiotic prophase has been divided into the several stages shown in Fig. 5.9 because it lasts longer and is more complex than mitotic prophase. The first important feature separating the two processes is that when the chromosomes become visible, they are seen to be single structures. Unlike the chromosomes in mitotic prophase, they are not divided longitudinally into two *chromatids* (Fig. 5.9). A second and very important feature is that following their appearance, homologous chromosomes pair in a highly specific manner, so that centromeres and homologous segments of the chromosomes lie in juxtaposition. The diploid number of chromosomes is thus resolved into a *haploid* number of pairs of homologous chromosomes, called *bivalents*. While pairing is taking place, the chromosomes twist around each other and become shorter and thicker.

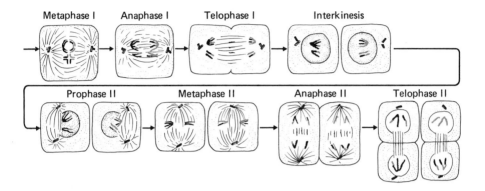

Some time after the process of pairing has been completed, the homologous members of each bivalent begin to separate. In the regions of separation, each homologue is now seen to be divided longitudinally into two chromatids. However, at one or more locations along the length of the bivalent, one chromatid from each homologue appears to have exchanged its partner so that two chromatids from opposite members of a bivalent lie criss-cross over each other (Fig. 5.10). Each of these cytologically visible points is known as a *chiasma* (plural, *chiasmata*). Chiasmata hold the homologous chromosomes together, and as chromosome separation proceeds, loops form between consecutive chiasmata, and half loops form between a chromosome end and its nearest chiasma. The process of separation forces the chiasmata to move toward the chromosome ends, a movement that is perhaps enhanced by the continuing coiling and contraction of the chromosomes. As prophase draws to a close, the centromeres of the homologous chromosomes of each bivalent orient themselves in positions equidistant above and below the spindle equator. In this way meiotic metaphase differs from mitotic metaphase, in which the centromeres of each chromosome align on the spindle equator (Fig. 5.8). At one point the association between sister chromatids lapses, the chiasmata slip apart, and each half of a bivalent moves polewards under the pull of its respective centromere.

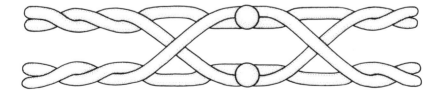

Figure 5.10

Paired chromosomes in meiosis. Frequently during prophase, a chromatid intertwines with a chromatid of the homologous chromosome. The points of contact are called chiasmata (singular, chiasma).

Since each bivalent contributes only one chromosome to each polar group, a haploid number of chromosomes collects at each pole. They differ from mitotic anaphase chromosomes, however, in that each chromosome consists of two chromatids. They shortly undergo a second division, mechanically similar to a mitotic division. Prior to the second division of meiosis, the chromosomes may form an interphase nucleus for a short period. The cytoplasm, too, may divide after the first meiotic division, and in plants even a cell wall may form. Alternatively, the second division of meiosis can follow immediately with no intervening interphase and either with or without a division of the cytoplasm.

The second meiotic division begins when all the chromosomes align themselves on the equator of the second division spindle. This step is followed by the simultaneous separation of the component chromatids of each chromosome. Four nuclei are thus produced by each meiotic sequence, and each nucleus contains a haploid number of single chromatids.

Meiotic division has been covered in great detail because Mendel's laws of segregation and independent assortment follow so directly from the behavior of the chromosomes at meiosis.

Consider an organism of the genetic constitution A_1A_2, each of the alleles inherited from a homozygous parent, one A_1A_1 and the other A_2A_2. Each allele will be carried in the exact same position on opposite homologous chromosomes. When the cell undergoes meiosis, homologous chromosomes separate during the first meiotic division, and they divide during the second meiotic division, giving rise to four haploid cells, two with the A_1 allele and two with the A_2 allele, corresponding to Mendel's principle of segregation.

If we now consider two pairs of alleles, A_1A_2 and B_1B_2, of two genes located on *different chromosomes*, we can see that because different bivalents arrange themselves on the metaphase I equatorial plate independently of each other, nonhomologous chromosomes undergo independent assortment during meiosis; that is, each of the four cells has equal probability of getting the paternal or maternal homologue. This is the physical basis of Mendel's second law of independent assortment.

However, not all the genes in a given *genome* (the full set of genes in an individual) assort independently, because those on the same chromosome are physically connected. However, the connection is not complete, because of the phenomenon of crossing over.

LINKAGE AND CROSSING OVER

It is essential that in each nuclear division, the daughter nuclei receive exact copies of the genetic complement of the mother nucleus. The aggregation of genes into chromosomes simplifies this process considerably. Indeed, if the more than 10,000 genes in a cell had to divide individually, and if each of the two halves had to move to the poles by itself, the traffic congestion thus created would undoubtedly result in unequal daughter nuclei. It is unlikely that complex multicellular organisms could have evolved under such circumstances. The stringing of the genes into chromosomes has made equal divisions of the chromosomes possible, but it has also had the effect of limiting the independent assortment of genes. In effect, at meiosis, all the genes in a chromosome have to move together to the same pole. This phenomenon, known as *linkage,* is of considerable genetic importance.

Genes in different chromosomes are not linked, whereas those in the same chromosome are. Given a certain number of genes for a particular species, it follows that the number of groups of mutually linked genes will be the same as the number of chromosomes of that species. The lower the number of linkage groups with a constant number of genes, the higher the number of genes in each linkage group, and the less operative the law of independent assortment. The phenomenon was first observed by two British geneticists, W. Bateson and R. C. Punnett, in 1906. When they crossed plants of sweet peas that were purple-flowered and had elongated ("long") pollen grains with plants that were red-flowered and had round pollen grains, they obtained plants that were all purple-flowered with elongated pollen grains, indicating that purple flowers and elongated pollen grains were dominant. In the F_2, they obtained the four combinations expected according to independent assortment of the genes (purple/long, purple/round, red/long, red/round), but not in the expected 9:3:3:1 ratio. Instead, the two parental combinations, purple/long and red/round, were present in excess over expectations. The proportion of the four classes was approximately 11:1:1:3. This surprising result contradicted Mendelian principles.

The answer to the riddle that Bateson and Punnett's experiments posed had to await work on the fruit fly *Drosophila* (1910–1915) by the American Nobel laureate geneticist Thomas H. Morgan and his students and close collaborators, Alfred Sturtevant, C. B. Bridges, and Hermann Müller. Bateson and Punnett had shown that the law of independent assortment did not apply to flower color and pollen shape in sweet peas. However, if the characters were on the same chromosome, only the parental types should appear. The presence of purple-flowered peas with round pollen and of red-flowered plants with elongated pollen is due to the exchange of corresponding segments of two chromatids of homologous chromosomes during meiosis. This phenomenon is called *crossing over* (Fig. 5.11), and its cytological manifestations are the chiasmata.

Figure 5.11 depicts crossing over. Note very specially that crossing over takes place between two chromatids at the four-strand stage of meiotic prophase and results at meiotic telophase in one chromosome each with the parental genetic combinations (A_1B_1 and A_2B_2) and two recombinant chromosomes (A_1B_2 and A_2B_1). Consequently, segments on each side of a crossover assort independently.

The number of chiasmata (and corresponding crossovers) that take place in each chromosomal arm is limited. It depends on a number of factors, such

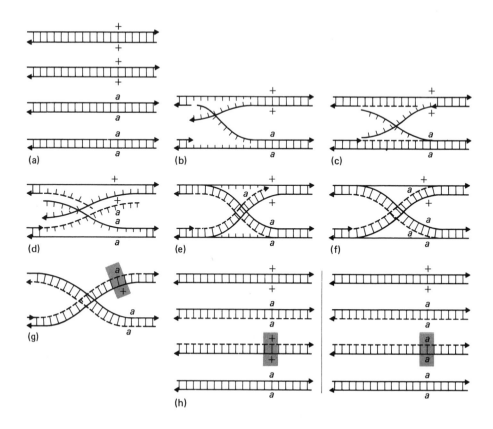

Figure 5.11

A breakage-synthesis-repair model to account for reciprocal and nonreciprocal crossing over. (a) DNA helices from two parents, representing tetrad formation. From this point until the last drawing in this figure, only the two central helices will be considered. The arrowheads indicate the polarity of the DNA strand in each segment of DNA. (b) Breakage of two of the DNA chains. (c) Nucleotide chains are synthesized along both unbroken chains. Newly made DNA is indicated by a dashed line. (d) Dissociation of newly made DNA strands from their templates. (e) Newly synthesized DNA associates with a strand from the other molecule. (f) Crossing over is completed. (g) The unpaired regions of the outer parental molecules are removed. Note the "heterozygous" region of DNA (shaded). This indicates a region of faulty base pairing. (h) The original, outer parental molecules of DNA are again shown, completing the tetrad after recombination has taken place. The "heterozygous" region is repaired. If the + strand is used as a template, reciprocal recombination is observed. If the *a* strand is used as a template for repair, nonreciprocal recombination is observed. (From D.I. Patt and G.R. Patt, 1975. *An Introduction to Modern Genetics.* Reading, Mass.: Addison-Wesley)

as the length of the chromosome, the genetic constitution of the species and individual, and the presence of other chiasmata, but it is usually on the order of one to three per arm. In some cases, such as in the males of *Drosophila*, no chiasmata occur, but such a situation is rare. Furthermore, the exact positions of the chiasmata vary from cell to cell during meiosis. Thus the probability of the occurrence of a chiasma between two genes is directly proportional to their physical separation along the chromosome and to the frequency with which crossovers occur in each chromosome.

This fact can be used to determine the order of the genes along a chromosome. In effect, since the frequency of crossovers between any two genes is proportional to their physical separation along the chromosomes, in crosses involving individuals heterozygous for three linked genes, the number of recombinants between them can be used to map their relative position on the chromosome (Fig. 5.12).

Linkage is a phenomenon that keeps genes on the same chromosome from assorting independently. As a result of crossing over, however, the proportion of chromosomes containing the various combinations of alleles at the two loci (A_1B_1, A_1B_2, A_2B_1, and A_2B_2) in the population will eventually coincide with the expected proportions under no linkage. However, although that equilibrium is reached in one generation of random breeding for nonlinked genes, the approach to equilibrium is slowed down considerably for linked genes. This may play an important role in evolution.

Linkage disequilibrium

Genes on the same chromosome cannot assort independently at meiosis because they are physically connected. However, crossing over breaks up these physically linked genes, so that in a population where selection is not operating, eventually the frequencies of the combinations of genes in the population correspond to what is expected by independent assortment. Let us look at why this is so and at the effect of linkage in greater detail.

Assume a population of organisms differing at two loci with two alleles at each locus, A_1, A_2, and B_1, B_2. Let us further assume that these two loci are on the same chromosome. There can be four genetically different chromosomes:

$$A_1B_1: \quad X_1$$
$$A_1B_2: \quad X_2$$
$$A_2B_1: \quad X_3$$
$$A_2B_2: \quad X_4$$

where X_1, X_2, X_3, and X_4 represent the frequencies of these chromosomes in the population. If we represent the frequency of each allele by p (A_1), q (A_2), s (B_1), and t (B_2), it follows that

$$p = X_1 + X_2, \qquad q = X_3 + X_4,$$
$$s = X_1 + X_3, \qquad t = X_2 + X_4.$$

Knowing the frequency of the four chromosomes, we can estimate the frequencies of the alleles. However, it is normally the allelic gene frequency that is known, not the chromosome frequency (which is equivalent to knowing the genotype frequency in this case, since X_1, X_2, X_3, and X_4 represent the four possible genotypes). The usual way to proceed in trying to estimate genotype frequency from allelic frequency is to multiply the value of the frequency of the first allele at the first gene by the value of the first allele at the second gene, thereby obtaining the value of the first genotype; and so on, allele by allele. This procedure assumes, however, that the frequencies of the alleles in the first locus are *independent* of the frequencies of the alleles in the second locus. For example, imagine a population of gametes where a frequency p of the gametes have $\mathbf{A}_1$ and q have $\mathbf{A}_2$, s have the allele $\mathbf{B}_1$ and t, $\mathbf{B}_2$. If we wish to know the proportion of gametes that have $\mathbf{A}_1\mathbf{B}_1$, we multiply q by s. This is always true under independent assortment. A corollary is that the product of the frequencies of chromosomes X_1X_4 minus the product of the frequencies of X_2X_3 must be zero ($X_1X_4 = ps \cdot qt$; $X_2X_3 = qs \cdot pt$).

However, in chromosomes where there is linkage, $X_1X_4 - X_2X_3$ will not be zero, because in cases of linkage certain gene combinations in a randomly breeding population are more frequent than expected according to independent assortment. Consequently, in cases of linkage the frequency of the chromosomes in the population cannot be calculated by simply multiplying the frequencies of the genes in them (that is, $X_1 = ps$). For example, if alleles $\mathbf{A}_1$ and $\mathbf{B}_1$ are completely linked in the population, all gametes having $\mathbf{A}_1$ will also have $\mathbf{B}_1$ rather than only a fraction, s. Therefore in cases of linkage, $X_1X_4 - X_2X_3$ will no longer be zero, because X_1 is no longer equal to the product ps, and similarly for the other chromosome types. The frequency X_1 will be equal to ps plus an excess or deficiency, which we will call *linkage disequilibrium* and will represent by the letter D. The value of D is given by the expression $X_1X_4 - X_2X_3$. Consequently, the chromosome frequency in a population with linkage as a function of gene frequency is

$$X_1 = ps + (X_1X_4 - X_2X_3) = ps + D,$$
$$X_2 = pt - (X_1X_4 - X_2X_3) = pt - D,$$
$$X_3 = qs - (X_1X_4 - X_2X_3) = qs - D,$$
$$X_4 = qt + (X_1X_4 - X_2X_3) = qt + D.$$

When $\mathbf{A}_1$ and $\mathbf{B}_1$ (as well as $\mathbf{A}_2$ and $\mathbf{B}_2$) are linked, a condition known as "coupled," D is positive. If $\mathbf{A}_1$ and $\mathbf{B}_2$ (as well as $\mathbf{A}_2$ and $\mathbf{B}_1$) are linked, a situation known as "repulsion," then the value of D is negative. To demonstrate these equations, we substitute for ps and D their chromosomal equivalents. For example,

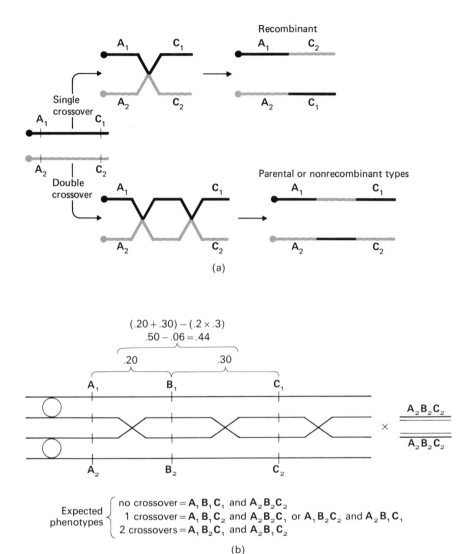

$$(.20 + .30) - (.2 \times .3)$$
$$.50 - .06 = .44$$

Expected phenotypes
{
no crossover = $A_1 B_1 C_1$ and $A_2 B_2 C_2$
1 crossover = $A_1 B_1 C_2$ and $A_2 B_2 C_1$ or $A_1 B_2 C_2$ and $A_2 B_1 C_1$
2 crossovers = $A_1 B_2 C_1$ and $A_2 B_1 C_2$
}

(b)

Figure 5.12

(a) Effect of single and double crossover events of two widely spaced loci. The single exchange will lead to a different and detectable gene arrangement, whereas the product of the double exchange is identical to the parental arrangement. Assume that A_1 and A_2 differ phenotypically; C_1 and C_2 are likewise different in their actions. (From D.I. Patt and G.R. Patt, 1975. *An Introduction to Modern Genetics.* Reading, Mass.: Addison-Wesley) (b) Establishment of the position of genes on the chromosome. A heterozygous individual is crossed to a homozygous recessive. The probability of a crossover increases with distance. If genes **A** and **B** show 20 percent crossovers, and **B** and **C** show 30 percent, then **A** and **C** show 50 percent if the position is **ABC**, but 10 percent if the position is **ACB**. In actuality, the percentage of crossovers between the end genes is less than the sum of the distances of **A** to **B** plus **B** to **C** because of double crossovers (in this case 6 percent).

$$ps + D = (X_1 + X_2)(X_1 + X_3) + (X_1X_4 - X_2X_3)$$
$$= X_1^2 + X_1X_3 + X_1X_2 + X_2X_3 + X_1X_4 - X_2X_3,$$

where $+X_2X_3$ and $-X_2X_3$ cancel each other, so that rearranging, we have

$$ps + D = X_1^2 + X_1X_2 + X_1X_3 + X_1X_4$$
$$= X_1 (X_1 + X_2 + X_3 + X_4).$$

The value within the parentheses is the sum of the frequencies of all possible chromosomes, which is 1, so that

$$ps + D = X_1,$$

which demonstrates the first equation. The other four are demonstrated in similar fashion.

In a diploid population where individuals differ at two loci with two alleles at each locus, there can be a total of nine genotypes. If the loci are linked, the genotype frequency cannot be predicted from gene frequencies under the assumptions of the Hardy-Weinberg theorem. Instead, the frequency of each *chromosome* (or gamete) has to be calculated, taking crossovers into account.

To predict the frequency of each chromosome in the next generation, we first enumerate the various ways by which the four kinds of chromosomes, A_1B_1, A_1B_2, A_2B_1, and A_2B_2, can be produced in the population. Starting with chromosome A_1B_1, we note that it can be generated in two different ways. First, it can be obtained from genotypes of the composition $A_1B_1/**$ (where * denotes any arbitrary allele), provided no crossover occurs between A_1 and B_1. If the recombination probability (probability of a crossover) is denoted by r, the probability of obtaining an X_1 chromosome from an $A_1B_1/**$ genotype is $1 - r$. A second way of obtaining an X_1 chromosome is from an $A_1*/*B_1$ genotype where a crossover does take place. The frequency of $A_1B_1/**$ genotypes in the population is equal to the frequency of the A_1B_1 chromosome, X_1. The frequency of $A_1*/*B_1$ genotypes is obviously the product of the gene frequencies, $p \times s$. Consequently, the frequency X_1 of the chromosome A_1B_1 in generation $t + 1$ is

$$X_{1_{t+1}} = X_{1_t} (1 - r) + ps(r), \tag{1}$$

and similarly,

$$X_{2_{t+1}} = X_{2_t} (1 - r) + pt(r),$$
$$X_{3_{t+1}} = X_{3_t} (1 - r) + qs(r),$$
$$X_{4_{t+1}} = X_{4_t} (1 - r) + qt(r).$$

But we know that $p = X_1 + X_2$ and $s = X_1 + X_3$, and therefore

$$ps = (X_1 + X_2)(X_1 + X_3) = X_1^2 + X_1X_2 + X_1X_3 + X_2X_3$$
$$= X_1 (X_1 + X_2 + X_3) + X_2X_3$$
$$= X_1 (1 - X_4) + X_2X_3.$$

Replacing in (1), we have

$$X_{1_{t+1}} = X_{1_t} (1 - r) + r [X_{1_t} (1 - X_{4_t}) + X_{2_t} X_{3_t}]$$
$$= X_{1_t} - rX_{1_t} + rX_{1_t} - rX_{1_t} X_{4_t} + rX_{2_t} X_{3_t}$$
$$= X_{1_t} - r_t (X_{1_t} X_{4_t} - X_{2_t} X_{3_t}).$$

But we saw that $X_1 X_4 - X_2 X_3 = D$, so that

$$X_{1_{t+1}} = X_{1_t} - rD_t, \tag{2}$$

and similarly,

$$X_{2_{t+1}} = X_{2_t} + rD_t,$$
$$X_{3_{t+1}} = X_{3_t} + rD_t,$$
$$X_{4_{t+1}} = X_{4_t} - rD_t.$$

That is, the frequency of the chromosome in generation $t + 1$ is equal to its frequency in the previous generation minus the recombination index between the two loci times the value of the linkage disequilibrium in generation t.

Since $X_{1_t} = ps + D_t$, and since $X_{1_{t+1}} = ps + D_{t+1}$, replacing in (2), we have

$$ps + D_{t+1} = (ps + D_t) - rD_t,$$
$$D_{t+1} = ps - ps + D_t - rD_t = D_t (1 - r),$$

and

$$D_t = D_{t-1} (1 - r),$$
$$D_{t+1} = D_{t-1} (1 - r) (1 - r) = D_{t-1} (1 - r)^2,$$

so that

$$D_{t+1} = D_0 (1 - r)^t,$$

where D_0 is the initial value of linkage disequilibrium in the population. Consequently, linkage disequilibrium decreases in the population at rate r, provided there is no selection and breeding is random. Therefore the expectation is that normally linkage disequilibrium will not be an important evolutionary force. For linkage disequilibrium to become a significant factor, *epistatic selection*, which is positive selection for a given gene combination, is required.

Evidence for linkage disequilibrium in nature was obtained in studies of population of species of *Drosophila*, primarily *D. pseudoobscura*. For example, Prakash and Lewontin showed that in the third chromosome of this species, the allele 1.04 of the gene **Pt-10** was always present with the so-called standard inversion, whereas three other alleles (0.94, 1.02, 1.06) of the same gene never were. T. Preston Webster found evidence for linkage disequilibrium in a salamander, *Plethodon cinereus*, that inhabits the forests of the eastern United States. In populations of this animal, the alleles of two esterase loci are in linkage disequilibrium. Among plants there is some evidence for linkage disequilibrium in barley and oats. There are also known cases in humans (Rh factor) and snails. The dearth of examples of gene disequilibrium in nature does not necessarily mean that it does not occur frequently, but rather reflects the technical problems involved in studying this phenomenon. Linkage disequilibrium is expected to occur most frequently between neighboring genes on a chromosome. At present, however, there is no systematic way of determining all the genes in a chromosomal sector, other than by a process of mapping each gene of the species. But homozygous genes are not identified this way, nor are those that have small effects on the phenotype. Consequently, many genes are missed, and at present we have not mapped all the genes of any species, and it is unlikely that we ever will do so. The precise assessment of the relative frequency of linkage disequilibrium in nature has to await the development of new techniques.

Linkage is a phenomenon that controls the rate at which nonallelic genes on the same chromosome will mix with other such genes in homologous chromosomes coming from different parents. If linkage is absolute, only the parental types will be produced, and no new combinations will be present. On the other hand, if there is no linkage, there will be random assortment, and the chances of the parental combinations surviving will be very low. In situations where parental combinations are very successful, natural selection will favor mechanisms that increase linkage. On the other hand, when variable offspring are advantageous, natural selection will favor less linkage and an increase in the number of recombinant types. Some of the mechanisms that increase linkage are a lowering of the rate of crossing over and a lower number of chromosomes. This last is the result of random transverse breakages and reunions of the broken ends of one or more chromosomes in novel ways, loss of centromeres, and selection of the new arrangements. How these particular changes come about we do not know exactly. But when they take place, they will be favored (and maintained) by natural selection in the populations where less recombination is advantageous, and they will be eliminated in the populations where less recombination is not desirable. The selective mechanisms and examples are presented in Chapter 8.

CHANGES IN CHROMOSOMAL STRUCTURE

The position of genes on a chromosome can be altered as a consequence of changes in the structure of the chromosome. Such changes can result in a modification of the number of chromosomes when the structural change produces gains or losses of centromeres, or they can take place without altering the chromosome number. In both cases the linkage relation of the genes in the modified chromosomes is affected. Some changes also result in an increase or loss of genes.

Changes in the morphology of the chromosome are a reflection of internal arrangements of the chromosomes. These changes are collectively referred to as *chromosomal aberrations*, although they are not aberrations in the usual sense of the word. Rather, they are phenomena that occur with a low frequency in most populations. The principal ones, shown in Fig. 5.13, are as follows:

Duplications. A part of the chromosome is present in double dose. The doubling probably occurs as a result of errors in replication.

Deficiencies. A part of the chromosome is missing. X-rays, ultraviolet, and other forms of high-energy radiation create deficiencies. All but very small deficiencies are lethal.

Shifts. A part of the chromosome is moved to another location as a result of breaks followed by fusion of the broken parts in an order different from the original one.

Inversions. Similar to shifts, inversions differ in that the segments are inverted in order without a shift. They are called *pericentric* when the centromere is included in the inverted segment and *paracentric* when the centromere is not included.

Translocations. A segment of a chromosome is shifted to another chromosome in a translocation. If a segment of chromosome A is moved to chromosome B and a segment of chromosome B is moved to A, the translocation is *reciprocal*. The segments need not be of the same size.

Chromosomal aberrations, especially translocation, can affect the length and aspect of the chromosome. Their effects are then observable in mitosis.

Normally, however, the existence of aberrations is deduced from the meiotic behavior of hybrids for the aberration. Figure 5.14 shows how the corresponding configurations look at pachytene. Chromosomal rearrangements play important evolutionary roles. An example is the duplications that gave rise to the various hemoglobin genes in higher primates.

Duplications can be divided into three groups. First, there are single gene duplications, in which a particular gene occurs in many copies, apparently in order to enable the cell to produce a large amount of a given substance at one time, such as ribosomal RNA. The second group consists of longer sequences of DNA on the order of 400 to 2000 nucleotide pairs, each repeated hundreds, thousands, even millions of times. Apparently these repetitive sequences of DNA, sometimes also called *redundant DNA,* are involved in the regulation of gene activity. Finally, there are occasional duplications that lead to the evolution of new biochemical functions. The best-known example of the last type of duplication is the evolution of the *globins,* proteins that are involved in oxygen transport: myoglobins in muscle tissue and hemoglobins in blood.

Myoglobin is a relatively simple protein, consisting of a single polypeptide chain and one single heme group (in this case a protoporphyrin ring with an iron atom), with a molecular weight of approximately 17,000. All vertebrates, with the exception of the very primitive jawless fishes, have a very similar myoglobin molecule. Consequently, all vertebrate myoglobins are thought to

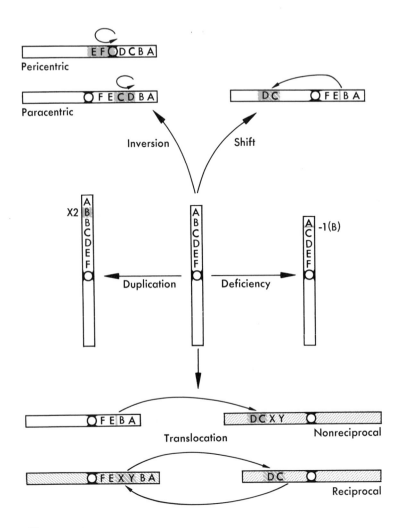

Figure 5.13

The various kinds of chromosomal aberrations and their effects on the morphology of the chromosomes. The size of the segment that is involved determines the magnitude of the change. (From O.T. Solbrig, 1970. *Principles and Methods of Plant Biosystematics*. New York: Macmillan)

be descendants with modification from a single ancestor gene. The *hemoglobins* are more complex. They consist of four polypeptide chains, each with a heme group, with a molecular weight of about 67,000, approximately four times the size of myoglobin. Hemoglobins are more efficient in unloading their oxygen than are myoglobins because of the interaction between the four hemes

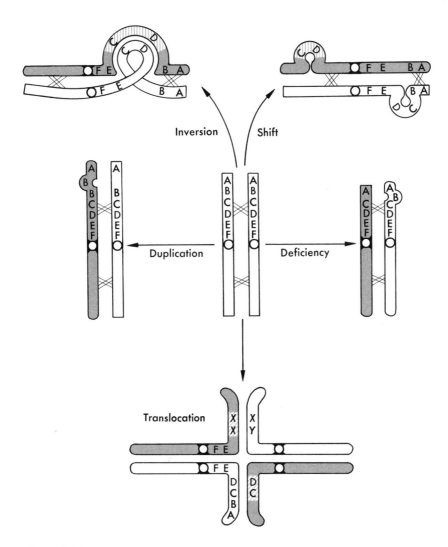

Figure 5.14

Effects of the various chromosomal aberrations on pairing of chromosomes at meiosis when in a heterozygous condition. (From O.T. Solbrig, 1970. *Principles and Methods of Plant Biosystematics.* New York: Macmillan)

in the molecule. Furthermore, the four chains of the hemoglobin molecule are not the same. In most vertebrates there are two each of two types of polypeptide chains, called α and β. In humans and in higher primates we find two types of hemoglobins: hemoglobin A_1, consisting of two alpha and two beta chains; and hemoglobin A_2, consisting of two alpha and two delta chains.

Finally, in the human embryo there is a third hemoglobin, fetal hemoglobin, consisting of two alpha and two gamma chains. Studies by a number of researchers, such as Fitch, Goodman, Margoliash, Moore, Pauling, and Zuckerkandl, have established that the alpha, beta, gamma, and delta chains are coded by different genes. They have also established that although these polypeptide chains differ in a number of amino acids, the basic sequence is such that they could have been derived from a common ancestor. This finding led the researchers to propose the following evolutionary scheme for the globins of the higher vertebrates (Figs. 5.15 and 5.16).

The ancestral gene was the one that coded for myoglobin. Some 500 million years ago it was duplicated, and there followed mutations that gave rise to a primitive beta chain that could aggregate into a primitive hemoglobin chain made up of four beta chains. The next step, around 380 million years ago, was a duplication of the primitive beta hemoglobin gene, followed by mutation that gave rise to a gene coding for the alpha chain. This allowed the formation of the more efficient A type hemoglobins ($\alpha_2\beta_2$). The next step took place some 150 million years ago, when the beta gene was duplicated once more, eventually giving rise to the gene that codes for the gamma chain. One more dupli-

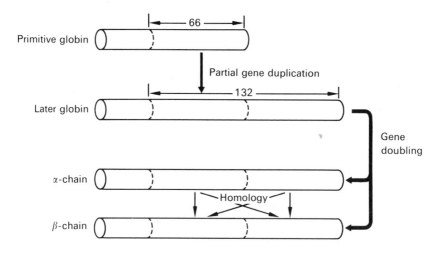

Figure 5.15

Evolution of the hemoglobins. Many amino acid sequences are the same in the α and β chains of hemoglobin. There are also amino acid sequences in the early part of the α chain that correlate strongly with sequences 66 amino acids further on in the β chain, and vice versa. From this finding it is deduced that a segment 66 amino acids long was duplicated in an ancestral molecule, and both the α and β chains are derived from that ancestor. (With permission from W.M. Fitch, 1966. Evidence suggesting a partial, internal duplication in the ancestral gene for heme-containing globins. *J. Mol. Biol.* 16:17–27. Copyright by Academic Press Inc. (London) Ltd.)

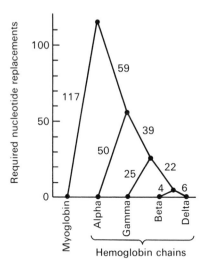

Figure 5.16

Evolution of globin molecules. The numbers represent nucleotide replacements necessary to derive the five present-day nucleotide sequences from common ancestors.
Myoglobin diverged earliest, with 117 substitutions. The different kinds of hemoglobin chains diverged later. An ancestral globin, from which all the present hemoglobin chains could be derived, is 59 substitutions from the molecule ancestral to
all the globins. (From W.M. Fitch and E. Margoliash, 1970. The usefulness of amino
acid and nucleotide sequences in evolutionary studies. *Evol. Biol.* 4:67–109)

cation occurred some 35 million years ago in the ancestral lineage of the higher
primates, giving rise to the gene coding for the delta chain. The globins illustrate how new genes (and new functions) evolve from ancestral ones. In this
case the ancestral molecule, myoglobin, is still being produced. In other cases
the original molecule probably got lost as a result of a deficiency, although
such an eventuality is difficult to demonstrate.

HAPLOIDY, DIPLOIDY, AND POLYPLOIDY

The first cell of the organism, the zygote, which is the product of the fusion of
two gametes, has two sets of chromosomes. Each is furnished by one of the
parents, and consequently the zygote is diploid. But not all organisms are
diploid. In some cases, particularly among plants, the spores produced at
meiosis form an organism without previous fusion. Such organisms, having
only one set of chromosomes, are haploid. Finally, because of accidents of development, a gamete, the zygote, or a cell produced by the zygote may double
its number of chromosomes. The resulting organisms will have more than two
sets of chromosomes and will be called *triploid, tetraploid, pentaploid, hexa-*

ploid, and so on, according to the number of sets of chromosomes present; a general term for them is *polyploids*.

Polyploidy is rare among most animals, but it is very frequent in plants, for which it has played an important evolutionary role. A plant with exactly twice the number of chromosomes of its progenitors is qualitatively like them, since no new genetic information has been added. But the additional sets of chromosomes usually have the effect of increasing the size of the cells, and at the same time of decreasing the total number of cells. In a tetraploid the first effect normally outweighs the second, and the plants are usually larger (a result that is of immediate application in horticulture, e.g., strawberries). With further doubling of the chromosomes, the net effect is often a decrease in the size of the plants.

Polyploids produced artificially by chemical treatment are often sterile. Since in tetraploids there are four homologous chromosomes, at meiosis they tend to pair together, and instead of bivalents, tetravalents are formed. These do not necessarily separate to the poles two by two, and the result is the formation of gametes that lack some chromosomes and have an excess of others (Fig. 5.17).

There is nevertheless a special kind of polyploid that is fertile. If two distantly related species cross, their chromosomes may not form bivalents at meiosis. Instead, *univalents* are formed (unpaired chromosomes that do not undergo regular meiosis), and consequently the hybrid plant will be sterile. If in such a plant the chromosomes are doubled, fertility is restored, since now

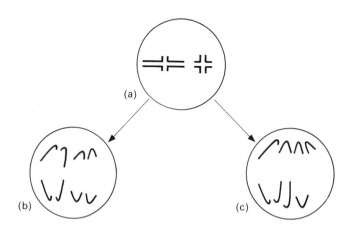

Figure 5.17

Meiosis in tetraploids. Tetraploids have four of each kind of chromosome and tend to group in fours at meiosis (a). When sister chromatids separate, they may migrate three to one pole and one to the other (c), instead of the expected two and two (b).

there will be two of each type of chromosome, and they will pair with each other at meiotic prophase. Such a hybrid is called an *allopolyploid*, to distinguish it from the former type, which is an *autopolyploid*. An example of an allopolyploid is the tobacco plant, *Nicotiana tabacum*. Tobacco is the result of the doubling of the chromosomes of the natural hybrid between *Nicotiana otophora* and *N. sylvestris*. This cross occurred naturally some hundreds or maybe thousands of years ago, but it has been successfully repeated under artificial conditions (Fig. 5.18).

A third type of polyploid is an intermediate between the two just noted. Some of the chromosomes of the two parents are sufficiently related so that they will pair, but others are not. This is a *segmental allopolyploid* (Fig. 5.19). Its fate depends largely on whether the allo- or the auto-components prevail (Fig. 5.20). Most natural polyploids are either allopolyploids or segmental allopolyploids. A few natural autopolyploids are also known.

Polyploidy is a mechanism by which new species can be formed. We will discuss it again in Chapter 11. Polyploidy also affects the pattern of inheritance.

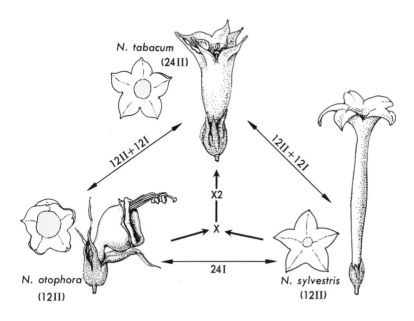

Figure 5.18

Probable origin of the tobacco plant, *Nicotiana tabacum*, as a result of hybridization between *N. otophora* and *N. sylvestris*, followed by chromosome doubling. (Data from T.H. Goodspeed, 1954. *The Genus Nicotiana*. Waltham, Mass.: Chronica Botanica)

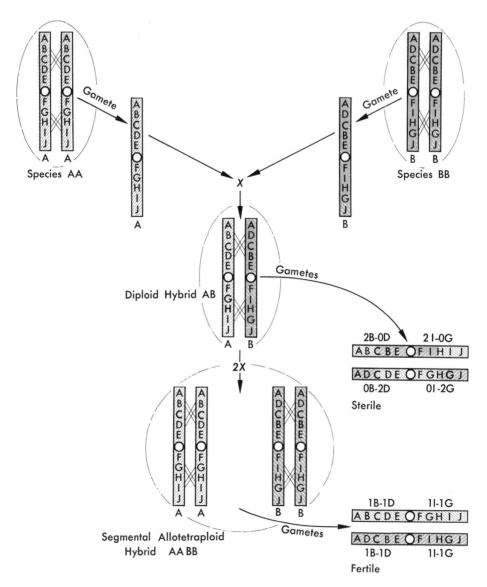

Figure 5.19

Segmental allopolyploid. The cross between two species, A and B, produces a diploid hybrid with good bivalent pairing, which nevertheless is sterile. The sterility can be explained as owing to small chromosomal inversions, which do not interfere with pairing but lead to duplicated and deficient chromosomal segments in the gametes when crossover takes place. If the number of chromosomes is doubled, however, pairing takes place preferentially among truly homologous chromosomes (those contributed by the same species). Pairing is again bivalent pairing, and the gametes are normal and consequently fertile. (From O.T. Solbrig, 1970. *Principles and Methods of Plant Biosystematics.* New York: Macmillan)

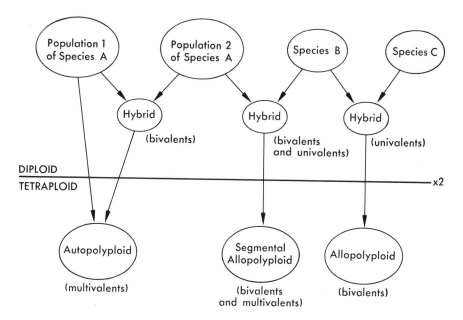

Figure 5.20

The various types of polyploidy. Autopolyploidy results from the doubling of the chromosomes of an organism. Because autopolyploids tend to form associations of four chromosomes at meiosis, they are expected to be sterile. Allopolypoids result from the doubling of the chromosomes of an interspecific hybrid between two relatively unrelated species. The diploid hybrid is sterile because the chromosomes of the two species are nonhomologous and will not pair; however, when the chromosomes are doubled, bivalents (sets of two) can again be formed, and the allopolyploid will be fertile. A segmental allopolyploid is halfway between an autopolyploid and an allopolyploid, in that the chromosomes of the two species are related enough to form bivalents in the diploid hybrid, but unrelated enough to form bivalents at the tetraploid level. (From O.T. Solbrig, 1970. *Principles and Methods of Plant Biosystematics.* New York: Macmillan)

Many of our crop plants are polyploid, for example, wheat, cotton, tobacco, and members of the cabbage family. The origin of their genomes can be traced by special genetic and cytological techniques. This knowledge enables the plant breeder to transfer to a cultivated species genes from a related wild species. An example of a superior polyploid is the cultivated strawberry.

The strawberry has been known in Europe probably as long as man has occupied that area. The European species, *Fragaria vesca,* is a diploid species with very small fruits. After the discovery of America, the New World *Fragaria chiloensis* (from Chile) and *Fragaria virginiana* (from Eastern North America) were introduced to Europe. These species are *octoploid* (eight sets of chromosomes) and have slightly larger fruits. These two species (and possibly

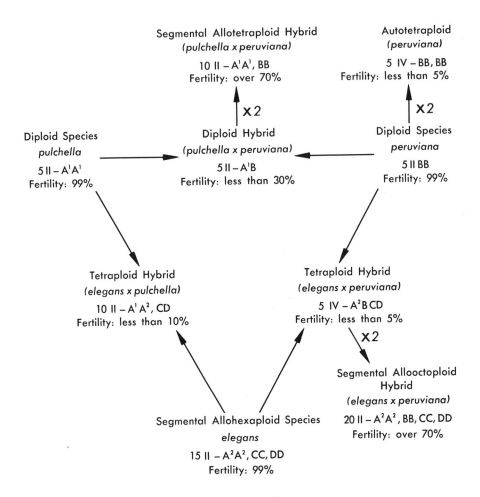

Figure 5.21

Polyploidy in *Glandularia*. Crosses were made between two diploid and one hexaploid (six sets of chromosomes) species of *Glandularia*. Artificial polyploids were produced by treating certain plants with colchicine. The pollen fertility, chromosomal pairing relationships, and inferred genomes are indicated. Note that fertility is restored when a segmental allopolyploid hybrid is produced, but reduced when an autopolyploid is produced. (From O.T. Solbrig, C. Passani, and R. Glass, 1968. Artificial hybridization between different polyploid levels in *Glandularia* (Verbenaceae). *Amer. J. Bot.* 55:1235–1239)

others) were hybridized, and varieties with larger fruits were selected, giving rise to the cultivated form *F.* x *ananassa* with fruits that are more than 10 times larger than those of the original species and that have other desirable horticultural traits as well. For another example, illustrated in Fig. 5.21, see the wild relatives of *Glandularia*, the garden verbena.

MENDELIAN INHERITANCE IN TETRAPLOIDS

Inheritance in tetraploids is complicated by the fact that each locus is represented four times in a somatic cell nucleus instead of twice. Consequently, when dealing with a one-locus–two-allele situation, there are five possible genotypes in a tetraploid ($A_1A_1A_1A_1$, $A_1A_1A_1A_2$, $A_1A_1A_2A_2$, $A_1A_2A_2A_2$, $A_2A_2A_2A_2$), rather than the three we are familiar with in a diploid. The same applies to dominance relations. Instead of one heterozygote, there are three. These can be identical to one of the homozygotes when there is absolute dominance, or there may be a quantitative effect, so that there is a gradient in phenotypic expression rather than five clear groups.

In describing Mendelian inheritance in tetraploids, we first have to consider the type of gametes that heterozygous genotypes produce. To improve communication, we will refer to the homozygous dominant genotype $A_1A_1A_1A_1$ as *quadruplex*, to the first heterozygote $A_1A_1A_1A_2$ as *triplex*, the second $A_1A_1A_2A_2$ as *duplex*, the third $A_1A_2A_2A_2$ as *simplex*, and the homozygote recessive as *nulliplex*.

A quadruplex and a nulliplex produce only one kind of gamete. A triplex, on the other hand, produces two kinds of gametes, A_1A_1 and A_1A_2. (Remember that in a tetraploid the gametes are diploid!) The gametes are produced in a 1:1 ratio as Table 5.3 shows. In a duplex there will be three kinds of gametes, A_1A_1, A_1A_2, and A_2A_2 in a 1:4:1 ratio (Table 5.4). In a simplex there will be two kinds of gametes, A_1A_2 and A_2A_2, in a 1:1 ratio (Table 5.5).

Table 5.3

Gametes produced by a triplex parent.

	Parental Genotype		
	A_1	A_1	A_1	A_2
A_1	x	x	x	x
A_1	A_1A_1	x	x	x
A_1	A_1A_1	A_1A_1	x	x
A_2	A_1A_2	A_1A_2	A_1A_2	x

Table 5.4

Gametes produced by a duplex parent.

	Parental Genotype		
	A_1	A_1	A_2	A_2
A_1	x	x	x	x
A_1	A_1A_1	x	x	x
A_2	A_1A_2	A_1A_2	x	x
A_2	A_1A_2	A_1A_2	A_2A_2	x

Table 5.5

Gametes produced by a simplex parent.

	Parental Genotype			
	A_1	A_2	A_2	A_2
A_1	x	x	x	x
A_2	A_1A_2	x	x	x
A_2	A_1A_2	A_2A_2	x	x
A_2	A_1A_2	A_2A_2	A_2A_2	x

Table 5.6

Progeny of a $A_1A_1A_2A_2 \times A_1A_1A_2A_2$ cross
($A_1A_1A_2A_2 : ♀$, $A_1A_1A_2A_2 : ♂$)

		Female Gametes		
		$1A_1A_1$	$4A_1A_2$	$1A_2A_2$
Male	$1A_1A_1$	$1A_1A_1A_1A_1$	$4A_1A_1A_1A_2$	$1A_1A_1A_2A_2$
Gametes	$4A_1A_2$	$4A_1A_1A_1A_2$	$16A_1A_1A_2A_2$	$4A_1A_2A_2A_2$
	$1A_2A_2$	$1A_1A_1A_2A_2$	$4A_1A_2A_2A_2$	$1A_2A_2A_2A_2$

Knowing the gametic ratio, we can easily deduce the genotype ratios of the progeny. So, for example, if we cross two homozygous tetraploids, $A_1A_1A_1A_1$ and $A_2A_2A_2A_2$ we can obtain a heterozygous duplex, $A_1A_1A_2A_2$. If we now self-fertilize it or cross it to a sibling, we obtain the F_2 given in Table 5.6. The ratio of the various genotypes obtained is as follows:

$$A_1A_1A_1A_1 = 1$$

$$A_1A_1A_1A_2 = 8$$

$$A_1A_1A_2A_2 = 18$$

$$A_1A_2A_2A_2 = 8$$

$$A_2A_2A_2A_2 = 1$$

In the corresponding diploid cross, the result would have been $1A_1A_1$, $2A_1A_2$, $1A_2A_2$. We see, then, that the proportion of heterozygotes has increased and that of the homozygotes has decreased in relation to the corresponding diploid. If we assume no dominance and intermediate phenotypes for the heterozygotes, and we graph them along an arbitrary phenotypic axis, we will discover that the proportion of homozygotes in the progeny of the tetraploid has decreased dramatically (Fig. 5.22).

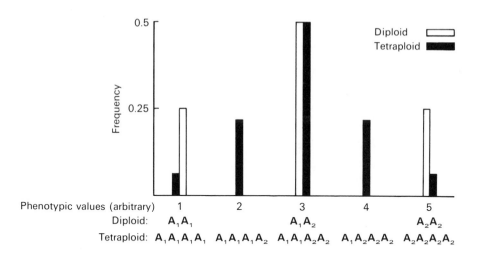

Figure 5.22

Distribution of offspring of diploid and tetraploid from a heterozygous cross. Imagine a locus **A** where the phenotypic value of individuals homozygous for A_1 is 1, and that of individuals homozygous for A_2 is 5. Heterozygous individuals have intermediate phenotypes, the value depending on the proportion of A_1 and A_2 alleles. In a cross between two heterozygous diploids (A_1A_2), three kinds of offspring are produced (white bars). In a cross between two $A_1A_1A_2A_2$ tetraploids, five kinds of offspring are produced (black bars), and the proportion of homozygotes is greatly reduced.

SUMMARY

Chromosomal cytology provides the physical foundation for the operation of the Mendelian laws of inheritance and the phenomenon of linkage. These laws are universal in their broad predictions. However, there is considerable diversity among species in details of chromosome characteristics. Even closely related species may have quite disparate chromosome lengths and may differ in chromosome number by a factor of two or three (if one species is polyploid). These differences in chromosome characteristics mean that differences exist in the pattern of inheritance, as we saw in the case of inheritance in polyploids as compared with diploids, and in the different degrees of linkage. There are advantages and disadvantages to each of these variations, and the variations are subject to natural selection, which is considered in the next chapter.

SUGGESTED FURTHER READING

General references

Dawson, G.W.P. 1962. *An Introduction to the Cytogenetics of Polyploids.* Oxford: Blackwell Scientific Publications.

Dyson, R.D. 1978. *Cell Biology: A Molecular Approach*, 2nd ed. Boston: Allyn and Bacon.

Lewis, K.R., and B. John. 1963. *Chromosome Marker*. London: Churchill.

Swanson, C.P. 1957. *Cytology and Cytogenetics*. Englewood Cliffs, N.J.: Prentice-Hall.

White, M.J.D. 1973. *Animal Cytology and Evolution*, 3rd ed. Cambridge: Cambridge University Press.

Specialized articles and reviews

Babcock, E.B. 1947. *The Genus Crepis*, Part 1. Berkeley: University of California Press, Botany series, vol. 21. (chromosome evolution)

Bateson, W., and R.C. Punnett. 1906. Experimental studies in the physiology of heredity. Reports to the Evolution Committee of the Royal Society (London). Reports 2, 3, 4. (linkage)

Casperson, T., and L. Zech (eds.). 1973. Chromosome identification: technique and application in biology and medicine (Nobel Symposium 23). Stockholm: Nobel Foundation. (chromosome staining)

Fitch, W.M. 1966. Evidence suggesting a partial, internal duplication in the ancestral gene for heme-containing globins. *J. Mol. Biol.* 16:17–27. (evolution of globins)

Fitch, W.M. 1973. Aspects of molecular evolution. *Ann. Rev. Genetics* 7:343–380. (protein evolution)

Fitch, W.M., and E. Margoliash. 1970. The usefulness of amino acid and nucleotide sequences in evolutionary studies. *Evol. Biol.* 4:67–109. (evolution of hemoglobins)

Gillies, C.B. 1975. Synaptonemal complex and chromosome structure. *Ann. Rev. Genetics* 9:91–110. (chromosome structure)

Goodspeed, T.H. 1954. *The genus Nicotiana*. Waltham, Mass.: Chronica Botanica. (evolution of tobacco)

Jackson, R.C. 1971. The karyotype in systematics. *Ann. Rev. Ecol. Syst.* 2:327–368. (chromosome evolution)

Kornberg, R.D. 1977. Structure of chromatin. *Ann. Rev. Biochem.* 46:931–954. (biochemical structure of chromosomes)

Lewis, H. 1957. Genetics and cytology in relation to taxonomy. *Taxon* 6:42–46. (use of chromosomes in systematics)

Paris Conference. 1972. Standardization in human cytogenetics. *Cytogenetics* 11:313–362. (identification of human chromosomes)

Renwick, J.H. 1971. The mapping of human chromosomes. *Ann. Rev. Genetics* 5:81–120.

Ris, H., and D.F. Kubai. 1970. Chromosome structure. *Ann. Rev. Genetics* 4:263–294.

Stebbins, G.L. 1947. Types of polyploids: their classification and significance. *Adv. in Genetics* 1:403–430.

Stebbins, G.L. 1960. The comparative evolution of genetic systems. In S. Tax (ed.), *Evolution after Darwin*, vol. 1. Chicago: University of Chicago Press.

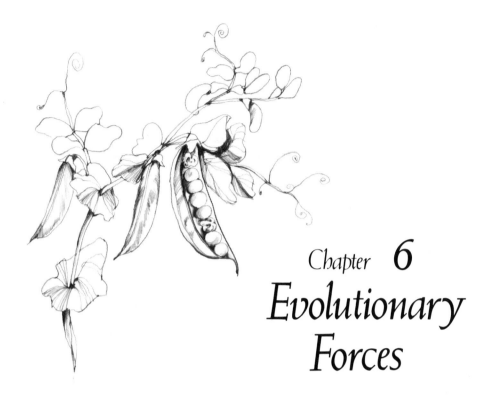

Chapter **6**
Evolutionary Forces

Population genetics is the branch of biology that deals with hereditary phenomena in whole groups of organisms. It is not directly concerned with the mechanism of inheritance but with its effects on the population. Population geneticists study the influence that mutation, recombination, and selection exert on heredity, as well as the effect of the size of the population, the type of mating, the longevity of individuals, and so on. Also considered are the effects that phenomena such as linkage, multiple alleles, sex-linked genes, and many other genetic processes have on the genetic composition of the population. Since all these phenomena can bring about changes in the population, the study of population genetics is thus inevitably related to that of organic evolution.

From the genetic point of view, evolution is the change in the frequency of genes in a population. The frequency of a gene (specifically an allele) is brought about by a number of different forces (Fig. 6.1). The principal ones are:

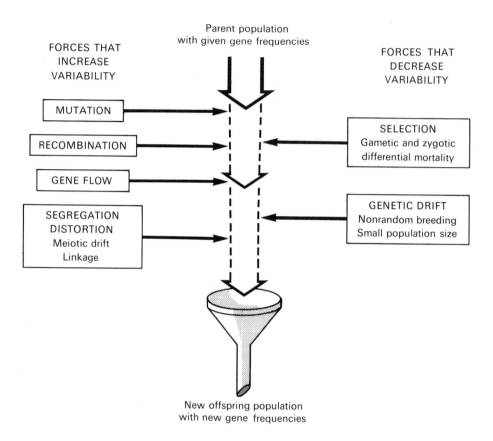

Figure 6.1

The forces of evolution. Forces such as mutation, recombination, gene flow, and segregation distortion increase the variability in a population. Other forces, selection and genetic drift, reduce the variability. The interaction among these forces can produce a new generation with different gene frequencies.

1. Mutation. Mutations are changes that affect the order of the nucleotides or their number in the DNA molecule and thereby "create" new alleles. Mutations can affect a single nucleotide or entire chromosomes.

2. Gene Flow. The gene frequency in the population can be altered by emigration of individuals from or immigration to the population. This phenomenon is called gene flow.

3. Recombination is the process by which the alleles in a population are mixed or recombined through crosses among individuals. Although recombination does not necessarily change the gene frequency in the population, it affects the action of natural selection and consequently has to be considered an important evolutionary force.

4. *Natural Selection,* or differential reproduction and survival of phenotypes, is a major way by which gene frequency is altered.

5. *Genetic Drift.* Gene frequency changes occur also as a result of chance alone. In small populations (less than about 100), genetic drift can play an important role in shaping evolutionary changes.

In this chapter and the next we review these forces in detail, trying to assess their relative importance. We start by considering frequency changes at one locus with two alleles. In later chapters we will consider multilocus changes.

MUTATION

A mutation is a change in the sequence, the type, or the number of nucleotides in the DNA molecule in the chromosome. This change can be very small, affecting only a single nucleotide pair, or it may be very large, affecting an entire chromosome segment or even a whole chromosome. In the previous chapter, under chromosomal aberrations, we described mutations that involve chromosomal segments. In this section we turn our attention to those mutations that affect a single gene locus. This kind of mutation is usually called *point mutation.*

Some point mutations are so rare as to be virtually unique. They do not have any permanent effect on the population, because they have an infinitely small chance of survival in a large population unless they give a very great and immediate advantage to their carriers, and that is highly improbable. Other mutations, called *recurrent mutations,* occur with a measurable frequency. The rate varies according to the gene and the organism (Table 6.1), but it is usually of the order of 10^{-4} to 10^{-8} per gene, per generation (that is, as many as one mutation at a given locus per 10,000 genes in a generation to as few as one per 100,000,000 genes in a generation). Recurrent mutations have a very important effect on gene frequency in the population, an effect that can be predicted mathematically.

Let us consider a population of organisms where all individuals have only the allele A_1 at a given locus. Allele A_1 mutates to allele A_2 at a constant rate, which we will call u. If the initial frequency of A_1 is p_0, the frequency q_1 of allele A_2 in the following generation will be

$$q_1 = u \times p_0$$

If we imagine that there were 1,000,000 gametes containing the A_1 allele, and that the rate at which they mutated to A_2 was $u = 10^{-4}$ (1/10,000), the following numerical values can be calculated:

$$p_0 = 1 \text{ (all alleles were } A_1\text{)};$$
$$u \times p_0 = 1 \times 10^{-4} = 0.0001;$$

and the new frequency of A_1,

$$p_1 = p_0 - (u \times p_0) = 1 - 0.0001 = 0.9999.$$

After one generation, there will be 999,900 gametes with the allele A_1 and 100 with the allele A_2.

But the new allele A_2 can mutate back to A_1, a phenomenon called *back mutation*. If we indicate by v the rate at which A_2 mutates to A_1 and apply the same reasoning used to calculate the changes accompanying the mutation of A_1 to A_2, we have

initial frequency of A_1 and A_2: p_0 and q_0;

mutation rates: $A_1 \underset{v}{\overset{u}{\rightleftharpoons}} A_2$;

new frequencies: $p_1 = p_0 - (u \times p_0) + (v \times q_0)$,
$$q_1 = q_0 - (v \times q_0) + (u \times p_0).$$

Considering forward and back mutation, the total change in the frequencies p and q are

$$\Delta p = p_1 - p_0 = v \times q_0 - u \times p_0,$$
$$\Delta q = q_1 - q_0 = u \times p_0 - v \times q_0.$$

As the frequency of one allele increases, fewer of the other are left to mutate in that direction, and more are available to mutate back to the original type. Eventually an equilibrium will be reached, at which point no further frequency changes will take place as a result of mutation. The point of equilibrium is found by equating the change of frequencies q and p to zero (when both rates are at equilibrium, no changes occur). Thus at equilibrium,

$$\Delta q = \Delta p = 0,$$

$$p \times u = q \times v, \qquad \text{or} \qquad \frac{p}{q} = \frac{v}{u},$$

and

$$p = \frac{v}{u} q.$$

Adding q to both sides of the equation and solving, we have

$$p + q = \frac{v}{u} q + q,$$

$$1 = \frac{vq + uq}{u} = \left(\frac{v + u}{u}\right) q,$$

$$q = \frac{u}{u + v}.$$

Table 6.1

Spontaneous mutation rates at specific loci for various organisms.

Organism	Trait	Mutation Rate per Gamete (or Cell)
Escherichia coli (a bacterium)	Streptomycin resistance	4×10^{-10}
	Phage T1 resistance	3×10^{-8}
Salmonella typhimurium (a bacterium)	Threonine resistance	4.1×10^{-6}
	Tryptophan independence	5×10^{-8}
Diplococcus pneumoniae (a bacterium)	Penicillin resistance	1×10^{-7}
Neurospora crassa (bread mold)	Adenine independence	3×10^{-8}
	Inositol independence	5×10^{-8}
Drosophila melanogaster (fruit fly)	Yellow body	1.2×10^{-4}
	Brown eyes	3×10^{-5}
	Ebony body	2×10^{-5}
	Eyeless	6×10^{-5}
Zea mays (corn)	Shrunken seed	1.2×10^{-6}
	Colorless aleurone	3×10^{-5}
	Sugary endosperm	2.4×10^{-6}
Mus musculus (house mouse)	Brown	8.5×10^{-6}
	Pink eye	8.5×10^{-6}
	Piebald	1.7×10^{-5}
Homo sapiens (man)	Epiloia	6×10^{-6}
	Retinoblastoma	$1.2–2.3 \times 10^{-5}$
	Aniridia	5×10^{-5}
	Huntington's Chorea	5×10^{-5}

After M.W. Strickberger, 1976. *Genetics*, 2nd ed. New York: Macmillan.

The frequency of allele A_1 depends exclusively (in the absence of selection) on the rate at which it mutates to A_2 and on the rate at which A_2 mutates back to A_1. Since these rates are very low (the rate of back mutations is usually only about one-tenth of the mutation rate), the change that mutation alone introduces into a population (at Hardy-Weinberg equilibrium) is very low. At the normal mutation rates (10^{-4} to 10^{-9}) it will take from a minimum of 5000 to a maximum of more than 50,000,000 generations to replace half of the A_1 alleles by A_2 alleles in a population. The effect of mutation is greater, the shorter the span of each generation. In the human species, 50,000 generations represents 1,000,000 to 2,000,000 years, but in a bacterium with a very fast generation time, on the order of half an hour, the same number of generations can elapse in less than three years. Whenever the rate of mutation is larger than that of back mutation, the mutant will become more numerous than the original gene unless selected against. An additional conclusion can be drawn. Increases in the rate of mutation will not produce any changes unless they affect the rate of mutation and back mutation in different ways. Otherwise the original equilibrium, as given by the formula, will not change.

The presentation above assumes a deterministic situation, with exactly 10^{-4} A_1 genes mutating to A_2, and with an offspring that precisely reflects the resultant frequency changes. But this is not how nature operates. The number 10^{-4} describes the average rate at which A_1 mutates to A_2. In any given generation the actual rate may be higher or lower than that value. In this context a relevant question is: Once a gamete contains the mutant allele, what is the probability of its giving rise to an adult organism?

The Australian geneticist and statistician Sir Ronald A. Fisher investigated this problem in 1930. He posed the following question: What is the probability of survival of a gene in a population if it is selectively neutral? Using a mathematical treatment known as branching processes (in the deterministic mode), Fisher showed that the probability of survival of such a gene is close to zero (Table 6.2). On the other hand, if the gene has a selective advantage of one percent, its probability of survival is only about two percent. Consequently, for a mutation to become established, it not only has to have some selective advantage but has to occur frequently!

GENE FLOW

Gene flow is the exchange of genes between populations in the form of occasional matings between individuals belonging to different breeding populations, incorporation into a population of an animal born elsewhere, or intro-

Table 6.2

Probability of survival and extinction of a neutral mutation.

Number of Generations	Probability of Extinction			Probability of Survival	
	No Advantage	1% Advantage	Difference	No Advantage	1% Advantage
1	0.3679	0.3642	0.0037	0.6321	0.6358
3	0.6259	0.6197	0.0062	0.3741	0.3803
7	0.7905	0.7825	0.0080	0.2095	0.2175
15	0.8873	0.8783	0.0090	0.1127	0.1217
31	0.9411	0.9313	0.0098	0.0589	0.0687
63	0.9698	0.9591	0.0107	0.0302	0.0409
127	0.9847	0.9729	0.0118	0.0153	0.0271
Limit	1.000	0.9803	0.0197	0.0000	0.0197

From R.A. Fisher, 1958. *The Genetical Theory of Natural Selection*, 2nd ed. New York: Dover.

duction of foreign seed or pollen. In all cases, genes derived by mutation and maintained by selection in a different environment are introduced into the population. The immediate effect is similar to the effect of mutation: A new source of variability is introduced. If the immigrant possesses an allele that is totally new in the population, the effect is identical to mutation. To calculate the effect of immigration of individuals, all possessing a new allele C_2, on a population initially possessing only the allele C_1, we use the same equations derived for mutation, replacing u (the mutation rate) by i, the immigration rate, and v (the rate of back mutation) by e, the rate of emigration of C_2 alleles from the population. If the frequency of C_1 is p_0 and that of C_2 is q_0, after one generation of gene flow we will have

$$\text{change in } C_1: \Delta p = e \times q_0 - i \times p_0$$

and

$$\text{change in } C_2: \Delta q = i \times p_0 - e \times q_0,$$

and at equilibrium

$$q = \frac{i}{i + e}.$$

However, the immigrants may possess the same alleles, C_1 and C_2, as the population under study, but the frequencies of these alleles in the two populations may be different. In that case we calculate the new frequencies in the following way. Let us consider a population that has at a given locus an allele C_2 with a frequency q_0. Each generation, the population begins to receive a number of immigrants that have C_2 at a different frequency, q_m. After a generation of gene flow, the population will have a proportion m of immigrants and $1 - m$ of nonimmigrant individuals. The frequency of the gene C_2 will now be equal to the sum of the proportion of the nonimmigrants times their frequency q_0 of C_2 plus the proportion of immigrants times their frequency q_m of C_2.

$$q_1 = mq_m + (1 - m)q_0$$
$$= m(q_m - q_0) + q_0.$$

The change Δq in the frequency of C_2 brought about by gene flow is the difference between the initial frequency q_0 and the frequency q_1 after gene flow has taken place:

$$\Delta q = q_1 - q_0$$
$$= m(q_m - q_0) + q_0 - q_0 = m(q_m - q_0).$$

Thus the change of gene frequency brought about by gene flow depends on the proportion of immigrants (m), the initial frequency q_0 of C_2, and the frequency of C_2 in the immigrants.

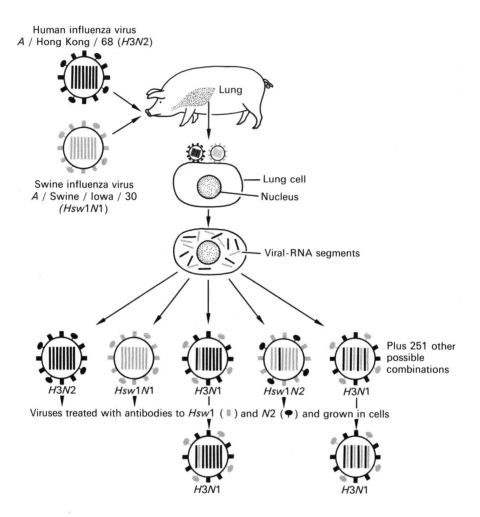

Figure 6.2

Recombination of the genetic material from two different type A influenza viruses. (The various strains of the virus are labeled, by convention among virologists, according to the sequence in which major variations in the two main antigenic components of the virion, hemagglutinin and neuraminidase, first make their appearance. In humans, for example, the variant H and N surface antigens are designated $H0$, $H1$, $H2$, $H3$, $N1$, $N2$, and so forth. For strains that appeared first in other animals, such as swine, appropriate identifying names or letters, in this case sw, are added to the coded label.) Here the two influenza viruses, one from the human ($H3N2$) and the other from swine ($Hsw1N1$), are shown being inoculated into the nose of a pig. The inoculation results in the simultaneous infection of a single lung cell with the eight separate RNA segments from each virus. Once inside the cell, the viruses multiply, and the 16 different RNA segments can be recombined in many ways during the "packaging" of the new virus particles. In the presence of antibodies to

An example of mutation and gene flow is provided by the influenza virus that attacks humans, causing the "flu." This virus contains two kinds of proteins that act as *antigens* that stimulate the human body to produce antibodies against them. They are *hemagglutinin,* a protein on the virus coat that forms projections capable of binding to red blood cells, and *neuraminidase,* an enzyme that dissociates the virus from the red blood cells. After an influenza virus infection, humans become immune against further attacks by virus containing hemagglutinins and neuraminidases identical to those in the virus that attacked them. However, they are still susceptible to attack by viruses that possess different hemagglutinins or neuraminidases. Almost every year there is an epidemic of influenza somewhere in the world. These local outbreaks are due to mutations in the virus protein (called *antigenic drift*), so that they become capable of reinfecting humans, since the antibodies that are specific to the unmutated protein no longer work. The human body can react quickly, however, and eventually produces antibodies against the mutated strain. Consequently, the epidemic tends to be local and relatively short-lived.

Occasionally there have been serious worldwide outbreaks of influenza that have caused a great number of deaths. These so-called influenza pandemics are known to have taken place in 1889–1890, 1918–1919, 1957–1958, and 1968–1969. Influenza pandemics seem to be caused by viruses radically different from the type B influenza virus that normally attacks humans. The difference is mainly in the hemagglutinins; the neuraminidases may or may not be the same as in type B virus. These viruses are called type A viruses. Because the hemagglutinin of type A is so radically different from that of type B, the human body is not able to produce antibodies against it with the flexibility that it can with type B, and morbidity of type A virus influenza is fairly high.

There are two hypotheses concerning the origin of the difference between the hemagglutinins of type A and type B viruses. The first hypothesis is that it is due to a very large mutation that drastically changes the protein. The second and more plausible one is that the new hemagglutinin is derived from a virus of another mammal or from a bird. Influenza viruses are known to undergo recombination readily (Fig. 6.2), and human influenza type A viruses have been isolated from other animals. The new hemagglutinins are very dif-

the neuraminidase of the human strain and to the hemagglutinin of the swine strain, all the resulting viruses will be neutralized except for those possessing the *H3N1* combination of surface antigens. The *H3N1* influenza virus may contain many different combinations of RNAs that include those coding for the two surface proteins, but their rearrangements with the remaining six segments can vary, and some of the recombinants may cause infection in pigs. *H3N2* strain was first detected in human beings in Hong Kong in 1968; *Hsw1N1* strain was first isolated from swine in Iowa in 1930. (From M.M. Kaplan and R.G. Webster, 1977. The epidemiology of influenza. *Scientific American* 237(6):88–106. Copyright © 1977 by Scientific American, Inc. All rights reserved)

ferent from other human influenza virus hemagglutinins in antigenic activity and in the kinds of polypeptides in their proteins, resembling the proteins normally found in the viruses of horses and ducks. Thus all the evidence seems to favor the second hypothesis, that of gene flow from viruses of other warm-blooded animals to the human influenza virus. Fear of a pandemic was the reason for so much concern with "swine-flu" virus in the winter of 1976–1977 (Fig. 6.3).

But gene flow has broader implications that are best understood if we consider the nature of the genes that are introduced. Inasmuch as the immigrants represent random samples of their original populations, they will carry in a high frequency alleles that are favorable there, and very infrequently they will introduce deleterious alleles, since these are kept at a low frequency by selection. Now, alleles that increase the fitness of individuals of population A do not necessarily increase the fitness of individuals of population B, because of different genetic backgrounds and different selection pressures in the two populations. Nevertheless, alleles introduced through gene flow are usually eliminated at a low rate because neighboring populations often have only small environmental differences. Furthermore, a relatively high proportion (compared with alleles arising by mutation) of the alleles acquired through gene flow have a beneficial effect, and they will tend to become established. In certain cases, they play a very important role, particularly when different species form hybrids, as we shall see further on.

The net effect of gene flow is that of a cohesive force that may work slightly counter to selection without affecting fitness to any great extent, but that hinders populations of the same species from becoming very different. Since the effect of gene flow is directly related to the number of immigrants, the farther apart two populations are physically, the less this cohesive force is felt. If two populations become completely isolated, selection may tend to push evolution in different directions, but as soon as contact is established, gene flow will again tend to unify the two populations.

The crucial data needed to assess the role of immigration are the rate at which gametes are dispersed with distance, the extent to which animals wander away from their population, and the distance seeds are blown or carried from where they were produced. There are not enough studies to make an overall assessment. However, the general rate of gene flow appears to decrease exponentially with distance, although it varies very much between species. Some species, such as humans, move around a lot; others, such as some butterflies (Fig. 6.4), appear to be very sedentary.

RECOMBINATION

The term *recombination* refers to the mixing in the offspring of the genes and chromosomes of their parents. Strictly speaking, recombination is not a force that changes the frequency of genes in the population. However, it is impor-

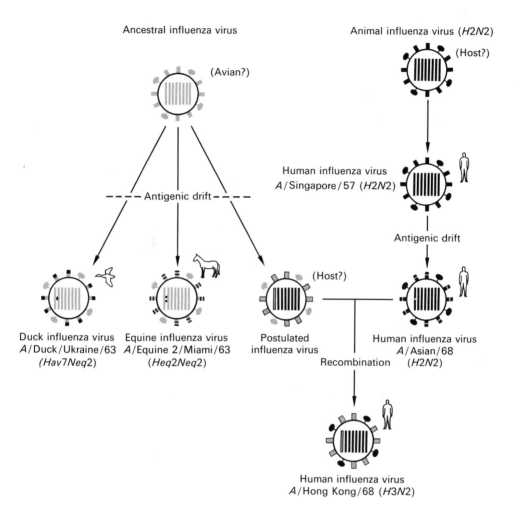

Figure 6.3

Proposed origin of the 1968 Hong Kong strain of human influenza virus. According to this hypothesis, an ancestral influenza virus (possibly avian) gave rise to a related series of animal viruses, including one virus isolated from horses in Miami in 1963 and another isolated from ducks in the Ukraine the same year. Five years later another virus in this family that was related to both the equine and the duck strains recombined with the then current Asian, or H2N2, strain of the human virus, giving rise to the Hong Kong, or H3N2, strain. Thus the new human strain acquired the hemagglutinin (H3) gene from the postulated intermediate animal strain and retained both the neuraminidase (N2) gene and the genes required to cause disease in humans from the human Asian strain. (From M.M. Kaplan and R.G. Webster, 1977. The epidemiology of influenza. *Scientific American* 237(6):88-106. Copyright © 1977 by Scientific American, Inc. All rights reserved)

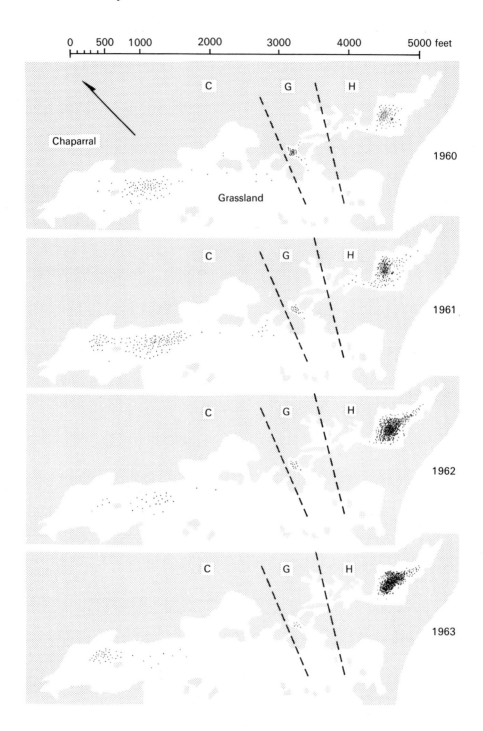

◀ *Figure 6.4*

Map of the density of the checkerspot butterfly, *Euphydryas editha*, in the years 1960–1963. Individuals in the various populations were marked and followed from year to year. Each dot represents the first capture of one individual. Note that the H population increased in size, the G population decreased very considerably, and the C population fluctuated in size. Individuals roamed relatively freely within one area, but marked individuals were seldom found outside the area where they were first captured. The populations are separated by not more than 4000 feet, well within the flight range of the butterflies, yet the populations remained distinct. (From P.R. Ehrlich, 1965. The population biology of the butterfly *Euphydryas editha*. II. The structure of the Jasper Ridge colony. *Evolution* 19:327–336)

tant for two main reasons. First, the different genes of a plant or animal inter-act, and certain combinations of genes are better than others. Second, the number of recombinants is infinitely larger than the possible number of muta-tions. Although the ultimate source of new genetic variation is mutation, most new types in a population arise by recombination. This process can produce more types and more varied types and at a faster rate than can mutation, and recombination rather than mutation is the immediate source of variability in a population. Sexuality is the outward manifestation of the modifications that have taken place in the phenotype to ensure recombination. Recombination takes place when the nuclei and the chromosomes of a male and a female gamete fuse to form the nucleus of the zygote, as we saw in Chapter 5.

If a heterozygous organism of the constitution $A_1A_2\ B_1B_2$ is crossed with another with the same gene combination, the offspring (in the absence of linkage between the two loci) will have the following genotypes (according to Mendel's second law):

$$A_1A_1B_1B_1;\ A_1A_1B_1B_2;\ A_1A_1B_2B_2;\ A_1A_2B_1B_1;\ A_1A_2B_1B_2;$$

$$A_1A_2B_2B_2;\ A_2A_2B_1B_1;\ A_2A_2B_1B_2;\ A_2A_2B_2B_2;$$

altogether nine different types. Each of these types can arise by mutation, but the chance that all nine types will arise by mutation in one generation is very small, so small that it is virtually impossible. (With the rate of mutation con-sidered to be equal to 10^{-4}, the probability in a population of $A_1A_2B_1B_2$ indi-viduals of obtaining these nine types in one generation by mutation is 10^{-24} $\times\ N$, where N is the number of individuals in the population.)

The number of diploid genotypes that can be assembled from any number of alleles of a gene is given by the formula

$$g = \frac{r(r+1)}{2},$$

where g is the number of diploid genotypes, and r is the number of alleles at any one locus.

If we consider two different loci, which are in different chromosomes so that they can be freely recombined, the total number of possible diploid genotypes that can be assembled for two genes considered together is

$$g_A \times g_B = \frac{r(r+1)}{2} \times \frac{r(r+1)}{2} = \left[\frac{r(r+1)}{2}\right]^2.$$

For example, if we consider that at each of the two loci there are two alleles, $r = 2$,

$$g_A \times g_B = \left[\frac{2(2+1)}{2}\right]^2 = 9,$$

which, as we saw above, is the number of possible combinations when we recombine two organisms heterozygous at two loci, each with two alleles.

If instead of two genes we are concerned with three, the number of recombinations is given by the product of $g_A \times g_B \times g_C$, and so on, so that for n genes (on different chromosomes) the number of recombinants will be

$$g_A \times g_B \times g_C \times \cdots \times g_n = \frac{r(r+1)}{2} \times \frac{r(r+1)}{2} \times \cdots \times \frac{r(r+1)}{2}$$

$$= \left[\frac{r(r+1)}{2}\right]^n.$$

The number of possible recombinations depends then on the number of alleles at a locus (r) and on the number of independent loci (n), and it is quite high whenever r and n are over 3, as is frequently true. For example, for $n = 10$ and $r = 3$, the value is $6^{10} = 60,466,176$. It is clear that the variability obtained by recombination is very high, so high that a series of mechanisms has evolved to control and restrict recombination (see Chapter 8).

An interesting example of the importance of recombination is provided by the wheat rust fungus, *Puccinia graminis*. Diploid (*dikaryotic*, or with two haploid nuclei per cell) spores of the fungus infect wheat plants (and oats, rye, barley, and many grasses) in the spring. The infection is externally evident because of the vertically elongate, reddish-brown or blackish, granular pustules on the stem and leaves, which give it its common name, "rust." Two kinds of spores are produced by the fungus on the wheat plant. One kind, called *uredospores*, are diploid spores that can reinfect other wheat plants and are produced throughout the summer. However, toward the end of the summer, the fungus produces *teleutospores* instead of uredospores. Each teleutospore undergoes meiosis and produces four haploid spores called *basidiospores*. Meiosis and basidiospore production often do not take place until the spring, with the teleutospores still attached to the old dead wheat plant or with the spores on the ground. Basidiospores are incapable of infecting wheat, and one can develop into a *mycelium* (fungus body) only if it falls on a barberry plant. A basidiospore falling on a barberry leaf or twig develops into a *haploid* mycelium. Eventually two haploid mycelia of different sexes (or mating types, as they normally are called in the fungi) will unite to form a *diploid* mycelium, still within the barberry plant. This diploid mycelium eventually forms diploid spores called *accidiospores,* which infect wheat plants (they are incapable of infecting barberry), and the entire cycle is repeated (Fig. 6.5).

The alternation of the life cycle of wheat rust in two hosts has interesting evolutionary and economic consequences. In evolutionary terms, recombination cannot take place if the barberry host is missing. Furthermore, in regions with severe winters all uredospores are winter-killed, and the rust cannot survive without barberry plants. Consequently in such areas—for example, the wheat fields of the Dakotas and the central Canadian provinces—eradication of barberry plants delays the infection of wheat in the spring. It does not

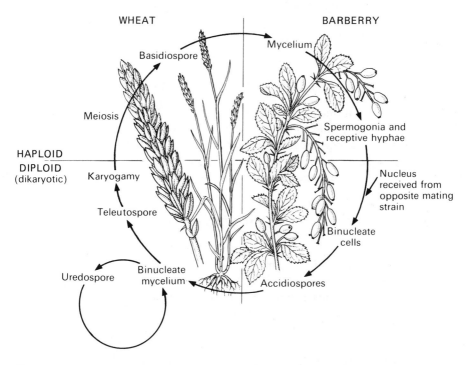

Figure 6.5

Life cycle of wheat rust (*Puccinia graminis*). Spores (accidiospores) germinate on wheat to form a binucleate mycelium. The mycelium gives rise to uredospores, which can infect other wheat plants. Later in the season, teleutospores are produced. The two nuclei in the teleutospore unite, and the fungus overwinters in that stage. In the spring, the teleutospores undergo meiosis, and haploid basidiospores are produced. These can germinate only on a barberry plant. The basidiospore forms a haploid mycelium, which develops spermogonia and receptive hyphae. Cells from the spermogonia are carried to the receptive hyphae, and binucleate cells are formed. Dikaryotic accidiospores, which can germinate only on wheat, are produced.

totally eliminate it, because uredospores are carried in by winds from regions with milder climates, where uredospores that have overwintered have already infected wheat plants and have begun the production of new spores. Most important, however, the eradication of the barberry plant eliminates sexual recombination in the rust. Modern wheat varieties are immune or highly resistant to rust infection. However, rust genotypes capable of breaking the resistance of the wheat plant are evolving constantly, necessitating the development of even more resistant wheat varieties. The genetic variation in the rust population is the result of mutation and recombination. The elimination of sexual recombination through eradication of barberry will slow evolutionary adjustments in the rust. This example is not very different from the situation of influenza virus and *Homo sapiens*. We can cope with the virus that lives in human populations and relies on mutation to break our defenses; we have a serious problem with the viruses derived from recombination of those of humans and other animals, such as monkeys or swine.

RECOMBINATIONAL DISTORTION

In Chapter 4 we mentioned that Mendel's second law (independent assortment of genes) is applicable only to genes in different chromosomes. Recombination of genes on the same chromosome is distorted by the combined effect of linkage and crossing over, which were discussed in detail in Chapter 5. Another phenomenon that distorts recombination is meiotic drive.

Meiotic drive

Two American geneticists, L. Sandler and E. Novitski, reported in 1957 that certain stocks of *Drosophila melanogaster* that were heterozygous for the recessive eye color genes cinnabar (**cn**) and brown (**bw**) located on the right arm of the second chromosome produced a deficiency of gametes with the **cn bw** genes and a corresponding excess of the wild-type chromosome. Consequently, when these flies were crossed to homozygous recessive flies of the composition **cn bw/cn bw** (which have white eyes), instead of the 1 : 1 ratio of red (wild-type) to white eyes, the result was a 25 wild-type : 1 white ratio. A similar case in plants is the tomato, both the cultivated *Lycopersicon esculentum* and its wild relative and presumed ancestor, *L. pimpinellifolium*. Charles M. Rick of the University of California at Davis found a gene in chromosome 4 that he called "Gamete eliminator" (**Ge**). Three alleles, Ge^n, Ge^c, and Ge^p, were found at this locus, but segregation distortion occurs only in Ge^c/Ge^p heterozygotes, which produce a ratio of approximately 5 Ge^c : 95 Ge^p gametes, rather than the expected 1 : 1 ratio.

Meiotic drive can be an important force in determining the survival of a gene or combination of genes. It is not entirely clear, however, whether it is a widespread phenomenon or, as it appears to be, an exceptional one. It is in any event an additional factor contributing to segregation distortion. As in linkage, if the beneficial chromosomes are produced in excess and detrimental ones in deficient numbers, the number of zygotes with low fitness is reduced and the costs of maintaining beneficial gene combinations in the population are decreased.

Recombination by itself does not produce any change in genetic information but uncovers arrangements of genes on which selection can act by favoring some and eliminating others. Let us now look at the way selection operates in a population to bring these changes about.

SELECTION

We have already seen that natural selection is the process that determines the differential contribution of offspring to the next generation. The proportionate contribution is called *fitness* or *adaptive value*. The number of offspring an organism contributes to the next generation depends on a series of factors. Some populations of plants and animals go through cycles of expansion and shrinkage. When the population is expanding, most individuals will probably produce some surviving descendants; when the population is shrinking, few will. But at every point some individuals will be producing a proportionately greater number of surviving offspring than others. To simplify the study of selection at all stages and over several generations, the geneticist considers frequencies rather than absolute numbers.

Absolute fitness consists in the contribution of offspring to the next generation; *relative fitness* consists in the contribution of offspring relative to some other genotype in the population, usually the one with the highest absolute fitness. We will use the notation m for the absolute fitness of an individual, w for the relative fitness. $\overline{M}$ and $\overline{W}$ will denote the average absolute and relative fitness in the population. Small letters will indicate individual fitness, capital letters values for the population. Furthermore, in comparisons of the effect of selection on different members of a population, selection is always considered to be acting against rather than in favor of the offspring of a certain organism.

In sexually reproducing plants and animals, the genic makeup of the offspring differs from that of either of its parents. Consequently, it is simpler to study the effect of selection on genes than on organisms. By convention, the fitness of the genotype transmitted to the next generation with the highest frequency is fixed at 1, regardless of the actual number of surviving offspring. Nevertheless, it should be understood clearly that selection operates on individual organisms, and only through them on genotypes and genes. Only when the differences in fitness between individuals are associated with the presence or absence of a particular allele or group of alleles in the individual's genotype does selection operate on that allele and the genotype that contains it. Consequently, the product of selection that affects evolution is determined by the way all genes together affect the fitness of an organism.

Most, if not all, new mutations reduce the fitness of the individuals that carry them in a homozygous condition. Many mutations are *lethal* when homozygous; that is, individuals with a double dose of that allele die. Other mutations are *semilethal;* most individuals carrying a double dose die, but some survive. And finally, other mutations are *subvital;* the fitness of the individuals homozygous for it is reduced but not drastically. In a heterozygous condition, on the other hand, lethal, sublethal, and subvital alleles can reduce the fitness of their carriers, be completely neutral (recessive), or even increase the fitness of their carriers. In the first case, there is no dominance for fitness (strictly speaking, dominance is said to be intermediate). In the second, recessiveness is complete, since the action of the mutant gene is completely masked. In the third case, it is an *overdominant* or *heterotic* gene (Fig. 6.6). Obviously the behavior of the allele in both the homozygous and heterozygous condition determines its selective advantage or disadvantage in the population. Remem-

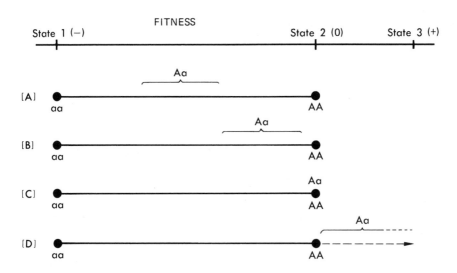

Figure 6.6

Dominance for fitness. The heterozygous phenotype may be (A) intermediate between the phenotype of the two homozygous parents (codominance); (B) similar to the dominant parental phenotype but not identical to it (incomplete dominance); (C) indistinguishable from the homozygous parental dominant phenotype (complete dominance); or (D) outside the variation of the homozygous parental phenotypes (heterosis or overdominance). The phenotype refers to any characteristic of the organism—morphological, physiological, or reproductive. (From O.T. Solbrig, 1970. *Principles and Methods of Plant Biosystematics.* New York: Macmillan)

ber that when a mutation first appears, it will be in the heterozygous condition; its early fate will depend on its effect in this condition. Only when it has become more frequent (as a result of favorable selection) will it become homozygous in some individuals.

Before proceeding to study these cases in detail, we must remember also that the genes in the chromosomes of an individual interact with one another, and that the fitness of a particular gene depends to a certain extent on other genes. For example, certain genes are completely recessive in certain genotypes but deleterious in other genotypes. Consequently, when we assign a certain fitness to a particular allele, we refer to the average fitness of that allele in the whole population.

Selection takes place whenever a particular phenotype experiences a higher rate of mortality than other phenotypes in the population, whenever it produces fewer offspring, or when the combination of mortality and reproduction rates results in fewer surviving offspring. It is unimportant from the point of view of the net effect on selection when and how these phenomena take place, as long as death occurs before the organism has reproduced. Death after reproduction has been completed has no effect on selection, unless there is parental care of the offspring after birth, such as in humans and many vertebrates. A large population of postreproductive adults can also affect selection indirectly by consuming resources needed by the young. The latter, however, is a special case, found primarily in humans and domestic animals. In nature, individuals of most, if not all, species die while still sexually active.

The evolution of resistance to pesticides in insects

A very interesting example of natural selection that combines many of the subjects discussed so far in this book is the resistance to pesticides that has developed over the last 30 years in more than 200 species of insects harmful to humans.

During World War II an insecticide, DDT, was developed that had a residual effect; i.e., once applied, it did not disappear by volatilization or degradation. Soldiers whose clothing was dusted with DDT were kept free of fleas and lice for a long time. After the war, a large number of compounds with residual effects were developed and became widely used in agriculture, medicine, and industry. However, as early as 1947 it was reported that houseflies in Sweden were not being killed by DDT. At first poor quality was suspected, but when the first report was followed by a series of others, it became apparent that insects were developing insecticide resistance. To combat this development, the pesticide dose was increased, and new kinds of insecticides were introduced. For example, in 1950 a dose of 0.25 lb of the insecticide parathion per 100 gallons of water was originally recommended against the walnut aphid in California. Seven years later the required dosage had increased to 1.5 lb of parathion per 100 gallons of water, a sixfold increase. There was of course a corresponding increase in the cost of insect control.

But the problem is even worse. Because these insecticides are usually toxic to all insects, they also kill the insects and the birds that normally eat the pest that we are trying to control. Consequently, the established ecological balance is upset. Destruction of beneficial insects and birds means that even more frequent applications of insecticides of greater potency are needed, since an increased number of pest species, including some formerly secondary pests, may now be involved in attacking the crops. In extreme but not unusual cases, insecticides must be applied at weekly intervals to afford protection, and this raises the cost of insect control tremendously. George Georghiou has reported that in Central America as many as 30 applications of pesticides during a six-month growing season are needed to grow cotton. This can render cotton production uneconomical and can lead to abandonment of cultivation, especially in the Third World, as was done with cotton production in the Cañete Valley of Peru.

Pesticide resistance has been studied extensively, and not surprisingly it involves several genetic changes, including biochemical, physiological, and behavioral characters. Evolution of pesticide resistance is best known in the common housefly (*Musca domestica*). Therefore we will concentrate on describing the evolution of insecticide resistance in that species.

Action of Pesticides. Unfortunately, the exact biochemical effect of an insecticide is not usually known. However, one class of compounds, the organophosphates (malathion, parathion, etc.) are known to kill the insect by binding with the neuromuscular enzyme acetylcholinesterase (AChE). When this enzyme is poisoned by the insecticide, it cannot function in its normal role of

Table 6.3

Bimolecular rate constants (k_i) for inhibition of acetylcholinesterase from resistant- and susceptible-housefly heads measured in the absence of substrate.

Insecticide	$10^{-3} \times k_i(\text{M}^{-1} \cdot \text{min}^{-1})$		
	Resistant	Susceptible	
		608	Cooper
Omethoate	1.77 ± 0.04	19.9 ± 0.32	20.5 ± 0.53
Methyl paraoxon	21.6 ± 1.0	—	90.1 ± 4.9
Paraoxon	139 ± 6.5	578 ± 22	—
Isopropyl paraoxon	9.23 ± 0.43	76.3 ± 3.5	80.6 ± 6.0
Malaoxon	47.3 ± 3.6	867 ± 35	601 ± 21
Ethyl malaoxon	94.3 ± 3.5	932 ± 12	—
Tetrachlorvinphos	24.7 ± 0.2	—	172 ± 26

From A.L. Devonshire, 1975. *Biochem. J.* 149:463–469.

breaking down acetylcholine to choline and acetic acid. The result is multiple muscle contractions when the motor nerve is stimulated. Consequently, the animal burns up all its reserves and eventually dies of exhaustion. Biochemical studies by Devonshire and others have shown that in resistant flies, changes in the enzyme take place that reduce significantly the rate of inhibition by organophosphates (Table 6.3). When resistant and susceptible strains of houseflies were crossed, the acetylcholinesterase of the progeny had the properties of a mixture of susceptible and resistant enzymes behaving independently, with activities in the ratio of 2 : 1 (Fig. 6.7). This finding can be explained if there were equal quantities of the resistant and the susceptible enzyme in the hybrids and if it is further assumed that the susceptible enzyme was twice as susceptible to the insecticide as the resistant one. This model was confirmed when it was determined that resistant flies possess a mutant allele, labeled

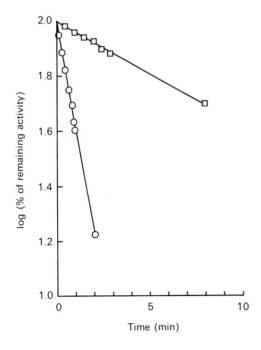

Figure 6.7

Effect of the insecticide malaoxon on the activities of the enzyme acetylcholinesterase of susceptible (circles) and resistant flies (squares). The enzyme was extracted from flies and assayed in vitro in the presence of malaoxon. Although the susceptible enzyme was inhibited drastically, the activity of the resistant one was affected much less. (*Note:* logarithmic scale on the ordinate.) (After A.L. Devonshire, 1975. Studies of the acetylcholinesterase from houseflies (*Musca domestica* L.) resistant and susceptible to organophosphorous insecticides. *Biochem. J.* 149:463–469)

AChE-R, of a single gene located in the second chromosome (the housefly has five chromosomes).

Although the allele **AChE-R** confers greater resistance to organophosphate insecticides, it does not make flies totally immune to the poison. Researchers soon found out that flies possess biochemical machinery that detoxifies the insecticide; that is, the fly changes the chemical composition of the insecticide, rendering it harmless. So far alleles at six different loci in the housefly have been identified that detoxify different insecticides (Table 6.4). Some of these, such as **Ox-2** and **Ses,** confer resistance to more than one insecticide. Consequently, when one is applying one kind of insecticide one can unwillingly select for resistance to another insecticide. It is also possible that there are more genes that have not been identified yet, since different resistant strains have only one or a few of the appropriate alleles.

In addition to these biochemical properties, researchers found that the housefly has a physiological mechanism that confers some resistance to insecticide by reducing its penetration through the integument into the animal. The significance of this mechanism of resistance lies in the fact that even a small reduction in the rate of entry of the insecticide per unit of time may permit the detoxification mechanisms to cope with the insecticide as it enters the metabolic

Table 6.4

Some of the genes that confer pesticide resistance in the housefly, *Musca domestica.*

Gene	Chromosome	Action
AChE-R	II	Produces a modified acetylcholinesterase that is resistant to organophosphates.
Ox-2	II	Detoxifies a variety of insecticides.
Deh	II	Produces DDT dehydrochlorinase activity.
Ox-2	II	Confers resistance to organophosphates.
py-ex	II	Confers resistance to pyrethrum.
y	II	Produces high levels of glutathione-dependent organophosphate metabolism.
Pen	III	Reduces insecticide absorption.
kdr	III	"Knock-down-resistance." Action unknown.
Dld-4	IV	Provides resistance to Cyclodienes.
Ses	V	Confer resistance to a variety of insecticides. May be allelic.
DDT-md	V	
Ox-5	V	
py-ses	V	Confers resistance to pyrethrum.

system of the fly. Two genes have been identified that affect the rate of entry. **Pen,** located in chromosome III, has actually been shown to reduce the rate of DDT penetration, and **kdr** (for "knock down resistance") also located in chromosome III, confers resistance to pyrethrum and DDT, although its actual mode of action is unknown.

Another way the insect can avoid being killed by a residual insecticide is by a behavioral change, such that the animal avoids coming into contact with the insecticide. Avoidance can be the result of increased irritability to the pesticide, so that the insect moves or hides when it perceives the pesticide. Alternatively, it can come by a process of selection; that is, only those animals survive that behaved in a certain way *before* the application of the insecticide. One such change is known to have taken place in a malaria mosquito, *Anopheles gambiae,* in Rhodesia. Before DDT was applied to human dwellings, the mosquito was found with equal frequency indoors and outdoors. After eight years of treating the houses and huts in the area with insecticides, the mosquito had become almost nonexistent in untreated as well as treated dwellings, and this condition did not change perceptibly even after insecticide treatments were discontinued (Table 6.5). Mosquitoes that fly into a dwelling, bite their victims, and then fly out without ever resting on the walls have been selected.

In summary, in a period of 30 years no less than nine mutations in the housefly (in different strains and different parts of the world) have occurred that confer some degree of resistance to their bearers against the new pesticides. Some of these mutations (especially the detoxifying enzymes) confer more resistance than others, so that when any allele is present alone, it decreases but does not eliminate mortality in the fly population. In combination, however, two or more genes enhance resistance multiplicatively (Fig. 6.8). In a very short period of time, a large number of insects the world over, through the combined action of mutation, recombination, selection, and possibly gene

Table 6.5

Numbers of hand-caught, resting *Anopheles gambiae* recorded in daytime catches indoors and outdoors in Shamva, Rhodesia.

	Before Treatment 1926–1928	After 8 Years of Spraying with Hexachlorocyclohexane, 1958
A. gambiae ♀ indoors	269	11*
A. gambiae ♀ outdoors	205	619

From R.C. Muirhead-Thomson, 1960. Significance of irritability, behavioristic avoidance, and allied phenomena in malaria eradication. *Bull. Wld. Hlth. Org.* 22:721–734.
* 96 untreated houses

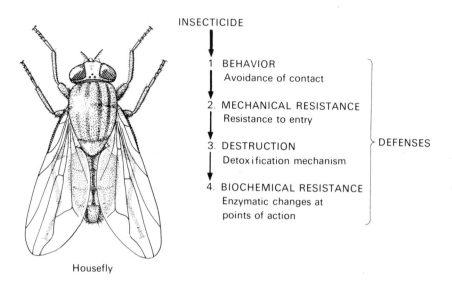

INSECTICIDE

1 BEHAVIOR
  Avoidance of contact

2. MECHANICAL RESISTANCE
   Resistance to entry

3. DESTRUCTION
   Detoxification mechanism

DEFENSES

4. BIOCHEMICAL RESISTANCE
   Enzymatic changes at
   points of action

Housefly

Figure 6.8

Resistance to insecticides as a combinations of different mechanisms, such as behavioral, anatomical, biochemical, and physiological mechanisms. No one mechanism alone provides total immunity, but together they combine to transform a potentially lethal effect (wide arrow at top) into an insignificant effect (narrow arrow at bottom).

flow, have acquired the ability to cope with a major change in the environment. The change threatened to make their environment and food source—agricultural crops—uninhabitable. As we will discuss in more detail in Chapter 15, at least some of these insects may have been pre-adapted to cope with the new barrage of chemical poisons, because plants produce naturally a large number of toxic compounds in their leaves and tissues. The changes in resistance that we have witnessed and documented in the last 30 years are remarkable, as are both the basic simplicity of the mechanism and the complexity of the interaction. The whole development has also been an important lesson for us. In trying to destroy harmful insects wholesale, we have probably succeeded only in destroying some useful insects and birds and in harming ourselves, since most of the compounds we have developed are poisonous to us. And unfortunately, because of our long generation and small reproductive capacity, we cannot evolve resistance with the same speed that flies can.

CALCULATION OF FITNESS IN A POPULATION

Survival Rate.   The survival rate, λ, for each genotype is simply the number of individuals of that genotype in the population after selection has taken place, divided by the initial number of individuals of that genotype. For example, given a starting population of 10,000 individuals, where the frequency p of A_1 is 0.6, and the frequency q of A_2 is 0.4, the initial numbers and a possible mortality due to selection are as shown in Table 6.6.

Table 6.6

Calculation of survival rate of three genotypes in a hypothetical population.

	A_1A_1	A_1A_2	A_2A_2	*Total*
Size of Initial Population	$p^2 \times N$	$2\,pq \times N$	$q^2 \times N$	N
Before Selection	$0.6^2 \times 10000$	$2 \times 0.6 \times 0.4 \times 10000$	$0.4^2 \times 10000$	
	$= 3600$	$= 4800$	$= 1600$	10000
After Selection	3300	2880	800	6980
Loss	-300	-1920	-800	-3020
Survival Rate	0.9	0.6	0.5	0.7

The survival rate, λ, of the three genotypes is as follows:

$$\lambda_{A_1A_1}: \frac{\text{number after selection}}{\text{number before selection}} = \frac{3300}{3600} = 0.9$$

$$\lambda_{A_1A_2}: \frac{\text{number after selection}}{\text{number before selection}} = \frac{2880}{4800} = 0.6$$

$$\lambda_{A_2A_2}: \frac{\text{number after selection}}{\text{number before selection}} = \frac{800}{1600} = 0.5$$

Although almost 2000 heterozygous individuals died, as contrasted with only 800 homozygous recessives, the rate of survival of the latter was lower than that of the former. By dealing with the *rate* of survival rather than with the absolute number, we can make meaningful comparisons of the performances of different phenotypes and genotypes.

Relative Fitness. The mortality in a population is not constant over time. For example, among many species of plants, mortality rates will be greater during a drought than in a year of normal rainfall. In order to assess the performance of a phenotype and genotype over time, we eliminate the yearly fluctuations by the common mathematical operation of dividing the survival rate of each genotype by the survival rate of the genotype with the highest survival rate.

The resultant quantity is the relative fitness, w, of the genotype. In our example,

$$w_{A_1A_1} = \frac{0.9}{0.9} = 1.00,$$

$$w_{A_1A_2} = \frac{0.6}{0.9} = 0.66,$$

$$w_{A_2A_2} = \frac{0.5}{0.9} = 0.55.$$

The relative fitness will not be affected by causes of mortality that affect all genotypes equally; it reflects only *differential* survival rates of the genotypes. Note that the genotype with the least mortality will have a fitness of 1. This does not mean that it is not affected by prereproductive mortality. It means only that it has the least mortality of all the genotypes.

Selection Coefficient. We can now define still another quantity, the selection coefficient, s. The selection coefficient is simply 1 minus the relative fitness, so that

$$s_{A_1A_1} = 1 - w_{A_1A_1} = 1 - 1.000 = 0,$$

$$s_{A_1A_2} = 1 - w_{A_1A_2} = 1 - 0.66 = 0.34,$$

$$s_{A_2A_2} = 1 - w_{A_2A_2} = 1 - 0.55 = 0.45.$$

Selection against double recessive

We will first study the changes in the frequency of allele A_2 brought about by selection when there is complete dominance—that is, when the mutant form is selected against only when it is in a homozygous state, since when it is in a heterozygous condition, it has no effect whatsoever on fitness (complete dominance). We know from Chapter 4 (Hardy-Weinberg law) that the frequencies of the three possible combinations between two alleles are

$$A_1A_1 = p^2; \quad A_1A_2 = 2pq; \quad A_2A_2 = q^2;$$

and

$$p^2 + 2pq + q^2 = 1.$$

The relative fitness of these combinations, by definition, is in this case

$$w_{A_1A_1} = 1, \quad w_{A_1A_2} = 1, \quad \text{and} \quad w_{A_2A_2} = 1 - s,$$

where s represents the coefficient of selection against homozygous recessives. The gametic contributions therefore will be

for A_1A_1, $p^2 \times 1$;   for A_1A_2, $2pq \times 1$;   for A_2A_2, $q^2 \times (1 - s)$;

since the gametic contribution is equal to the product of the frequency times the coefficient of selection. The total contribution will no longer be unity; it will be $1 - sq^2$. In order to find out the frequency of A_2 genes in the next generation, we have to add one-half of the contribution of A_1A_2 plus the contribution of A_2A_2 and divide by the new total, so that the frequency of A_2 in generation 1 is

$$q_1 = \frac{q^2 (1 - s) + pq}{1 - sq^2}.$$

The change that has resulted in one generation as the result of selection is

$$\Delta q = q_1 - q$$
$$= \frac{q^2 (1 - s) + pq}{1 - sq^2} - q,$$

which on simplification reduces to

$$\Delta q = - \frac{sq^2 (1 - q)}{1 - sq^2}.$$

The same type of reasoning is applied to calculate the effects of selection when selection works against a dominant gene, A_1, rather than the recessive A_2; when selection affects A_2 both when homozygous and when heterozygous; or when there is overdominance and both genes A_1 and A_2 are selected against when homozygous. The resulting formulas are shown in Table 6.7.

From the formulas in Table 6.7 several conclusions can be drawn. First of all, we see that selection depends on the coefficient of selection, s, and on the initial frequencies, q, or p and q, when overdominance is present. The relationship is a rather complex one. From the formulas in Table 6.7 and the graphs in Fig. 6.9 we can see that selection is most effective when the gene is at an intermediate frequency, and it becomes least effective when the mutant is at a very high or a very low frequency. Furthermore, selection against a recessive mutant is highly ineffective when that mutant is at a low frequency. This is due to the fact that in such circumstances it will be found largely or entirely in heterozygous combinations, where it is "shielded" from selection.

SELECTION ON PHENOTYPIC TRAITS

Natural selection acts on phenotypes and only indirectly on genotypes. Changes in gene frequency obtain only when there is a correlation between genotype and phenotype. Studies of embryology and development of individual plants and animals show that different genotypes give rise to different phenotypes, but that the relation is very complex, being affected by the environment in many and diverse ways (Chapter 4). Consequently, it is often simpler to observe the effect of selection directly on the phenotype than to observe it on the under-

Table 6.7

Change of gene frequency, Δq, after one generation of selection under different conditions of dominance.*

Conditions of Dominance and Selection	Initial Frequencies and Fitness of the Genotypes			Change of Frequency, Δq, of Allele A_2
	A_1A_1 p^2	A_1A_2 $2pq$	A_2A_2 q^2	
No dominance selection against A_2	1	$1 - \frac{1}{2}s$	$1 - s$	$\dfrac{-\frac{1}{2}sq\,(1-q)}{1 - sq}$
Complete dominance selection against A_2A_2	1	1	$1 - s$	$\dfrac{-sq^2\,(1-q)}{1 - sq^2}$
Complete dominance selection against A_1	$1 - s$	$1 - s$	1	$\dfrac{+sq^2\,(1-q)}{1 - s(1 - q^2)}$
Overdominance selection against A_1A_1 and A_2A_2	$1 - s_1$	1	$1 - s_2$	$\dfrac{+pq\,(s_1 p - s_2 q)}{1 - s_1 p^2 - s_2 q^2}$ †

From D.S. Falconer, 1960. *Introduction to Quantitative Genetics.* New York: Ronald Press.

* When s is small, the denominators differ little from 1, and the numerators alone can be taken to represent Δq accurately enough for most purposes.

† For derivation of this expression, see Chapter 8.

lying genotype. Furthermore, in artificial selection the breeder is primarily interested in the phenotypic character, such as milk production by a herd of cows, number of eggs laid per hen, or yield of grain per hectare.

Selection on phenotypic characters can be divided into three types: stabilizing selection, disruptive selection, and directional or progressive selection. These three types of selection represent ways by which the population remains adjusted to an environment that may be (1) constant and stable, (2) becoming more variable and breaking up into different subenvironments (disruptive selection), or (3) changing constantly in a single direction (directional selection) (Fig. 6.10).

The consequence of selection is usually thought of as phenotypic change. This is not necessarily so and, as a matter of fact, is often not so. Actually, a very common type of selection is *stabilizing selection* (Fig. 6.10). In stabilizing selection, phenotypes with characteristics near the mean of the population are more fit than those at the extreme because of the existence of a stable environment. As a result of this kind of selection, the mean of the population will not vary although its variability may do so, depending on the genetic determination of the character under consideration.

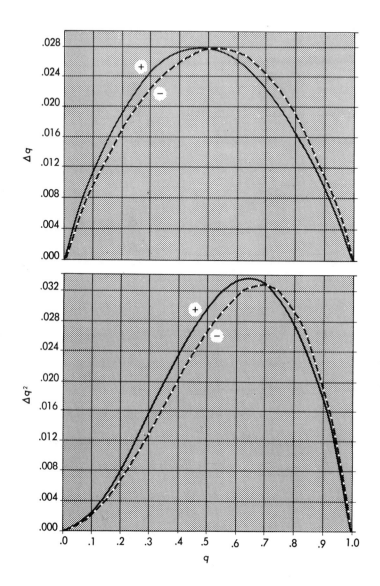

Figure 6.9

Change of gene frequency, Δq, under selection of intensity $s = 0.2$, at different values of initial frequency, q. Upper figure shows a gene with no dominance; lower figure shows a gene with complete dominance. The graphs marked ($-$) refer to selection against the gene whose frequency is q, so that Δq is negative. The graphs marked ($+$) refer to selection in favor of the gene, so that Δq is positive. (From D.S. Falconer, 1954. Asymmetrical responses in selection experiments. In *Symposium on Genetics of Population Structure*, Instituto di Genetics, Universita di Pavia, Italy, Aug. 20–23, 1953. *Un. Int. Sci. Biol.* Series B, no. 15:16–41)

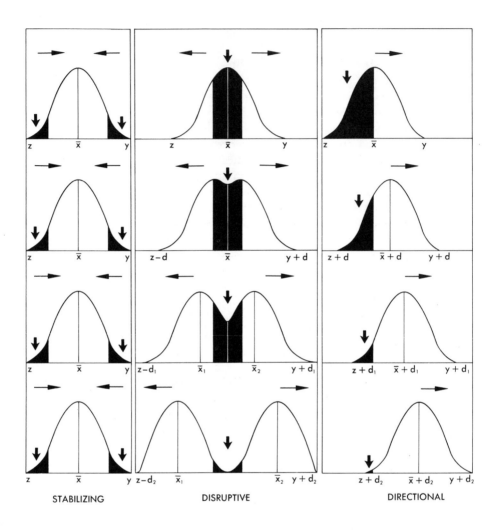

STABILIZING   DISRUPTIVE   DIRECTIONAL

Figure 6.10

The three main kinds of selection. In stabilizing selection the environment favors the organisms with values close to the populational mean; consequently, little or no change is produced in the population. Disruptive selection favors the extremes and will tend to divide the population in two. Directional selection favors one extreme, and it will tend to move the mean of the popul. tion toward that extreme. Directional selection accounts for most of the change observed during evolution. The curves represent the frequency of organisms with a certain range of values between x and y; the shaded areas, those phenotypes that are being eliminated by selection; the long arrows, the direction of evolutionary change; d, amount of change. (From O.T. Solbrig, 1970. *Principles and Methods of Plant Biosystematics*. New York: Macmillan)

Blooming period in many spring blooming plants is under stabilizing selection: Members of species that bloom too early in the season will, on the average, produce fewer seeds than will those emerging later, because pollinating insects have not yet emerged; plants blooming late in the season will, on the average, produce fewer seeds than those that bloomed before, because water is lacking or there is more competition with other plants.

Another possible kind of selection is one that favors the extremes over the center because of an environment that is breaking up into two or more types (Fig. 6.10). However, because the extremes breed with one another, in each generation a large number of organisms having the intermediate characteristics will be produced. In an environment favoring the extremes, any mechanisms that would keep opposite extremes from breeding with one another would be highly advantageous, leading to the second type of selection, called *disruptive selection*. Disruptive selection requires some kind of mechanism ensuring that like breeds with like—in other words, some kind of isolating mechanism.

The mechanics of selection that favor the extremes—that is, disruptive selection—are not well understood. However, an interesting experiment was performed by the British geneticists Thoday and Boam in 1959 with *Drosophila melanogaster*. Starting with eight females and eight males that were left to mate at will, Thoday and Boam selected in each generation the four males and four females with the largest number of sternopleural chaetae and the four males and four females with the smallest number. The selected animals were allowed to mate together at will. Because both extremes were equally favored, selection at the beginning was of the stabilizing type, with extremes favored. By the time the experiment was terminated, however, no more intermediate types were present in the population, apparently because the cross between the flies with high bristle number and the flies with low bristle number produced sterile offspring. Selection had become disruptive rather than stabilizing.

If situations such as the one experimentally produced by Thoday and Boam exist in nature, disruptive selection could be a mechanism leading to speciation.

A final kind of selection, *directional selection*, is the type in which one extreme phenotypic expression of a character is favored, as happens in nature when the environment is changing constantly in one direction (Fig. 6.10). Directional selection produces phenotypic and genotypic changes more rapidly than any other kind of selection. Artificial selection is mostly of this kind.

At the Illinois Agricultural Experimental Station, several selection experiments have been performed with corn, *Zea mays*, over the last 60 years. In one of the experiments, plants with ears borne close to the ground were selected in each generation. In the original population the average height of the ears above the ground ranged from 43 to 56 inches in different individuals. After 24 generations, all the plants produced ears only eight inches above the ground, and no more change was produced by selection, presumably because there was no more genetic variability for the determining character. At the same time, in another line derived from the same foundation stock, plants with ears high off the ground were selected. At the end of 24 generations the plants produced their ears an average of 10 feet above the ground!

In another set of experiments, plants with seeds with high protein content were selected in one line and with low protein content in another. The high line after 76 generations of selection was producing up to 15 percent more protein than the original strain, and the low line was producing about 9 percent less. In still another experiment, lines were selected for high and low oil content. Starting with plants producing about 5 percent of oil, a line that produced almost 19 percent of oil was selected and one that produced less than 1 percent (Fig. 6.11).

Directional selection probably accounts for the majority of the phenotypic changes that occur during evolution. However, any kind of selection produces genotypic changes by definition. If no phenotypic and/or genotypic changes occur, or if such changes as occur are truly random, no natural selection exists.

MUTATION AND SELECTION

As we have seen before, when the frequency of the original gene is high (and that of the mutant low), mutation is most effective, since there are more genes that can mutate ($p \times u$). This situation coincides with the conditions at which selection is most ineffective. But as the frequency of the mutant form increases, selection becomes more effective and mutation less so, until an equilibrium is reached, when no more changes in gene frequency take place.

Let us consider the situation where the mutant is selected against only in the homozygous situation. At equilibrium the changes produced in one generation by mutation should equal those brought about by selection, so that

$$u \times p - v \times q = \frac{sq^2 (1 - q)}{1 - sq^2},$$

$$u (1 - q) - v \times q = \frac{sq^2 (1 - q)}{1 - sq^2}.$$

We can simplify this equation, sacrificing some accuracy but not too much, in order to get a better understanding of the relationships. First, we can simplify the equation by disregarding the back mutations ($-v \times q$), which are almost negligible when the frequency of A_2 is low, and we can further disregard the denominator $1 - sq^2$, since its value is near unity when the frequency of A_2 is low. The simplified equation then becomes

$$u (1 - q) = sq^2 (1 - q),$$
$$u = sq^2,$$
$$q = \sqrt{\frac{u}{s}} \qquad \text{(approximate)}.$$

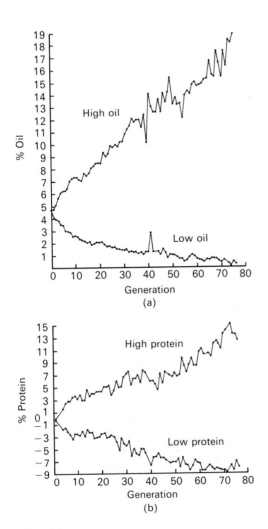

Figure 6.11

Directional selection in *Zea mays* over 76 generations. (a) Strains were selected for high oil content and for low oil content. (b) Strains were selected for high and low protein content. Because the protein content varies in response to the weather, the percent protein figures for each year were adjusted by subtracting the mean protein content of the high and low oil strains for that year. (After J.W. Dudley, 1977. 76 generations of selection for oil and protein percentage in maize. In E. Pollak and others (eds.), *Proceedings of the International Conference on Quantitative Genetics*, Aug. 16–21, 1976. Ames: Iowa State University Press)

This approximate expression for the gene frequency of the mutant gene A_2 when mutation and selection are at equilibrium indicates that the equilibrium frequency of the gene A_2 will depend on the relative values of the mutation rate and the coefficient of selection, and theoretically at least, the equilibrium frequency could have many different values. But given the normally low values of mutation (on the order of 10^{-4} to 10^{-8}), the frequency of the mutant will be kept at low values even without too great a selection against it. Therefore the joint action of selection and mutation should keep deleterious mutants that have no heterotic effects at a low frequency, a fact confirmed by observations on natural populations. A further conclusion is that mutation alone is not a likely cause of evolutionary change, and it is selection that chiefly determines whether a gene spreads through the population or remains a rarity.

A fine example of the interaction of selection with mutation is the evolution of myxomatosis in Australian rabbits. European rabbits were introduced into Australia in 1859. They thrived and became pests. In 1950 the myxoma virus, which causes myxomatosis, was introduced in an attempt to control the rabbit population. This acute rabbit disease kills its host a short time after infection. The treatment was very effective: A large proportion, more than 99 percent, of the Australian rabbits died the first year the myxoma virus was introduced.

The virus is spread by mosquitos. They are plentiful in the summer, and the disease spreads rapidly. However, there are no adult mosquitos in the winter, so the virus (which has no intermediate host) also dies out in the winter. Consequently, farmers must reintroduce the virus each spring. But within the virus population, selection favors mutant strains that are less virulent and therefore allow the sick rabbit and its parasitic virus to survive over the winter, since such strains can spread year after year naturally. The original strain killed its host quickly and was not available to spread in the following summer (except where it was reintroduced). A less virulent strain, on the other hand, would be accessible to mosquitos in the summer following its introduction and therefore could spread naturally. In Australia, within three to four years, the strains collected from wild rabbits were indeed less virulent than the original.

At the same time, there was strong selection among the rabbits for resistance to myxomatosis. In populations with continuing exposure to the virus, genetic resistance increased, presumably as a result of mutation followed by selection. In one population, over a period of seven years, deaths from a particular strain dropped from 90 percent to 25 percent.

Thus there is coevolution between the virus and the rabbits. Changes in either affect the selective forces acting on the other. In highly susceptible rabbit populations, selection on the viruses favors less virulent strains. With exposure to myxomatosis, resistant strains of rabbits are favored. As genetic resistance builds up in a rabbit population, increasingly virulent strains of virus are favored. Selection operates independently on both populations to maximize their fitness. The eventual result will be a large population of rabbits that is host to a large population of viruses.

COMPLETE ELIMINATION OF RECESSIVES

Many mutations are lethal when homozygous, but they show little or no effect in a heterozygous condition. The calculation of the change in gene frequency in such situations is relatively simple and gives us further insights into the operation of selection.

The initial frequency in the population will again be assumed to be $p^2 + 2pq + q^2 = 1$. The relative fitness of these combinations will be:

$$w_{A_1A_1} = 1; \quad w_{A_1A_2} = 1 \quad \text{and} \quad w_{A_2A_2} = 0.$$

Consequently, after selection we will have the following genotype frequencies:

$$A_1A_1: p^2 \times w_{A_1A_1} = p_0^2,$$

$$A_1A_2: 2pq \times w_{A_1A_2} = 2p_0q_0,$$

$$A_2A_2: q^2 \times w_{A_2A_2} = 0,$$

$$\text{Total}: p_0^2 + 2p_0q_0 = p_0(p_0 + q_0 + q_0) = p_0(1 + q_0),$$

and the frequency

$$q_1 = \frac{p_0q_0}{p_0(1 + q_0)} = \frac{q_0}{1 + q_0}.$$

After one generation of breeding, Hardy-Weinberg equilibrium will be restored (at least on the zygote populations), selection will act again, and by the same reasoning, we will find that

$$q_2 = \frac{q_1}{1 + q_1} = \frac{\dfrac{q_0}{1 + q_0}}{1 + \dfrac{q_0}{1 + q_0}} = \frac{q_0}{1 + 2q_0},$$

and so on, so that

$$q_n = \frac{q_0}{1 + nq_0}, \qquad \text{(Harmonic series)}$$

and the change Δq is

$$\Delta q = q_1 - q_0 = \frac{q_0}{1 + q_0} - q_0 = \frac{q_0 - q_0 - q_0^2}{1 + q_0} = \frac{-q_0^2}{1 + q_0}.$$

Complete elimination of recessives is a very drastic kind of selection. However, the intensity depends on the frequency of the gene. When the gene is at a very high frequency (close to 1), the numerator will have a value close to 1 and the denominator a value approaching 2. Therefore, close to half the genes will be eliminated. When the frequency of the deleterious gene is 0.5, the numerator of the equation will be 0.25 and the denominator 1.5, and only about one-sixth of the A_2 genes will be eliminated from the population. As the frequency of the A_2 genes continues to decrease, the numerator will decrease proportional to the square of q^2, the denominator in proportion to $1 + q_0$, so that Δq will become smaller and smaller. We see then that selection against recessives is very powerful when they are abundant in the population but very ineffective when they are at a low frequency.

From the formula for the frequency of A_2 in the nth generation,

$$q_n = \frac{q_0}{1 + nq_0} \text{ ,}$$

we can calculate the number of generations needed to produce a given gene frequency change, as follows:

$$1 + nq_0 = \frac{q_0}{q_n} \text{ ,}$$

$$nq_0 = \frac{q_0}{q_n} - 1 = \frac{q_0 - q_n}{q_n} \text{ ,}$$

$$n = \frac{q_0 - q_n}{q_n q_0} = \frac{1}{q_n} - \frac{1}{q_0} \text{ .}$$

From this we see that if we want to reduce the gene frequency of A_2 from 0.99 to 0.5, we need

$$n = \frac{1}{0.5} - \frac{1}{0.99} = 2 - 1 = 1 \text{ generation (approximate),}$$

but if we want to reduce it from 0.5 to 0.25, we need

$$n = \frac{1}{0.25} - \frac{1}{0.5} = 4 - 2 = 2 \text{ generations,}$$

and from 0.25 to 0.125,

$$n = \frac{1}{0.125} - \frac{1}{0.25} = 8 - 4 = 4 \text{ generations.}$$

Complete elimination is the most drastic type of selection. If selection is less drastic, the rate of change will be slower, but the general form of the process will be the same. That is, selection will still be very effective when the gene is at a high frequency and very ineffective when the gene is at a low frequency. Furthermore, as recurrent mutation is constantly incorporating the defective gene into the population, a recessive gene can never be eliminated, no matter how drastic its harmful effect.

An example is the genetic disease in humans called phenylketonuria, or PKU. This disease is due to a recessive allele that interferes with the transformation of the amino acid phenylalanine into tyrosine. People with this disease excrete abnormal quantities of phenylpyruvic acid in their urine. Nearly all the individuals who have so far been found to excrete phenylpyruvic acid continuously in their urine have exhibited some degree of intellectual impairment. It is usually severe, amounting to idiocy or imbecility, so reproduction is seldom achieved.

About two to 10 children in 100,000 (or 0.2 to 1 in 10,000) exhibit this syndrome. For the following calculations we will accept the higher frequency of $1/10,000 = 10^{-4}$:

$$q^2 = 10^{-4},$$
$$q = \sqrt{10^{-4}} = 10^{-2} = 0.01.$$

That is, on the average about one person in 100 carries the recessive gene for PKU. If two persons with the recessive gene marry, the expectation is that one-fourth of their children will have the disease. The incidence of the disease is equivalent to the mutation rate.

Recently it has been discovered that if the disease can be diagnosed in the newborn (through analysis of the urine), the harmful effects of the disease can be prevented by feeding the child a diet that does not contain phenylalanine, and these people can be normal and can reproduce. Consequently, in each generation there will be an increase in the frequency of the defective allele, which will be equivalent to the mutation rate (if all cases are identified). The increase can be calculated.

$q_0 = 0.01,$

$q_1 = q_0 + (p_0 \times u) = 0.01 + [0.99 \times (2 \times 10^{-4})] = 0.01 + 0.000199$
  $= 0.010199,$

$q_2 = q_1 + (p_1 \times u) = 0.010199 + (0.9898 \times 2 \times 10^{-4}) = 0.010395,$

$q_3 = q_2 \times (p_2 \times u) = 0.010395 + (0.9896 + 2 \times 10^{-4}) = 0.010594.$

Thus after three generations, or roughly 100 years, the incidence of the people with the gene will increase from roughly 100 out of every 10,000 individuals to about 106, and people with the disease from 10 in 10,000 to 11 in 10,000. However, once the frequency increases to higher values, such as 10 percent (0.1) of the population, people with the disease will be much more commonly found (one out of every 100), creating serious problems since phenylalanine is present in most foods, and removing it from the diet is difficult and expensive. Preventing the creation of a population where almost everybody has some severe genetic disease requiring continuing medical treatment is a major moral and social problem.

THE FUNDAMENTAL THEOREM OF
NATURAL SELECTION

The essence of the theory of evolution through natural selection is that genotypes with a higher fitness (w) leave a proportionally greater number of offspring, and that consequently their genes will be present in a higher frequency
in the next generation. It is intuitively clear that for any change to occur there
must be some genetic variation in the population, and that the greater the
variation, the greater the possibilities for change. But is there any precise relation between genetic variation and the maximum possible rate of selection?

The existence of such a relationship was demonstrated in 1930 by Ronald
Fisher, who called it "The Fundamental Theorem of Natural Selection." This
theorem says: "The rate of increase in fitness of any population at any time
is equal to its genetic variance in fitness at that time." The theorem can be
demonstrated for one locus with two or more alleles and random mating.
When it is extended to more than one locus and to situations where mating is
not random, it does not necessarily hold in all situations.

Consider a population where fitness is determined at one locus, and consider further that there are only two alleles, A_1 and A_2, at the locus in question,
with frequencies p and q ($q = 1 - p$), respectively. Let w_{11} be the fitness of
A_1A_1, let w_{12} be the fitness of A_1A_2, and let w_{22} be that of A_2A_2. The frequency
of A_1 after one generation of selection is

$$p_1 = \frac{w_{11}p_0^2 + w_{12}p_0q_0}{w_{11}p_0^2 + 2w_{12}p_0q_0 + w_{22}q_0^2} = \frac{w_{11}p_0^2 + w_{12}p_0q_0}{\overline{W}},$$

where

$$\overline{W} = w_{11}p^2 + 2w_{12}pq + w_{22}q^2,$$

which is the average fitness of the population.

We now can ask what is the change in $\overline{W}$ after one generation of selection.

$$\Delta W = \overline{W}_1 - \overline{W}_0$$
$$= (w_{11}p_1^2 + 2w_{12}p_1q_1 + w_{22}q_1^2) - (w_{11}p_0^2 + 2w_{12}p_0q_0 + w_{22}q_0^2)$$
$$= w_{11}(p_1^2 - p_0^2) + 2w_{12}(p_1q_1 - p_0q_0) + w_{22}(q_1^2 - q_0^2).$$

We can now simplify this equation by a few algebraic manipulations.

$$\Delta\overline{W} = w_{11}[(p_1 + p_0)(p_1 - p_0)] + 2w_{12}[p_1(1 - p_1) - p_0(1 - p_0)] +$$
$$\quad w_{22}[(q_1 + q_0)(q_1 - q_0)]$$
$$= w_{11}[(p_1 + p_0)(p_1 - p_0)] + 2w_{12}[(p_1 - p_1^2) - (p_0 - p_0^2)] +$$
$$\quad w_{22}\{[(1 - p_1) + (1 - p_0)][(1 - p_1) - (1 - p_0)]\}$$
$$= w_{11}[(p_1 + p_0)(p_1 - p_0)] + 2w_{12}[(p_1 - p_0) + (p_0^2 - p_1^2)] +$$
$$\quad w_{22}[(1 - p_1 + 1 - p_0)(1 - p_1 - 1 + p_0)]$$

$$= w_{11}[(p_1 + p_0)(p_1 - p_0)] + 2w_{12}[(p_1 - p_0) + (p_0 - p_1)(p_0 + p_1)] +$$
$$w_{22}(2 - p_1 - p_0)(-p_1 + p_0)$$
$$= w_{11}[(p_1 - p_0)(p_1 + p_0)] + 2w_{12}[(p_1 - p_0)(1 - p_0 - p_1)] +$$
$$w_{22}[(p_1 - p_0)(p_1 + p_0 - 2)].$$

And factoring out $(p_1 - p_0)$, we have

$$\Delta\overline{W} = (p_1 - p_0)[w_{11}(p_1 + p_0) + 2w_{12}(1 - p_0 - p_1) + w_{22}(p_1 + p_0 - 2)].$$

But $p_1 - p_0 = \Delta p$, and $p_1 = \dfrac{w_{11}p_0{}^2 + w_{12}p_0q_0}{\overline{W}}$.

Replacing in the equation above and with some algebraic manipulations, we arrive at

$$\Delta\overline{W} = (\Delta p)^2 \left(w_{11} - 2w_{12} + w_{22} + \frac{2\overline{W}}{p_0 q_0} \right)$$
$$= 2p_0 q_0 [w_{11}p_0 + w_{12}(1 - 2p_0) - w_{22}q_0]^2$$
$$= 2p_0 q_0 [w_{11}p_0 + w_{12}q_0 - w_{12}p_0 - w_{22}q_0]^2.$$

If we now define E_1 and E_2 as

$$E_1 = w_{11}p_0 + w_{12}q_0 - \overline{W},$$
$$E_2 = w_{12}p_0 + w_{22}q_0 - \overline{W},$$

we can write the equation above as

$$\Delta\overline{W} = 2p_0 q_0 (E_1 - E_2)^2.$$

Now, the positive part of E_1, $w_{11}p_0 + w_{12}q_0$, represents the fitness of A_1A_1 genotypes and the half of A_1A_2 genotypes containing A_1, and the positive part of E_2, $w_{12}p_0 + w_{22}q_0$, represents the fitness of the half of A_1A_2 genotypes containing A_2 plus the fitness of the A_2A_2 genotypes. Then E_1 and E_2 represent the deviations from the mean population fitness of the two groups. The quantity $E_1 - E_2$ Fisher called the "average excess" of A_1.

Stated in words, what we have just shown is that the change in fitness ($\Delta\overline{W}$) is equal to the product of the frequencies of the two alleles multiplied by the square of the difference between the deviations in mean population fitness of the two alleles. By definition of a variance, this is the additive variance in fitness of the two alleles (p. 73). That the value of $\overline{W}$ will be zero whenever $p = 0$ or $p = 1$ can be easily shown by substituting in the appropriate places of the equations just presented. In the case of heterozygote superiority, where $w_{12} > w_{22}$ and $w_{12} > w_{11}$, the proof is more elaborate, but it can be shown by the use of regression statistics that irrespective of the value of w_{11}, w_{12}, and w_{22}, the additive genetic variance will be zero when $\overline{W} = 0$.

The significance of Fisher's theorem is twofold. First, it formally demonstrates that no selection is possible in a uniform population, and second, the theorem points out that it is not the fitness value of an allele that determines the rate of change but the degree of variation in fitness at that locus in the population. For example, a population where $w_{11} = 0.9$, $p = 0.5$, and $w_{22} = 1$, and $q = 0.5$ will change more in the next generation than one where $w_{11} = 0.1$, $p = 0.01$, and $w_{22} = 1$, and $q = 0.99$, because there is more total genetic variance in fitness in the first population.

RANDOM PHENOMENA

So far we have assumed that the population we are dealing with is a very large one (theoretically infinitely large) and that mating is strictly random. Neither of these two situations is ever met in nature. All populations are finite, and most are medium-sized (10^2–10^3 individuals) or small (fewer than 10^2 members) and go through periods when their numbers are much decreased, particularly in times of hardship (lack of food, extreme temperatures, droughts, and so on). Furthermore, mating is not random; such simple factors as physical proximity in plant populations and mate preferences among animals make random mating impossible.

The effect of a finite population size (Chapter 7) and of nonrandomness of mating (Chapter 8) is to make each generation to a certain extent a random sample of the previous one. This means that genes may be eliminated or maintained in a population in part by chance. The probability of a gene staying or being eliminated in a population is determined by its frequency, which is the result of mutation, selection, and gene flow. But there can occur in populations from year to year random fluctuations that are not the result of selection. These will seldom have long-lasting evolutionary effects, but they will keep the population from reaching its theoretical selective peak. Random phenomena can be of evolutionary significance only when a certain number of important gene arrangements are completely eliminated from the population. The probability that this will occur, even in populations as small as 100 individuals, is low. It is doubtful that random genetic drift, as this phenomenon is called, has ever been a major positive evolutionary force. We will take up these points in the next chapter.

SUMMARY

In this chapter we analyzed briefly the simplest situation in a population, that of two autosomal alleles. Such a situation probably never obtains in nature. Organisms usually differ at many loci. Furthermore, seldom does only one allele affect fitness independently of the other genes of the organism. However, if we are to understand the more complex situation found in nature, we must understand the simpler one first. It is therefore justifiable and of value to study what is a rather abstract situation.

The population geneticist studies gene frequency changes in populations and the phenomena that cause these frequencies to change. The tools are statistics and mathematics, as well as experimentation with artificial and natural phenomena. In the last two decades, two new tools, computer simulation and gel electrophoresis have been added to that arsenal. On the basis of theoretical and experimental studies, the population geneticist sets up models to explain general evolutionary phenomena. These can be simulated in a digital computer, which can test the logic and consistency of the model. However, the real test is experimentation and observation in field and laboratory with actual populations of plants, animals, and microorganisms. Through measurement of enzyme frequencies using gel electrophoresis—a tool derived from knowledge acquired by the molecular biologist in the last two decades—the population geneticist estimates allele frequencies in nature. In this manner the actual changes in gene frequencies that take place in nature can be documented. The results of some of these studies and the controversies they have produced will be the topic of Chapter 9.

SUGGESTED FURTHER READING

General references

Crow, J.F., and M. Kimura. 1970. *An Introduction to Population Genetics Theory.* New York: Harper and Row.

Dobzhansky, T. 1970. *Genetics of the Evolutionary Process.* New York: Columbia University Press.

Falconer, D.S. 1960. *Introduction to Quantitative Genetics.* New York: Ronald Press.

Grant, V. 1963. *The Origin of Adaptations.* New York: Columbia University Press.

Grant, V. 1975. *Genetics of Flowering Plants.* New York: Columbia University Press.

Li, C.C. 1976. *First Course in Population Genetics.* Pacific Grove, Cal.: Boxwood Press.

Spiess, E.B. 1977. *Genes in Populations.* New York: Wiley.

Specialized articles and reviews

Beckwith, J., and P. Rossow. 1974. Analysis of genetic regulatory mechanisms. *Ann. Rev. Genetics* 8:1–14. (genetic regulation)

Brussard, P.F., P.R. Ehrlich, and M.C. Singer. 1974. Adult movements and population structure in *Euphydryas editha. Evolution* 28:408–415. (gene flow)

Devonshire, A.L. 1975. Studies of the acetylcholinesterase from houseflies (*Musca domestica* L.) resistant and susceptible to organophosphorus insecticides. *Biochem. J.* 149:463–469. (evolution of insecticide resistance)

Ehrlich, P.R., and P.H. Raven. 1969. Differentiation of populations. *Science* 165: 1228–1232. (gene flow)

Fenner, F.J. 1974. *The Biology of Animal Viruses,* 2nd ed. New York: Academic Press. (evolution of myxomatosis)

Fisher, R.A. 1958. *The Genetic Theory of Natural Selection.* New York: Dover. (Fisher's fundamental theorem)

Georghiou, G.P. 1972. The evolution of resistance to pesticides. *Ann. Rev. Ecol. Syst.* 3:133–168. (evolution of insecticide resistance)

Kaplan, M.M., and R.G. Webster. 1977. The epidemiology of influenza. *Scientific American* 237 (6):88–106. (influenza epidemics)

Levin, D.A., and H.W. Kerster. 1974. Gene flow in seed plants. *Evol. Biol.* 7:139–220. (gene flow)

Muirhead-Thomson, R.C. 1960. The significance of irritability, behavioristic avoidance, and allied phenomena in malarial eradication. *Bull. WHO* 22:721–734. (insecticide resistance)

Thoday, J.M., and T.B. Boam. 1959. Effects of disruptive selection: III. Polymorphism and divergence without isolation. *Heredity* 13:205–218. (disruptive selection)

Woodworth, C.M., E.R. Leng, and R.W. Jugenheimer. 1952. Fifty generations of selection for protein and oil in corn. *Agron. J.* 44:60–65. (directional selection)

Wright, S. 1931. Evolution in Mendelian populations. *Genetics* 16:97–159. (action of natural selection)

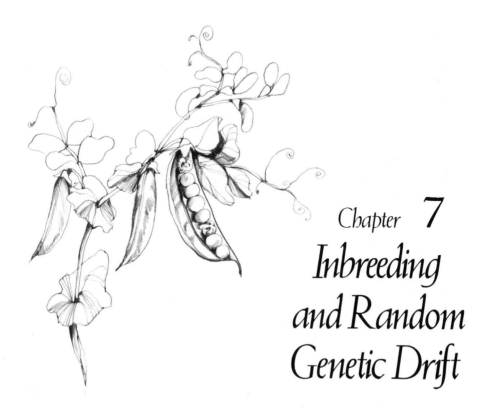

Chapter 7

Inbreeding and Random Genetic Drift

In the previous discussions on evolutionary forces (Chapter 6) we have assumed that the population is very large and that breeding is random. But a population with such characteristics probably does not exist in nature. Instead, all populations are limited in size, and most important, breeding is usually not random. In this chapter we investigate the effect that nonrandom breeding and finite population size have on the genetic structure of the population, briefly mentioned in Chapter 6.

RANDOM BREEDING

What precisely is meant by random breeding? It means that the probability that individual *a* in the population will mate with *b* is the same as the probability that *a* will mate with *c*, with *d*, with *x*, and so on. Let us analyze this statement in practical terms. Let us assume we are dealing with a species that has well-defined, discrete populations that can be delimited clearly in space (which is the exception). If we further assume this to be a population of plants, for breeding to be perfectly random, the probability that a pollen grain from plant *a* will fertilize an ovule of its neighbor plant *b* must be the same as the probability that the same pollen grain will reach and fertilize an ovule of plant *s* at the other end of the field. However, this is not true. The probability that a pollen grain from plant *a* will reach *b* is greater than the probability that it will reach *s*. The same applies to a population of animals. The probability of encounters (and consequently matings) with individuals in close physical proximity is greater than with those that are physically removed.

If we plot the probability of mating against distance, we obtain a curve such as the one depicted in Fig. 7.1. The probability of mating decreases very steeply with distance, but it does not go to zero. Instead, a small probability always exists that two very distant individuals will mate. Such a curve is described as *leptokurtic* (it deviates from the normal, bell-shaped curve in favor of the center, or in other words, the standard deviation is smaller than for the corresponding normal curve).

Imagine now the probability distribution in three dimensions. It will look like a cone (Fig. 7.2), and if we picture all the organisms in that population, the mating probability for all of them in three dimensions looks like a "bed of nails," according to Bruce Wallace (see Fig. 7.3).

In addition to the spatial constraint on random breeding, the finite size of the population and details of the breeding mechanism result in additional deviations from random breeding.

So far, in discussing the genetic mechanism, we have assumed that the gene frequency in the population of gametes could be predicted precisely from the parent. For example, we assumed that a heterozygote A_1A_2 would produce

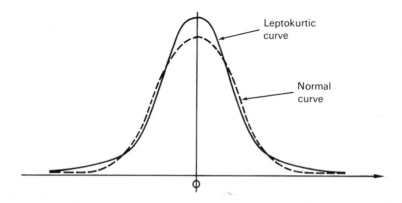

Figure 7.1

Leptokurtic curve and normal curve. The leptokurtic curve has more values close to the mean than the normal curve. A graph of probability of mating (*y*-axis) against distance (*x*-axis) gives a leptokurtic curve.

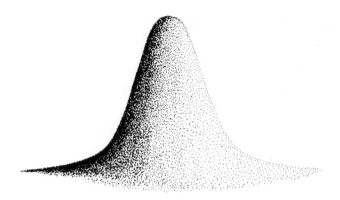

Figure 7.2

Leptokurtic dispersal curve in three dimensions.

two kinds of gametes, A_1 and A_2, in a 1 : 1 ratio. We further assumed that these gametes would unite to give three types of zygotes, A_1A_1, A_1A_2, and A_2A_2, in a 1 : 2 : 1 ratio. We now review this assumption and show that the ratios mentioned are probabilities, and that it is very unlikely that these *exact* ratios will ever be encountered.

Normally individual organisms produce a large number of gametes. In pollen grains and sperm, the number can be in the thousands; in some animal eggs and plant ovules, the number is smaller, but even then it tends to be more

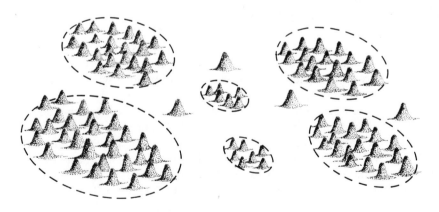

Figure 7.3

Mating probabilities in a population. Each cone represents the probability of mating of an individual plant or sedentary animal in two-dimensional space. The spacing of cones reflects the distribution of individuals in the population. The closer two individuals are together, the greater the chance that they will cross.

than a hundred per individual. If meiosis proceeds normally and there is no systematic distortion of the segregation ratios (Chapter 5), the allelic composition of the gametes and the frequencies should be predictable from the genotype composition of the parents, provided every gamete is viable. However, occasional pollen grains or sperm are malformed. Since it is highly unlikely that these malformed gametes will occur proportionately among the various kinds of gametes, the composition of gametes and their frequency will no longer be exactly like that of the parent. Furthermore, other factors can distort the ratios in the filial generation. In the production of female egg cells in both the higher plants and animals, only one out of four products of meiosis results in the formation of a gamete. In addition, not all individuals in the population produce exactly the same number of gametes. Of the large number of gametes produced by the members of the population, only a reduced number become part of a successful mating that results in offspring. Consequently, the pool of zygotes produced by the population is a *sample* of the pool of zygotes the population could have produced if each meiotic product had resulted in a viable gamete and each gamete had resulted in a cross. As a sample of the theoretical population, it obeys all the rules of probability that apply to sampling. In other words, the process of zygote formation has aspects that are similar to picking a sample of red and black balls from a bag of many such black and red balls. The proportion of red balls in the sample will approximate the proportion in the bag, but the probability that the sample will have exactly the same proportion of red and black as the bag is very low. Probability theory tells us that the larger the sample (i.e., the larger the population in our case), the smaller the deviations; the smaller the sample (i.e., the smaller the population), the greater the deviation.

That a perfect ratio is hardly ever obtained is illustrated in Table 7.1, which presents the F_2 segregation ratios obtained by Mendel for seven characters in the garden pea (*Pisum sativum*). A 3 : 1 ratio was expected in each case, and although the observed ratios are remarkably close to expectation, they never were exactly as predicted. The closest fit was obtained when the population was largest.

Finite population size and deviations from random breeding are characteristics of all populations. They act as constraints imposed by nature. These two characteristics of all populations introduce a stochastic element into evolution. By that we mean that the outcome of the interplay of the basic evolutionary forces, mutation, recombination and selection, cannot be predicted with absolute certainty; it can be predicted only in a statistical sense. That is, if the same evolutionary event were to take place repeatedly, we could predict only the most probable outcome. It also means, if exactly the same evolutionary event were to take place repeatedly, that the result would be slightly different each time. But evolutionary events take place only once. In the majority of cases they will proceed according to what is most likely to happen, but occasionally the outcome will be a very unlikely event. This is a very important concept to remember.

Table 7.1

Segregation ratios observed by Mendel in the F_2 of garden pea crosses.

Character	Numbers Observed in F_2	Ratio
Round seed	5474	2.96 : 1
Wrinkled seed	1850	
Yellow cotyledon	6022	3.01 : 1
Green cotyledon	2001	
Grayish-brown seed coat	705	3.15 : 1
White seed coat	224	
Inflated pod	882	2.95 : 1
Constricted pod	299	
Green pod	428	2.82 : 1
Yellow pod	152	
Axial flower	651	3.14 : 1
Terminal flower	207	
Long stems	787	2.84 : 1
Short stems	277	

From G. Mendel, 1965. *Experiments in Plant Hybridization.* Translation and reprint from the original German publication of 1866, edited by J. H. Bennett. Edinburgh and London: Oliver.

CHANCE LOSS OF ALLELES IN A POPULATION

We have emphasized that the Hardy-Weinberg equilibrium, $p^2 + 2pq + q^2 = 1$, holds only (1) in an infinitely large population, and (2) where breeding is random. We see now that neither of these assumptions is strictly true. How does this affect the Hardy-Weinberg equilibrium?

Let us assume that we start with a finite population of x individuals. Let us further assume that at a locus A half of the individuals in the population have allele A_1 and the other half A_2; that is, the frequency of A_1 is $p = 0.5$, and that of A_2 is $q = 0.5$. Let us still further assume that there is no mutation, segregation distortion, or selection at the A locus. We expect that half the gametes will be A_1 and the other half A_2. However, for the reasons given above, there will be some deviation from the exact 1 : 1 ratio. Furthermore, as a result of some nonrandom breeding and the finite size of the zygote populations, further statistical deviations will take place during fertilization. These may be in opposite directions to those that took place during gamete formation (e.g., there may be an excess of A_2 genes in the population of gametes, but a higher proportion of A_1 gametes may form zygotes), so that their effects cancel each other somewhat, or they may be in the same direction and reinforce each other. In any case, the probability is that in the next generation the frequency of A_1 will not be exactly $p = 0.5$, as predicted by the Hardy-Weinberg theorem, but that a new equilibrium will be established. Let us assume that the new frequencies are $A_1 : p = 0.55$ and $A_2 : q = 0.45$. In the next generation the same kind of sampling errors will take place, but sampling will be in a population where there are slightly more A_1 genes than A_2. The probability that A_1 will again increase slightly or that it will decrease in frequency is exactly 1/2. That is, the fact that A_1 increased by chance in one generation does not mean that it will necessarily decrease in the next generation. Conceivably the value may be $p = 0.6$ and $q = 0.4$, or it may be $p = 0.51$ and $q = 0.49$. Let us assume that the new value is $p = 0.6$. Again we expect a deviation from the gene frequency of the parent population, either toward a lower value of p or toward a higher value of p, the probability of these two events being equal. Conceivably the new value could be $p = 0.7$. If these deviations alternate in increasing and decreasing the gene frequency of A_1, the population will be oscillating between values slightly higher or slightly lower than the original $p = 0.5$. However, it is also possible that for a number of consecutive generations most of the changes will be in the same direction. If that happens, one of the alleles can eventually be lost (Fig. 7.4). The probability of such an occurrence is a function of population size and of time. Given enough time in a finite population, one of a pair of heterozygous alleles will be lost eventually by chance alone if there is no selection or mutation. This is an important point.

Population subdivision and the Wahlund formulas

We now proceed to discuss another factor that affects gene frequency in a stochastic way, the spatial distribution of individuals.

Figure 7.5 depicts a map of the distribution of a species of plants, *Clematis fremontii* var. *riehlii*, in Missouri. This type of distribution, though not universal, is fairly common. The principal aspect to note is that the density of individuals is not equal across the area; in some glades plants are close together whereas in others they are separated. Since plants that are close to each other have a greater probability of mating than those that are far apart, the population of *Clematis fremontii* depicted in Fig. 7.5 is effectively subdivided into subpopulations.

Although crosses are possible between all the plants in the glade, most crosses will take place within a subpopulation, and mostly between neighboring plants. Let us now follow gene frequency changes in such a population.

Let us assume once more that at generation 0, the frequency of A_1 equals the frequency of A_2—that is, $p = q = 0.5$ across the population and that there is no selection or mutation. Now, in generation 1 there will be deviations from this initial frequency that are due to chance. However, since the events in each subpopulation are independent, the deviations observed in each of the subpopulations will be different: In some A_1 will be larger than 0.5, in others less. The probability is that in about one-half it will increase, and in the other half

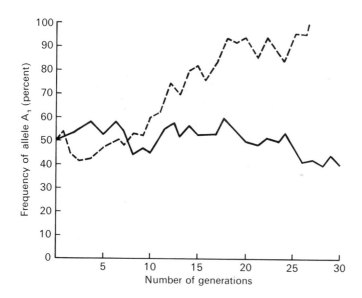

Figure 7.4

Random sampling can cause the frequency of a gene to vary markedly from one population to another. In this calculation performed with random numbers, allele A₁ appears in 50 percent of the members of two populations. Only 27 generations later, allele A₁ has become "fixed" in one population (dashed line), and in the other its frequency fluctuates from 40 to 60 percent. (From L.L. Cavalli-Sforza, 1969. "Genetic drift" in an Italian population. *Scientific American* 221(2):30–37. Copyright © 1969 by Scientific American, Inc. All rights reserved)

it will decrease. The same is true for the next generation and every one thereafter. However, because the deviations cancel each other out, the overall gene frequency will remain about the same, that is, 0.5 in our example. This is so even after one of the alleles gets lost in each of the subpopulations. In effect, the probability that A₁ will become fixed in a subpopulation is equal to its initial frequency. If the initial frequencies were $p = q = 0.5$, then approximately half of the subpopulations should eventually have A₁ fixed and the other half A₂, and consequently the gene frequency of A₁ and A₂ in the population will not have changed (Table 7.2).

However, although the gene frequency has not changed, the genotype frequency has. In effect, when all subpopulations have either A₁ or A₂ fixed only the two homozygous genotypes will be present. From this we conclude that the main effects of nonrandom breeding and finite population size is 1) to subdivide the population into smaller subpopulations, 2) to increase the probability of gene fixation within a subpopulation, and 3) to increase the frequency of homozygous genotypes, without necessarily changing the gene frequency

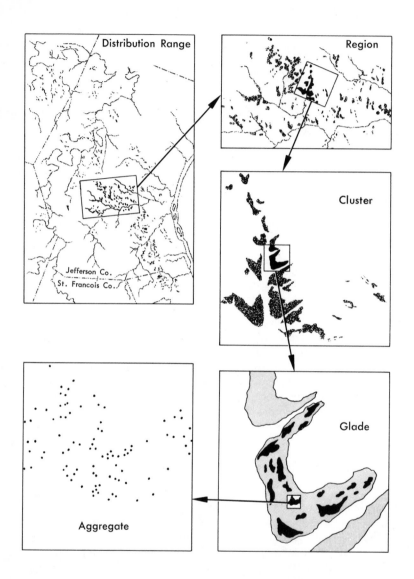

Figure 7.5

Distribution of *Clematis fremontii* var. *riehlii*. The variety is restricted to glades within an area of about 400 square miles in east-central Missouri. Within that range, there is a hierarchy of natural subdivisions: regions of glade concentration; clusters of glades; colonies of the plant, which correspond approximately with glades; and aggregates of very few to perhaps a thousand individuals in each glade. (From R.O. Erickson, 1945. *The Clematis fremontii* var. *riehlii* population in the Ozarks. *Ann. Mo. Bot. Gard.* 32:413–460)

Table 7.2

Estimation of actual and expected heterozygosity in a hypothetical subdivided population.

Subpopulation	N	Gene Frequency		Genotype Number		
		p	q	p^2	$2pq$	q^2
1	450	0.6	0.4	162	216	72
2	50	0.2	0.8	2	16	32
3	100	1	0	100	—	—
4	500	0.3	0.7	45	210	245
5	100	0	1	—	—	100
6	350	0.7	0.3	172	147	31
	1550	0.5	0.5	481	589	480
			Expected	387.5	775	387.5

in the overall population. This represents a deviation from the expected Hardy-Weinberg equilibrium.

Let us now present the same concept in mathematical terms.

Suppose we take k subpopulations each with different gene frequencies p for allele A_1 and q for allele A_2, each subpopulation being in Hardy-Weinberg equilibrium. Let us assume too that all subpopulations have the same numbers of individuals. The gene and genotype frequencies are given in Table 7.2.

The average gene frequency in the total population will be

$$\bar{p} = \frac{\sum\limits_{i=1}^{i=k} p_i}{k},$$

and the variance (p. 73) of the gene frequencies between subpopulations will be

$$\sigma^2 = \frac{\sum p_i^2}{k} - \bar{p}^2 = \overline{p^2} - \bar{p}^2;$$

that is, the variance is equal to the average of the squared gene frequencies minus the square of the average gene frequency.

Now, the average frequency of heterozygotes A_1A_2 in the total population will be

$$\bar{H} = \frac{\sum(2p_iq_i)}{k} = \frac{2\sum p_i}{k} - \frac{2\sum p_i^2}{k}$$

(remember that $q_i = 1 - p_i$)

$$\bar{H} = 2\left(\bar{p} - \frac{\Sigma p_i^2}{k}\right) = 2\left(\bar{p} - \bar{p}^2 + \bar{p}^2 - \frac{\Sigma p_i^2}{k}\right)$$

$$= 2\left[\bar{p} - \bar{p}^2 - \left(\frac{\Sigma p_i^2}{k} - \bar{p}^2\right)\right] = 2(\bar{p} - \bar{p}^2 - \sigma^2)$$

$$= 2(\bar{p}\bar{q} - \sigma^2),$$

since $\bar{p}\bar{q} = \bar{p}(1 - \bar{p}) = \bar{p} - \bar{p}^2$,

and therefore

$$\bar{H} = 2\bar{p}\bar{q}\left(1 - \frac{\sigma^2}{\bar{p}\bar{q}}\right),$$

where σ^2/pq = *standardized variance of gene frequencies*, also known as *Wahlund's variance of gene frequencies*.

The average frequency of heterozygotes in a population where breeding is nonrandom but takes place randomly within subpopulations of equal size is equal to twice the product of the average allele frequencies minus twice the variance of the gene frequency between subpopulations.

Similarly, the mean frequencies for homozygotes are

for A_1A_1: $\bar{D} = \dfrac{\Sigma p_i^2}{k} = \bar{p}^2 + \sigma^2;$

for A_2A_2: $\bar{R} = \dfrac{\Sigma q_i^2}{k} = \bar{q}^2 + \sigma^2.$

These are known as *Wahlund's formulas*.

INBREEDING

So far we have assumed that each individual in the population finds a mate without bias. We have seen that perfect random mating is not possible because of finite population size and neighborhood structure. We now relax the random mating assumption even further by acknowledging that preferential mating between relatives is common in many species.

Breeding with oneself is possible only in hermaphroditic organisms, such as plants and earthworms. However, few hermaphroditic species self-fertilize. Breeding with close relatives, on the other hand, is much more common. Close proximity, ecological preference, morphological resemblance, and biases of many sorts, increase the likelihood of breeding between relatives. Even among humans, family, tribe, geography, national origin, culture, social considerations, and economic factors, bring about stratification in the population. Marriages within each of these strata are much more common than between strata, increasing the likelihood of marriages between relatives. Whenever mates are on the average more closely related than they would be if they had been chosen at random from the population, we say that *inbreeding* takes place. Inbreeding also affects the genotypic composition of the population. This can be seen when an extreme example, selfing (self-fertilizing) in a hermaphroditic organism, is considered. Species that are exclusively selfing are probably non-existent. However, there are a number of plant species that habitually self-pollinate, such as cultivated wheat (*Triticum aestivum*), where about 99 percent of all seeds are the result of selfing, tobacco (*Nicotiana tabacum*), cotton (*Gossypium barbadense*), and many of the common garden weeds, such as crabgrass (*Digitaria sanguinalis*), shepherd's purse (*Capsella bursa-pastoris*), and chickweed (*Stellaria media*).

Let us start with a population that is in Hardy-Weinberg equilibrium, where the frequency p of A_1 is equal to the frequency q of A_2, and both are equal to 0.5, so that $p^2 = 0.25$; $2pq = 0.50$; and $q^2 = 0.25$. All breeding is strictly by selfing from this point on. Consequently, all homozygotes will breed true, whereas the heterozygous individuals will segregate 1/2 homozygotes and 1/2 heterozygotes in each generation. As Table 7.3 shows, the proportion of heterozygotes is halved in each generation. Eventually the population will be formed exclusively of homozygous individuals.

The same result is obtained if the initial population has different gene frequencies; let us say the frequency of A_1 is $p = 0.3$ and that of A_2 is $q = 0.7$. The proportion of heterozygotes is halved in each generation, adding equal numbers of homozygotes to the population, until the population consists of a frequency p of A_1A_1 homozygotes and a frequency q of A_2A_2 of homozygotes. In general, the expected frequency of heterozygotes in a population that is mating exclusively by selfing at any generation n is

$$2pq \times (1/2)^n,$$

where $2pq$ was the initial frequency of heterozygotes at generation 0, and n is the number of consecutive generations of selfing.

Table 7.3

Proportion of homozygous and heterozygous genotypes in a self-fertilizing population.

	Genotypes		
Generation	A_1A_1	A_1A_2	A_2A_2
0	—	1	—
1	1	2	1
2	3	2	3
3	7	2	7
4	15	2	15
5	31	2	31
10	1023	2	1023
n	$2^n - 1$	2	$2^n - 1$

If only some individuals in the population self, and others breed randomly, the proportion of homozygotes will be higher and that of heterozygotes lower than predicted by the Hardy-Weinberg equilibrium, but the relationship will no longer be simple. At this point we must introduce a new concept, the inbreeding coefficient F.

Inbreeding coefficient

This very useful and much used statistic was first introduced in 1923 by the great American pioneer of population genetics, Professor Sewall Wright, who called it the "fixation" index.

Before defining F, we have to define what is meant by the terms "identical by descent" and "alike in state." We say that a homozygous individual has two genes that are *identical by descent* when the genes in question can be traced to one *common ancestor*. We also refer to that type of homozygote as *autozygous*. However, if the two alleles in a homozygous individual are identical because they are the result of *two independent mutations*, we say that those genes are *alike in state* and that the individual is *allozygous*. This may seem a very subtle distinction, and indeed it is, but it is necessary because, as we shall see, *inbreeding* increases the frequency of *autozygous* individuals. The coefficient of inbreeding, F, measures identity by descent through the degree of relationships of mating pairs. We now will define the *coefficient of inbreeding* as "the probability that an individual receives at a given locus two genes that are identical by descent."

Let us assume for a moment that we are dealing with an idealized population of self-compatible plants, where mating takes place strictly at random. We further assume that in the base population each of N_0 individuals has two alleles for the locus under study, which are different from each other and from every other allele in the population. Being heterozygous, each individual will produce two kinds of gametes (one kind with one allele, the other kind with the other allele). Since no two alleles are alike, the total number of kinds of gametes that will be produced in the population is $2N_0$. When these gametes unite (strictly at random, as we stated) to form zygotes of the next generation, there is a probability of $1/2N_0$ that two identical gametes will unite to form a homozygote. Since we started at generation 0 with a population where all gametes were different, the probability that in the first filial generation an individual will have its two alleles *identical by descent* is the same as the probability of having two identical alleles as defined above, namely, $1/2N_0$. This is the value of the inbreeding coefficient, F, for the first generation.

In the next generation there will be N_1 individuals. Some of them will be homozygous and some heterozygous. Let us assume, however, that all the alleles are different, as was true in the previous generation. There would then be a probability $1/2N_1$ that two identical gametes would unite to form a homozygote. Although these are not the only homozygotes in the second generation, they are the only ones homozygous in this generation by virtue of the union of two gametes *identical by descent* of an allele in generation 1. But not all alleles are different; some are identical, having descended from a common "ancestor" in generation 0. Consequently, homozygotes will be produced other than those identical by descent from an allele in generation 1, for example, those produced by the union of two gametes from a homozygous individual, or the homozygous individuals that are produced by the union of two gametes from different individuals in F_1 but descending from the same ancestor in generation 0. What is the probability of this second type of autozygote being produced? If $1/2N_1$ is the proportion of zygotes that are autozygous by virtue of the union of two gametes identical by descent of an allele in generation 1, there obviously remains $(1 - 1/2N_1)$ zygotes where this did not occur. The probability that among the remainder a homozygote will be produced by random mating of two identical gametes descended from an ancestor allele in generation 0 has already been calculated as the F for the first generation $(= 1/2N_0)$, which is the probability that any two alleles in a zygote are copies of a single allele from generation 0. Consequently, we have

First generation: $\quad F_{(1)} = \dfrac{1}{2N_0}$

Second generation: $F_{(2)} = \dfrac{1}{2N_1} + \left(1 - \dfrac{1}{2N_1}\right) F_{(1)}$

Third generation: $\quad F_{(3)} = \dfrac{1}{2N_2} + \left(1 - \dfrac{1}{2N_2}\right) F_{(2)}$

$$\cdots \qquad \cdots \quad \cdots$$

nth generation: $\quad F_{(n)} = \dfrac{1}{2N_{n-1}} + \left(1 - \dfrac{1}{2N_{n-1}}\right) F_{(n-1)}$

Thus we see that the inbreeding coefficient is made of two parts: a part $1/2N_{(n-1)}$ derived from new inbreeding and a part which can be attributed to previous inbreeding and which has the inbreeding coefficient of the previous generation.

An important corollary is that if the population suddenly increased in size so that the new inbreeding was negligible because $1/2N_{(n-1)}$ became a very small number, the previous inbreeding would not be eliminated but would remain where it was before the increase in population size.

The value $1/2N_{(n-1)}$ applies of course only to the idealized population. We can remove that constraint if we call $1/2N_{(n-1)}$ "Delta F," that is, ΔF, the change in the inbreeding coefficient. Then

$$F_n = \Delta F + (1 - \Delta F) F_{(n-1)}$$
$$= \Delta F - (\Delta F) F_{(n-1)} + F_{(n-1)},$$
$$F_n - F_{(n-1)} = \Delta F(1 - F_{(n-1)}),$$
$$\Delta F = \frac{F_n - F_{(n-1)}}{1 - F_{(n-1)}},$$

where ΔF is the rate of inbreeding and is equal to the part of the population that inbred in the last generation divided by the part that did not. When the inbreeding coefficient is expressed in terms of ΔF, the equation for F_n is valid for any breeding system and is not restricted to the original population.

Let us now return to the original population and assume that all individuals self. Let us again assume that in the base population each individual has two alleles for the locus under study, which are different from each other and from every other allele in the population. Half of the offspring of each heterozygote are homozygotes identical by descent in each generation. Consequently, $\Delta F = \frac{1}{2}$ for each generation. Consequently, we have

First generation: $\quad F_{(1)} = \frac{1}{2}$

Second generation: $F_{(2)} = \frac{1}{2} + (1 - \frac{1}{2}) F_{(1)} = \frac{3}{4}$

Third generation: $\quad F_{(3)} = \frac{1}{2} + (1 - \frac{1}{2}) F_{(2)} = \frac{7}{8}$

$\cdots \qquad\qquad \cdots \quad\cdots$

nth generation: $\quad F_{(n)} = \frac{1}{2} + (1 - \frac{1}{2}) F_{(n-1)}$

To calculate the number of heterozygotes (H) in the nth generation, we proceed as follows:

First generation: $\quad H_1 = \frac{1}{2} H_0 = H_0 - \frac{1}{2} H_0$

Second generation: $H_2 = H_1 - \frac{1}{2} H_1$

and substituting,

$$H_2 = H_0 - \frac{1}{2} H_0 - \frac{1}{2} (H_0 - \frac{1}{2} H_0)$$
$$= H_0 - \frac{3}{4} H_0 = H_0(1 - \frac{3}{4})$$

or, in other words, the number of heterozygotes equals the number of original heterozygotes multiplied by 1 minus the fraction of heterozygotes that has been made homozygous by recurrent selfing, which is F. Consequently,

nth generation: $H_n = H_0(1 - F_n)$

This relationship between the number of heterozygotes and the inbreeding coefficient F is always true whatever the F values may be.

Equilibrium with Selfing and Random Mating Combined. Under continuous selfing the proportion of heterozygous genotypes is halved in each generation. However, if selfing is interrupted by one generation of perfect random breeding, the initial Hardy-Weinberg equilibrium of genotypes ($p^2 + 2pq + q^2$) is restored (Chapter 4). In nature neither continuous selfing nor perfect random breeding takes place, but populations show mixtures of some selfing (or inbreeding with close relatives) by some individuals and more or less random breeding by others. Such populations can be considered to consist of a fraction (s) of selfed individuals and a fraction ($1 - s$) of randomly mating individuals.

Table 7.4

Genotype frequencies and F values for six genic loci in a population of *Avena barbata* from CSA locality (Napa County, California). (Genotypes 11 and 22 are the homozygotes; 12 is the heterozygote.)

Gene	Genotype	Frequencies	Sample Size	F
E_1	11	0		
	12	0	54	1.00
	22	1		
E_4	11	0.30		
	12	0.11	54	0.76
	22	0.59		
E_7	11	0		
	12	0	54	1.00
	22	1		
E_{10}	11	0.46		
	12	0.13	85	0.70
	22	0.41		
P_5	11	0.40		
	12	0.15	86	0.70
	22	0.45		
APX_5	11	0.48		
	12	0.11	86	0.78
	22	0.41		

From D.R. Marshall and R.W. Allard, 1970. Maintenance of isozyme polymorphisms in natural populations of *Avena barbata*. *Genetics* 66:393–399.

The value of F in a population at equilibrium with selfing and random mating combined is

$$F = \frac{s}{2-s}.^*$$

Two geneticists, D.R. Marshall and R.W. Allard, determined that the average degree of outcrossing in a California population of wild oats, *Avena barbata*, was only 0.014; that is, 98.6 percent of the plants in the population produced seeds by selfing. The inbreeding coefficient is

$$F = \frac{s}{2-s} = \frac{0.986}{1.014} = 0.972.$$

* See E.B. Spiess, *Genes in Populations*, p. 246, for a derivation of this formula.

They then determined the genotype frequencies for six loci in the same population (Table 7.4). From these values an independent estimate of F can be obtained as follows:

$$H_n = H_0 (1 - F), \qquad \text{(p. 172)}$$

and therefore

$$1 - F = \frac{H_n}{H_0},$$

and

$$F = 1 - \frac{H_n}{H_0},$$

where H_0 is obtained from the expected number of heterozygotes under random breeding ($2pq$).

For the four heterozygous loci the F values calculated from the number of heterozygotes in the population was consistently lower (Table 7.3) than was estimated from the degree of outcrossing. Consequently, either there must be a greater degree of outcrossing or there is selection in favor of the heterozygous genotypes. Marshall and Allard favored the second hypothesis. We will return to it in Chapter 9.

Inbreeding in human populations

Each of us has two parents, four grandparents, eight great grandparents, and so on, so that for n generations back, each of us has 2^n ancestors. This means that each of us had more than 700,000 ancestors 20 generations back. If we calculate four generations per century, 20 generations span five centuries. If all the inhabitants of the United States that are not immediately related (let us assume one out of every 20) had an entirely independent set of ancestors, three trillion ancestors in the fifteenth century would have been required. This is several orders of magnitude greater than the population of the world today. Humans must be more interrelated than most are aware of.

In human populations it is customary to measure the average identity by descent, now called *consanguinity*, in a population as the average inbreeding coefficient of its individuals. This quantity, called α, is defined as

$$\alpha = \Sigma p_i F_i,$$

where p_i = frequency of individuals in the population with inbreeding coefficients of F_i.

We can see from the estimates of Table 7.5 that α varies by more than two orders of magnitude. Large social or religious groups with strict rules against marriages between close relatives, such as Catholics, have low values of alpha, whereas small groups (such as Samaritans) or groups that habitually

Table 7.5

Some estimates of inbreeding, α.

U.S., Roman Catholics	0.00009
Argentina, Roman Catholics	0.00058
U.S., Mormons	0.00038
Japan	0.0046
Guinea	0.026
India, Andra-Pradesh	0.032
Israel and Jordan, Samaritans	0.0434
Third-cousin marriage	0.0019
Second-cousin marriage	0.0078
First-cousin marriage	0.0625
Uncle/niece marriage	0.125

Adapted from L.L. Cavalli-Sforza and W.F. Bodmer, 1971. *The Genetics of Human Populations.* San Francisco: Freeman.

allow first-cousin marriages (such as some East Indian societies) have much higher values of α.

Inbreeding can have several consequences: (1) It leads to an increase in homozygosity. (2) Recessive genotypes become expressed. (3) Since many recessives are deleterious, inbreeding usually leads to a decrease in size, fertility, vigor, yield, and fitness. (However, the degree of this effect is in proportion to the number of deleterious homozygous genes in the population. In species that normally inbreed, such as wheat and many of our crops, the number of deleterious genes in the population is very low.) (4) There is an increase of phenotypic variability, since similar genes tend to be concentrated in the same individual, which is likely to differ from other individuals.

EFFECTIVE POPULATION NUMBER

So far in this chapter we have assumed a finite and constant population of N individuals produced each generation from the union of male and female gametes randomly sampled from the previous generation. We showed how heterozygosity tends to decrease and homozygosity to increase in such a population.

Real populations seldom keep their size fixed from year to year, since circumstances dictate that some years favor growth and reproduction and others do not. In populations of stable habitats, the yearly fluctuations will be less drastic than those experienced by species that exploit temporary habitats, such as weeds or species of temporary ponds. Furthermore, the population is made up of a mixture of breeders and nonbreeders, and among the breeders, reproductive success varies with age and size, as well as with other factors.

We saw (p. 171) that in an idealized population of self-compatible plants, where mating takes place strictly at random, there is a probability of $1/2N_0$ that two identical gametes will unite to form a homozygote, and in general in generation n there is a $1/2N_{(n-1)}$ probability of such an event.

If instead of dealing with a population of hermaphroditic, self-compatible plants, we are dealing with a population of animals consisting of N_m males and N_f females, the formulas have to be revised slightly, but the reasoning will still be the same. Since each zygote is the result of the union of *one male* and *one female* gamete, the probability that two genes in different individuals in generation n are both derived from males in generation $n-1$ is $1/4$ $(1/2 \times 1/2)$, and the probability that they came from the *same* male is $1/4N_m$. Likewise, the probability of their coming from the *same female* is $1/4N_f$.

Whenever the number of males is equal to the number of females in the population ($N_m = N_f$),

$$\frac{1}{4N_m} + \frac{1}{4N_f} = \frac{1}{4N_m} + \frac{1}{4N_m} = \frac{2}{4N_m} = \frac{1}{2N_m} = \frac{1}{N}.$$

However, when $N_m \neq N_f$,

$$\frac{1}{4N_m} + \frac{1}{4N_f} = \frac{1}{N_e} > \frac{1}{N},$$

where N_e is the *effective population number*, which is equal to the size of an ideally behaving population that has the same increase in homozygosity as the observed population number.

Any deviation from the ideal population (where $N_m = N_f$) will make the effective population number less than the actual population number. We have already mentioned two factors that lower the effective population number: inequality of the numbers of males and females, and population size fluctuations from generation to generation. Other factors are uneven contribution of gametes by different individuals in the population, nonrandom breeding, and subdivision of the population into subpopulations. Since all these factors are present in real populations, it follows that the effective population size is always less than the actual number of individuals. There are very few precise studies of the effective population size, but in studies of *Drosophila* it varied from 0.48 to 0.71 $\times N$; in humans it was about 0.69 to 0.95 $\times N$; and in the snail *Lymnaea* it was 0.75 $\times N$. In plants the values appear to be still lower. The effective population size is a measure of the degree to which an actual population deviates from the ideal population. The smaller the ratio of N_e/N, the greater the tendency of the population to inbreed and increase its level of homozygosity.

RANDOM GENETIC DRIFT

The gene pool of each succeeding generation in a population represents a sample of the parental generation that gave rise to it. Consequently, it is subject to sampling error variations. These fluctuations are greater, the smaller the sample size, i.e., the smaller the effective population number. All populations are at least occasionally faced with times of hardship when their numbers are drastically decreased. Consequently, changes in gene frequency occur in populations that are *not* the direct result of natural selection. This process is known as *random genetic drift.*

The importance of random genetic drift as an evolutionary force has been the object of much discussion. When a population is being subjected to selection, any random change in the *same direction* as selection will have the result of reinforcing or speeding up selection. Conversely, any random change in the *opposite direction* of selection retards selection. In this sense, random genetic drift is equivalent to "static noise" in the system. However, occasionally these perturbations can lead to the loss of the fittest allele and the fixation of the less fit. When that occurs, a nonreversible change has taken place (unless the lost allele is restored by mutation). The probability of such changes occurring is indirectly related to the intensity of selection (the stronger the selection, the less likely the nonreversible change) and to the size of the population (the larger the population, the less likely the change).

If the changes brought about by random drift appreciably lower the fitness of the population, the population probably will become extinct. Therefore not all the changes brought about by random drift will survive. However, it is possible that occasionally a number of genetic changes can take place that alter the genetic structure at many loci, producing an organism that is fairly distinct genetically and also of high fitness. Such a process Ernst Mayr called a "genetic revolution." A genetic revolution is a highly improbable event, and we have only indirect and limited evidence for its occurrence. According to some biologists, every new species originates as a result of a genetic revolution. However, this contention, which is questioned by many geneticists, still has to be demonstrated. When a new population is established as a result of the immigration of a very few individuals, many genes from the mother population are lost. This process, known as the *founder principle,* may also play an important evolutionary role.

The Italian-American geneticist Luigi Cavalli-Sforza studied the distribution of blood groups in the populations of the cities and villages in the Parma valley in Italy (Fig. 7.6), mainly to test the effects of population size on drift. The geology of the valley of the Parma River determines that in the upper steep reaches of the valley, villages are small (200–300 people), but further downstream the villages become increasingly larger, with the city of Parma the largest (175,000). There has been no major immigration into the Parma valley since the seventh century B.C., and consequently demographic and genetic equilibrium has presumably been reached. However, within the region, the mountain villages are more stable and have fewer migratory exchanges than the villages of the lower valley. Cavalli-Sforza studied the frequency of the blood types as a function of population density. As Fig. 7.7 shows, the degree of genetic variation between populations decreases as the population density

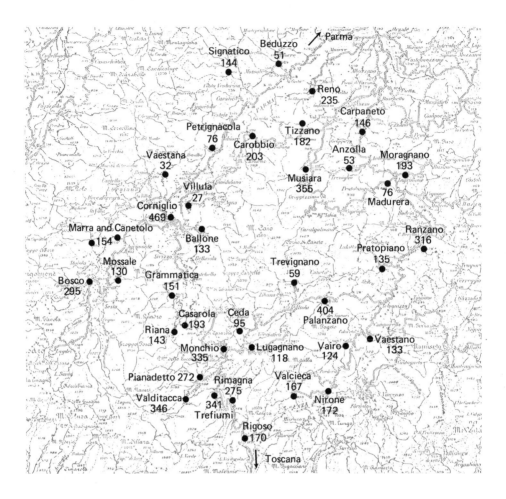

Figure 7.6

Map of villages of the Upper Parma valley (northern Italy) showing main connecting roads and village sizes in 1950. (Adapted from *The Genetics of Human Populations* by L.L. Cavalli-Sforza and W.F. Bodmer. W.H. Freeman and Company. Copyright © 1971)

increases, but it is significant only in the small hamlets. Inbreeding is also a factor in the mountain villages (Table 7.6), but much less so in the city of Parma ($F = 0.002$).

In summary, not all changes in a population are the direct result of natural selection. Some beneficial changes may occur by chance alone. The smaller the population, the greater the effects of random sampling and the more likely that changes are the result of chance. Although these random effects may at times be important events in the evolution of a lineage, this is not often so.

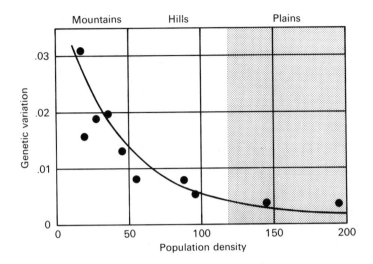

Figure 7.7

Variation in the frequency of a blood type between one village and another was greatest in the isolated upland hamlets and declined as population density increased farther
down the valley in the hill towns, on the plain, and in the city of Parma. The measure
of genetic variation used is the F value. (From L.L. Cavalli-Sforza, 1969. "Genetic
drift" in an Italian population. *Scientific American* 221(2):30–37. Copyright © 1969
by Scientific American, Inc. All rights reserved)

Table 7.6

F values for three blood group loci
among 37 parishes in the upper Parma
valley (66 villages with 15,000 persons).

Blood Group Alleles	$F = \dfrac{q^2}{\bar{p}\bar{q}}$
M-N	0.048
A_1	0.052
A_2	—
B	0.034
O	0.061
r	0.038
R_1	0.033
R_2	0.017
rh remainder	0.037
Average	0.0356 ± 0.006

From L.L. Cavalli-Sforza, I. Barrai, and
A.W.F. Edwards, 1964. Analysis of human evolution under random genetic drift.
*Cold Spring Harbor Symposia on Quanti-
tative Biology* 29:9–20.

On the other hand, some change in gene frequency due to random effects is a recurrent aspect of the evolutionary process.

The tendency to inbreed and increase the level of homozygosity can be countered by mutation, recombination, and selection, but not without some genetic consequences. In the next chapter we explore these consequences.

SUGGESTED FURTHER READING

General references

Cavalli-Sforza, L.L., and W.F. Bodmer. 1971. *The Genetics of Human Populations.* San Francisco: Freeman.

Crow, J.F., and M. Kimura. 1970. *An Introduction to Population Genetics Theory.* New York: Harper and Row.

Nei, M. 1975. *Molecular Population Genetics and Evolution.* Amsterdam: North-Holland.

Wallace, B. 1968. *Topics in Population Genetics.* New York: Norton.

Specialized articles and reviews

Buri, P. 1956. Gene frequency in small populations of mutant *Drosophila. Evolution* 10:367–402.

Carson, H.L. 1967. Inbreeding and gene fixation in natural populations. In R.A. Brink (ed.), *Heritage from Mendel.* Madison: University of Wisconsin Press.

Cavalli-Sforza, L.L. 1969. Genetic drift in an Italian population. *Scientific American* 221(2):30–37.

Crow, J.F. 1954. Breeding structure of populations II. Effective population number. In O. Kempthorne, T.A. Bancroft, J.W. Gowen, and J.L. Lush (eds.), *Statistics and Mathematics.* Ames, Iowa: Iowa State College Press.

Crow, J.F., and T. Maruyama. 1971. The number of neutral alleles maintained in a finite geographically structured population. *Theoret. Pop. Biol.* 2:437–453.

Erickson, R.O. 1945. The *Clematis fremontii* var. *richlii* population in the Ozarks. *Ann. Mo. Bot. Gard.* 32:413–460.

Felsenstein, J. 1971. Inbreeding and variance effective numbers in populations with overlapping generations. *Genetics* 68:581–597.

Greenwood, J.J.D. 1974. Effective population numbers in the snail (*Cepaea nemoralis*). *Evolution* 24:513–526.

Kimura, M. 1962. On the probability of fixation of mutant genes in a population. *Genetics* 47:713–719.

Kimura, M., and J.F. Crow. 1964. The number of alleles that can be maintained in a finite population. *Genetics* 49:725–738.

Kimura, M., and T. Ohta. 1969. The average number of generations until fixation of a mutant gene in a finite population. *Genetics* 61:763–771.

Nei, M., and M.W. Feldman, 1972. Identity of genes by descent within and between populations under mutation and migration pressures. *Theoret. Pop. Biol.* 3: 460–465.

Wright, S. 1923. Mendelian analysis of the pure breeds of livestock: I. The measurement of inbreeding and relationship. *J. Heredity* 14:339–348.

Wright, S. 1951. The genetical structure of populations. *Ann. Eugenics* 15:323–354.

Chapter **8**
Recombination

In the previous chapter we saw that nonrandom breeding and finite population size can profoundly affect the genetic structure of a population, primarily by decreasing the proportion of heterozygous genotypes and by causing a loss of alleles. Consequently, if these forces were to act unopposed by other factors in a relatively uniform environment, sooner or later populations would become totally homozygous at all loci. However, there are forces that oppose the loss of alleles (Chapter 6). The primary force is mutation, but recombination and certain types of selection are also very important. Also, environments are seldom entirely uniform. We will now discuss recombination in greater detail than in Chapter 6.

FACTORS AFFECTING RECOMBINATION

Table 8.1 shows the major factors that regulate the amount of recombination. We have already discussed the role of chromosome number and of crossing over in affecting recombination (Chapter 5) and the effect of gene flow (Chapter 6).

Table 8.1

Factors that regulate recombination.

---

1. Chromosome number
2. Frequency of crossing over
3. Gene flow
4. Length of generation
5. Population size
6. Breeding system
7. Incompatibility system
8. Crossability barriers and external isolating mechanisms

Adapted from V. Grant, *Cold Spring Harbor Symposia on Quantitative Biology*, 23:329, 1958.

Length of Generation. The effect of the length of generation can be simply stated. If the number of recombinations is given by the formula (p. 130)

$$\left[\frac{r(r+1)}{2} \right]^n,$$

the number of recombinants in a unit of time (that is, per year) is obtained by multiplying the number of recombinations by the number of generations (N) in a year:

$$\left[\frac{r(r+1)}{2} \right]^n \times N.$$

Other factors being equal, an animal with two generations a year will produce twice as many recombinants in a year as an animal with only one generation a year.

The length of generation has another important effect. Not all the offspring will survive to adulthood. In most natural situations, on the average the number of offspring that survive is roughly equivalent to the number of adults that die. If the species is long-lived, the rate of turnover is slow. But long-lived organisms produce more offspring over a lifetime than do short-lived ones of equal fertility. It follows, then, that a much smaller proportion of the offspring survives to adulthood each year in a long-lived organism than in a short-lived one. For an extreme example, consider the sequoias. If we assume that there are as many sequoias as possible in the area now occupied by them, the rate of replacement is about one offspring per adult sequoia every 100 to 500 years, although the number of seeds produced by one plant in that span of time is several million. On the other hand, in a population of annual plants, such as wild lettuce, each plant has to be replaced each year, and yet the number of seeds produced by each individual is much less than the number produced by the sequoias. The life span is very short in most animals, exceeding two years in only a very small number of species. Among plants extremely great ages are known, not only among trees but even in many kinds of herbs that reproduce vegetatively.

Population Size. The larger the population, the greater the possibility of storing genetic variability. In the extreme case of a population consisting of two organisms, there can be at most four different alleles for each gene and the potential of forming by sexual reproduction 10 different combinations for each gene in the following generations. In a population with three individuals, there is a maximum of six alleles for each gene and a potential of forming 21 different combinations; with four individuals, eight alleles and 36 combinations; with five, 10 and 55; with 10, 20 and 210; and with n individuals, there can be a maximum of $2n$ alleles and a potential of forming $n(2n + 1)$ recombinants (see Chapter 6 for the derivation of this formula). Other effects of population size were presented in Chapter 7.

In no known case is a population heterozygous for each genic locus, but the larger the population, other factors being equal, the greater the probability of forming new combinations, and the greater the variability that may be present.

Breeding System and Incompatibility System. Not all organisms have the same mode of reproduction. The different mechanisms employed are referred to as breeding systems.

Table 8.2 lists the major breeding systems of plants and animals. Asexual systems are those where no recombination takes place. Since in this type of reproduction each plant or animal reproduces its own kind without dependence on others, the reproductive mechanism itself is very efficient. In obligate *apomicts* (plants that produce seed without fertilization) and in parthenogenetic animals, the only new source of variability is gene mutation. Obligate asexuality is very rare, probably because the highly uniform populations that result are eliminated in the course of time when environmental conditions change. *Vegetative reproduction* is the ability of many plants to produce new physiological individuals through the production of runners, bulbs, or other such structures.

In certain insects, particularly the aphids, there is an alternation of a sexual generation (usually in the fall) with asexual ones (in the spring and summer). Also, in some plants that are usually asexual, occasional sexual crosses occur (not necessarily tied to the seasons, as in animals). In this way there is still the possibility of genetic recombination, coupled with the advantages of asexuality. Anybody who has to fight aphids on roses knows how fast and efficiently aphids can reproduce!

Several mechanisms exist that regulate the amount of self-fertilization in sexually reproducing organisms. True selfing is possible only in hermaphroditic organisms. Obligatory self-fertilizing hermaphrodites are very rare in the animal kingdom, and though more frequent in the plant kingdom, they are

Table 8.2

Major breeding systems in plants and animals.

A. *Asexual:*

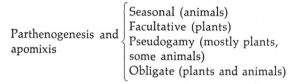

Parthenogenesis and apomixis
- Seasonal (animals)
- Facultative (plants)
- Pseudogamy (mostly plants, some animals)
- Obligate (plants and animals)

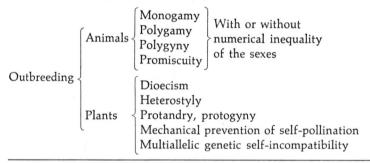

Vegetative reproduction
- With sexual reproduction — mostly plants, some animals
- Without sexual reproduction — Some animals and plants

B. *Sexual:*

Complete inbreeding (plants and a few animals)
Close inbreeding (plants and animals)

Outbreeding
- Animals
  - Monogamy
  - Polygamy
  - Polygyny
  - Promiscuity
  — With or without numerical inequality of the sexes
- Plants
  - Dioecism
  - Heterostyly
  - Protandry, protogyny
  - Mechanical prevention of self-pollination
  - Multiallelic genetic self-incompatibility

---

relatively rare among plants, too. Most animals have individuals of different sexes, as do *dioecious* plants, a fact that ensures outbreeding. Although most plants are hermaphroditic, many mechanisms exist that prevent the sperm from fertilizing the egg of the same plant. Some of the major ones are listed at the end of Table 8.2. In *heterostylous* plants there are two types of flowers: In some the styles are longer than the stamens, and in others the stamens are longer than the styles (Fig. 8.1). Pollen of short stamens is normally viable only on short styles, and pollen of long stamens on long styles, thereby ensuring outbreeding.

Protandrous flowers are those in which the pollen matures ahead of the female organs, and *protogynous* flowers are those in which the female organs ripen before the pollen of the same flower. Here again there is a reduced possibility of self-fertilization. In addition, there are various mechanical contrivances that keep pollen away from the stigmas.

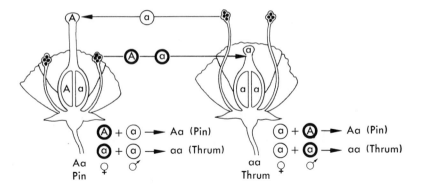

Figure 8.1

Crossing relationships between heterostylous plants. A pollen grain produced by the pin plant (with a long style) will be viable only on the style of a thrum plant (with anthers longer than styles). Both pin and thrum plants will have one-half pins and one-half thrums in their progeny. (From O.T. Solbrig, 1966. *Evolution and Systematics.* New York: Macmillan)

The most common mechanism to prevent self-fertilization in plants is *genetic incompatibility* (Fig. 8.2). A self-incompatibility gene, called the S gene, is the controlling factor. The S gene has many alleles, referred to as S^1, $S^2, S^3, \ldots, S^n$, of which there are often more than 30, 50, or even 100 in a population. A plant will not ordinarily accept pollen with the same allele as itself. If a plant is of the constitution S^1S^2, it will not be fertilized by S^1 or S^2 pollen, which are the two types of pollen the plant produces (remember that the pollen is haploid, and meiosis has preceded its formation—see Chapter 5). An S^1S^2 plant resists fertilization by any other S^1S^2 plant in the population, and when an S^1S^2 plant is crossed with an S^1S^3 plant, it is the S^3 pollen that will be viable on the S^1S^2 plant. The alleles determining self-incompatibility apparently produce their effect by controlling the rate of pollen tube growth or even pollen germination. The stylar tissue produces a substance (probably an antigenlike product) that inhibits or drastically slows down the growth of the pollen tubes of the same genetic constitution.

Crossability barriers and external isolating mechanisms are the subjects of Chapters 11 and 12.

Recombination is then the result of a chain of processes that begins with gene mutation and continues through the sexual cycle. Genes are associated in chromosomes that restrict recombination. Individual organisms are associated in breeding populations and species that set a limit to recombination. The genes of a whale cannot be recombined with those of a walrus. Consequently, only a fraction of the potentialities of recombination are ever realized in nature.

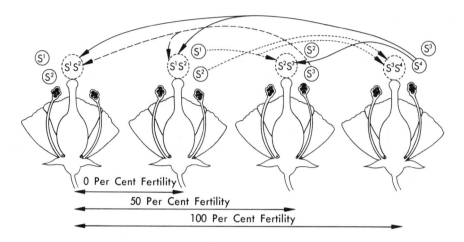

Figure 8.2

Crossing relationships between self-incompatible plants. Unbroken circles show pollen grains; broken circles show genotypes of styles. A pollen grain will fertilize only a plant that does not have the incompatibility allele the pollen grain carries. Thus only 50 percent of the pollen from an S^1S^2 plant (the pollen with the S^1 allele) is capable of fertilizing an S^2S^3 plant. (From O.T. Solbrig, 1966. *Evolution and Systematics*. New York: Macmillan)

CLASSIFICATION OF RECOMBINATION SYSTEMS

The fraction of attainable recombination varies, depending primarily on the breeding system but also on such factors as the size of the population, frequency of crossing over, and chromosome number. Following a proposal of the American geneticist Hampton Carson, we can classify organisms according to their recombination system into those with a relatively open system, those with a restricted system, and those with a closed system.

A *closed recombination system*, exemplified by obligate parthenogenetic or apomictic organisms, such as the common dandelion (*Taraxacum officinale*), is one where no recombination takes place. A *restricted recombination system* is one that imposes strong barriers to recombination, such as in facultative apomicts (normally asexual plants that nevertheless can breed sexually) or close inbreeders, for example, species of aphids that normally reproduce asexually but have a sexual generation in the autumn. An *open recombination system* is one where a large number of the potential recombinants are normally produced, such as in most vertebrates.

We may ask ourselves why these variations in recombination systems exist. Why do certain organisms have a closed recombination system and others an open one, sometimes even in closely related species? The question has not yet been fully answered. One way of approaching the problem is to view it as an economic problem. Each type of recombination system brings certain benefits to the organism and carries different "costs." The "benefits" minus the "costs" translate into a fitness increase (or decrease), which we call "profit." Natural selection should favor within each population the individuals with the recombination system that shows the greatest "profit," i.e., the highest fitness.

Open recombination systems have greater "costs" and presumably also show greater "benefits." The costs of an open recombination system can be arbitrarily divided into three types: (1) the genetic load, (2) the energy cost of out-breeding, and (3) the meiotic costs. The benefits presumably consist in the gain in fitness an organism achieves by producing offspring with a higher probability of surviving to maturity because it has a special genetic constitution resulting from a cross. We now take up each of these "costs" and "benefits" in detail.

GENETIC COSTS AND THE CONCEPT OF GENETIC LOAD

In the previous chapter we showed that all populations tend to inbreed; that is, the number of homozygote genotypes tends to increase, and alleles tend to be lost. We showed that this is a consequence of finite population size, non-random breeding, and the semistochastic nature of the reproductive system. Counteracting this tendency means favoring heterozygotes over homozygotes, and rare alleles over common ones. Consequently, there is a wastage of alleles, or "genetic cost," which we call the *genetic load*.

The term "genetic load" was first used by the great American geneticist and Nobel laureate Hermann J. Müller in 1950. His use of the term was very specific and much more restricted than its present use. Specifically, Müller used it to refer to the fact that in human populations recurrent mutations lead to the accumulation in heterozygous combinations of alleles that are deleterious or debilitating in a homozygous state. As the frequency of these alleles increases, homozygous zygotes are produced that either die before birth (resulting in spontaneous abortion) or produce defective individuals.

The situation alluded to by Müller has broader implications. Let us imagine a population of genetically dissimilar organisms living in a perfectly uniform environment, and let us further assume that we can measure precisely the fitness of each individual. Some individuals will necessarily be more fit than others, and one will have the highest fitness, W_{max}. Obviously, the average fitness of this population is lower than one made up of genetically uniform individuals, each with W_{max} fitness. The loss in fitness, L, of the first population can be represented by

$$L = \frac{W_{max} - \overline{W}}{W_{max}},$$

where $\overline{W}$ is the average fitness of the population. The loss in fitness, L, is called the *genetic load*. Note that the load applies to the *population* and not to the individual. Assuming that the W_{max} individual is heterozygous, L is a way of measuring the cost of counteracting the forces that decrease heterozygosity.

Any factor that tends to decrease the fitness of an individual contributes to the genetic load of the population. However, two phenomena are paramount: mutation and segregation.

The Mutational Load. In Chapter 6 we showed that the equilibrium frequency of a recessive allele under adverse selection is

$$q = \sqrt{\frac{u}{s}},$$

where $u = $ the mutation rate and $s = $ the selection coefficient against the homozygote. If the allele is dominant, the equilibrium frequency is

$$q = \frac{u}{hs},$$

where h is the degree of dominance.

In a population consisting of three genotypes, A_1A_1, A_1A_2, and A_2A_2, where A_2 is a completely recessive deleterious allele, the average fitness of the population is the sum of the products of the genotypic frequencies and their individual fitnesses, as in the following table (see Chapter 6).

	Frequency		Fitness (W)	
$A_1A_1 =$	p^2	$\times$	1	$= p^2$
$A_1A_2 =$	$2pq$	$\times$	1	$= 2pq$
$A_2A_2 =$	q^2	$\times$	$1 - s$	$= q^2 - sq^2$
			Average fitness	$= p^2 + 2pq + q^2 - sq^2$
				$= 1 - sq^2$

Consequently, in this case the value of the genetic load is sq^2. By substituting $\sqrt{u/s}$ for q at equilibrium, we can calculate the fraction of the genetic load that owes its existence to recurrent mutation.

$$L = sq^2 = s\left(\sqrt{\frac{u}{s}}\right)^2 = s\left(\frac{u}{s}\right) = u.$$

That is, the mutational load at equilibrium due to a completely recessive allele equals the rate at which it is introduced into the population by mutation and is independent of the effect of the allele on fitness. This is due to the fact that the frequency at equilibrium of the allele is inversely proportional to its deleterious effect.

By the same reasoning we can show that for semidominant alleles

$$W = 1 - 2pqhs - sq^2$$
$$\simeq 1 - 2hsq,$$

since p in the case of populations with deleterious genes is close to unity and sq^2 is very small.

Substituting u/hs for q gives

$$L = hsq = \frac{2hsu}{hs} = 2u.$$

Consequently, the mutational load of a deleterious allele in a population lies between its mutation rate and twice that value, depending on the degree of dominance of the allele, regardless of how harmful it is.

The Segregational Load. In Chapter 6 we saw that sometimes the heterozygote has a higher fitness than either homozygote. We will now consider this situation in greater detail.

Imagine a population differing only at one locus with two alleles, A_1 and A_2, with frequencies p and q, respectively, with initial frequencies and relative fitness of the three genotypes, as follows:

Genotype	A_1A_1	A_1A_2	A_2A_2
Initial frequency	p_0^2	$2p_0q_0$	q_0^2
Fitness	$1 - s$	1	$1 - t$

where s and t are the selection coefficients against A_1A_1 and A_2A_2, respectively; the genotypic frequencies are

$$A_1A_1: p^2; \qquad A_1A_2: 2pq; \qquad A_2A_2: q^2.$$

The individual gametic contributions of the various genotypes will therefore be

$$A_1A_1: p^2 \times (1 - s); \qquad A_1A_2: 2pq \times 1; \qquad A_2A_2: q^2 \times (1 - t);$$

since the gametic contribution is equal to the product of the frequency of each genotype and the coefficient of selection for that genotype. The total gametic contribution of the population will be

$$p^2(1 - s) + 2pq + q^2(1 - t) = p^2 - sp^2 + 2pq + q^2 - tq^2$$
$$= 1 - sp^2 - tq^2.$$

In order to find out the frequency of A_2 genes in the next generation, we have to add one-half the contribution of A_1A_2 to the contribution of A_2A_2 and divide by the new total, so that the frequency of A_2 in generation 1 is

$$q_1 = \frac{q^2(1 - t) + pq}{1 - sp^2 - tq^2},$$

and the change that has resulted in one generation as the result of selection is

$$\Delta q = q_1 - q$$
$$= \frac{q^2(1-t) + pq}{1 - sp^2 - tq^2} - q = \frac{q^2 - tq^2 + pq - q + sp^2q + tq^3}{1 - sp^2 - tq^2},$$

which on simplification reduces to

$$\Delta q = \frac{pq(sp - tq)}{1 - sp^2 - tq^2}.$$

Whenever the product sp is larger than tq, their difference will be greater than zero and Δq will be positive; that is, A_2 will increase. When sp is smaller than tq, the numerator and Δq will be negative and A_2 will decrease. When $sp - tq = 0$, no more change in gene frequency takes place in the population. At equilibrium, then,

$$sp = tq,$$

but

$$p = 1 - q.$$

Therefore $s(1 - q) = tq$, and $s - sq = tq$, or

$$s = sq + tq = q(s + t),$$

and rearranging,

$$q = \frac{s}{s + t}.$$

Likewise,

$$p = \frac{t}{s + t}.$$

Consequently, when heterozygotes have higher fitness than either homozygote, no genotype is fixed at the expense of the other, but both alleles remain in the population at intermediate frequencies. This situation is called a *balanced polymorphism*, or *single gene heterosis*.

The average fitness of the population at equilibrium, when no more change in gene frequency takes place and $\Delta q = \Delta p = 0$, is

$$\overline{W} = 1 - sp^2 - tq^2$$
$$= 1 - s\left(\frac{t}{s + t}\right)^2 - t\left(\frac{s}{s + t}\right)^2$$
$$= 1 - \frac{st}{s + t}.$$

Consequently, the genetic load resulting from overdominance is

$$L = \frac{st}{s + t}.$$

This kind of load is called the *segregational load*, and it is fairly substantial whenever s and t have values above 0.1. In the extreme case when both homozygotes are lethals and $s = t = 1$, $L = 0.5$; that is, half of the offspring die every year.

Other Loads. The same reasoning can be applied to calculate the fitness loss to a population when some crosses in the population fail because of incompatibility between the partners (incompatibility load), or when a selectively superior mutation appears in the population (substitution load).

Is the concept of genetic load valid?

The analysis just presented shows that a genetic "cost" can be assigned to phenomena that increase genetic diversity, such as mutation and heterozygote superiority. However, such costs can easily be overemphasized. The reader is reminded of the constraint of the model, namely, an environment uniform in time for all members of the population. Only in such a situation does W_{max} have a fixed value, and only in such a situation can a fixed fitness value be given to each genotype. Furthermore, the genetic load applies to the population and not to the individual. Let us consider the more common situation, where the physical environment changes in both space and time.

When we introduced the concept of relative fitness, we showed that fitness of a genotype is measured in relation to all the other genotypes in the population. Therefore, in a variable environment, the fitness of a genotype in relation to the others in the population changes with the environment. Let us imagine a population made up of two phenotypes, each uniquely corresponding to one genotype. One phenotype is adapted to drought; the other functions best under wet conditions. Consequently, the fitness of each genotype will depend on the environment it is in. In swampy places, the second genotype will be favored, in dry places the first one. Furthermore, in extremely rainy years, the "wet" adapted phenotype may do better than the "dry" adapted phenotype even in dry places, and vice versa. Natural selection will be constantly eliminating certain phenotypes from the population, but the adverse selection will be against both phenotypes, depending on the place and the particular year. If we apply the corresponding formulas, the average fitness of the population will be below the theoretical maximum value of a uniform population restricted to either wet or dry conditions.

Let us now imagine the same physical setting but with the population's having only one phenotype, let us say the "wet" adapted one. In this case the fitness of all genotypes will be fairly uniform, and if we use the formulas, we will see that the population will be at or close to W_{max}. However, it will be restricted to swampy areas, and in dry years it will be in danger of being eliminated!

This example shows that a relative fitness value of 1, that is, W_{max}, with a load value of zero is not a measure of population success in a variable environment. What the existence of a load indicates is that there is a genetic cost associated with diversity, namely, the production of genotypes that prove inferior at given points in time and space. (See also Chapter 9.)

But selection operates primarily through individuals. If an individual can produce a larger proportion of surviving offspring, that individual will be selected, regardless of the fate of the nonsurviving offspring. Whether they die because of unfit genotypes or because of competition matters little. There are situations, however, when the cost in lost zygotes due to a high genetic load can be detrimental, so the concept of "load" has some practical value.

Energy costs of outbreeding

Whenever an individual has to search for a potential mate, it expends some energy in the process. Organisms that discharge their gametes into the water do not have that cost, and it is minimal in species that live in flocks of mixed sexes. The cost is highest in sedentary species that rely on an outside agent to carry the gametes from one individual to another, such as plants. In addition, outbreeding always involves some wastage of gametes.

There are actually few good studies of the energy cost of outbreeding. In a variety of species of plants it has been established that the number of pollen grains necessary to fertilize one ovule is greater by several orders of magnitude in cross-fertilized than in self-fertilized plants (Table 8.3). Likewise, outcrossing species have larger and showier flowers and produce greater amounts of nectar in order to attract insects. Energy diverted into those structures results in either fewer flowers or a smaller plant, or in both.

Meiotic cost

Offspring produced by self-fertilization inherit all their genes from their parent; those that are the result of cross-fertilization inherit half their genes from

Table 8.3

Pollen to ovule ratio in flowers of plants with different breeding systems.

Breeding System	Number of Species Investigated	Average Pollen/Ovule
Cleistogamy	6	4.7
Obligate autogamy	7	27.7
Facultative autogamy	20	168.5
Facultative outcrossing	38	796.6
Obligate outcrossing	25	5859.2

From R.W. Cruden, 1977. Pollen-ovule ratios: A conservative indicator of breeding systems in flowering plants. *Evolution* 31:32–66.

each parent. Consequently, a parent producing offspring by outcrossing transmits one-half as many genes to its offspring as one that produces offspring by self-fertilization. This difference has been termed the *meiotic cost* by G.C. Williams. The meiotic cost is 50 percent if the two individuals that cross have no genes in common by descent ($F = 0$, Chapter 7), but it is less than that if there is some degree of relationship, as is often true (Fig. 8.3).

Benefits from outcrossing

Since the great majority of organisms are cross-fertilizing, and since the cost of outbreeding is higher than that of inbreeding, it must follow that there is a clear advantage in the production of a variable offspring, and that this advantage is greater than the combined costs of the genetic load, the energy costs of outcrossing and the meiotic cost.

The meiotic cost depends on the degree of inbreeding, and it is zero in a totally homozygous and homogeneous population and maximum when no two individuals in the population are genetically related ($F = 0$, Chapter 7). The mutational genetic load depends, as we have seen, on the mutation rate; the segregational load, on the other hand, is related to the amount of heterozygosity, and it will therefore increase with outbreeding. Finally, the cost of ran-

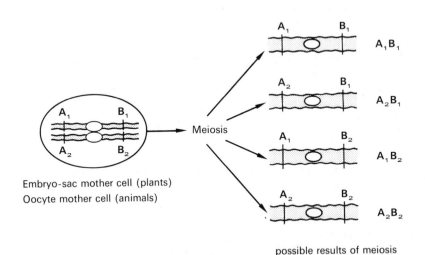

possible results of meiosis

Figure 8.3

Cost of meiosis. In female gametogenesis, half of the alleles in the ovule mother cells are not "used." This has been called the "cost of meiosis" by G. C. Williams. Since only one of the four possible gametes becomes an egg, half of the alleles are lost. (For instance, if the egg is A_2B_1, the alleles A_1 and B_2 are lost. (Of course, in other eggs produced by the same individual, different alleles may be "lost.")

dom breeding—that is, searching for a mate—depends on the size and density of the population and the degree to which the individual searches for a mate outside the circle of relatives and neighboring individuals. Consequently, costs go generally up with open recombination systems.

Although there is general agreement regarding the cost of outcrossing, there is no consensus regarding the benefits derived from a genetically variable progeny. However, there are two mutually nonexclusive benefits that can be pointed out.

First, there are the apparent advantages from producing heterozygous offspring that will be discussed in Chapter 9. The second advantage is related to life in an environment that is spatially and temporally variable. Individual organisms exist in a milieu that includes many other individuals. Some belong to the same species and can be potential competitors; others belong to different species and may also be competitors for resources, or potential predators, or potential prey. This "biological environment" is very complex and constantly changing, often at rapid rates. Furthermore, this biological environment responds to any adaptive change in any one species. For example, we saw that flies responded to the introduction of DDT into the environment by developing immunity. In the same manner, herbivorous insects will often respond to the appearance of a toxic compound in a plant species by evolving a mechanism of detoxification. In this context the immediate advantages of producing variable offspring by recombination are clear. An organism that always relies on the exact same mechanism of defense or attack, or of competition, to interact with the biological environment risks losing all its offspring if and when a predator or pathogen breaks the defense, a potential prey develops a defense, or a competitor develops a new mechanism of competition. Therefore the organism will have a greater probability of producing viable offspring if each one has a slightly different kind or degree of morphological, physiological, chemical, and behavioral mechanism of interaction. Homozygosity leads to the former situation, heterozygosity to the latter. Given that producing offspring that survive to reproduction is in part determined by chance, the probability of winning (i.e., producing such offspring) is usually increased by the equivalent of having many different lottery numbers (i.e., variable offspring) rather than many copies of the same number (i.e., homozygous and invariable offspring).

The preceding discussion shows that there are energy and genetic advantages and disadvantages associated with open and closed recombination systems. Consequently, it is very likely that the recombination system that shows the greater "profit" (benefit-cost) is selective in each different environment. However, there are no studies in which all the costs and benefits of the recombination system of a species have been analyzed. Not only is such a study laborious and time-consuming, but at present there is no general way to translate genetic costs (i.e., meiotic cost, or mutational load) into fitness costs (i.e., loss of offspring). There are some partial studies that indicate that this hypothesis may be correct.

RECOMBINATIONAL DISTORTION

In Chapter 4 we mentioned that Mendel's second law (independent assortment of genes) is applicable only to genes on different chromosomes. The reasons that the second law is not applicable to genes on the same chromosome are linkage (discussed in Chapter 5) and meiotic drive (discussed in Chapter 6). We now show that recombinational distortion can be beneficial in certain cases and therefore may be maintained by selection.

The reason that linkage may be beneficial is that it is a way of maintaining certain gene combinations in a population without a wastage of gametes. Let us assume that in a population the genotypes $A_1A_1B_1B_1$ and $A_2A_2B_2B_2$ have the same high fitness value, and the genotypes that have A_1 and B_2 or A_2 and B_1 have the same low fitness value. If A and B are two loci on different chromosomes and if we assume that the frequencies of A_1 and A_2 are $p = q = 0.5$, and of B_1 and B_2 are $m = n = 0.5$, then the genotype frequencies will be as follows.

$$\text{for } A_1A_1B_1B_1, \, p^2 \times m^2 = (0.5)^2 \times (0.5)^2 = 0.0625$$
$$\text{for } A_2A_2B_2B_2, \, q^2 \times n^2 = (0.5)^2 \times (0.5)^2 = \underline{0.0625}$$
$$\text{Total} \qquad 0.125$$

Thus only $\frac{1}{8}$ of the population will have the superior genotype.

However, if A and B are two linked loci on the same chromosome with an initial chromosome frequency for A_1B_1 of $X_1 = 0.5$ and for A_2B_2 of $X_4 = 0.5$, with a recombination probability r of 0.01, then, by the equation on p. 100,

$$X_{1_{t+1}} = X_{1_t}(1 - r) + pm(r)$$
$$= 0.5(1 - 0.01) + (0.5)(0.5)(0.01)$$
$$= 0.495 + 0.0025 = 0.4975$$

and equally,

$$X_{4_{t+1}} = \underline{0.4975}$$
$$\text{Total} \qquad 0.995$$

whereas the chromosome frequency X_2 of A_1B_2 is

$$X_{2_{t+1}} = X_{2_t}(1 - r) + pn(r)$$
$$= 0(1 - 0.01) + (0.5)(0.5)(0.01)$$
$$= 0 + 0.0025$$

and equally, for the chromosome A_2B_1,

$$X_{3_{t+1}} = \underline{0.0025}$$
$$\text{Total} \qquad 0.005$$

and the genotypic composition will be

$$A_1A_1B_1B_1 = \overline{A_1B_1} \times \overline{A_1B_1} : (X_{1_{t+1}}) \times (X_{1_{t+1}}) = (0.4975)^2 = 0.2475$$
$$A_2A_2B_2B_2 = \overline{A_2B_2} \times \overline{A_2B_2} : (X_{4_{t+1}}) \times (X_{4_{t+1}}) = (0.4975)^2 = \underline{0.2475}$$
$$\text{Total} \qquad 0.495$$

where $\overline{A_1B_1}$ indicates that A_1 and B_1 are on the same chromosome. That is, 49.5 percent or almost half of the population will have the superior genotype, compared with only ⅛ without linkage. Since a loss of half the offspring is not as drastic as a loss of ⅞ (87.5%), linkage is a mechanism that can reduce the costs of maintaining superior gene combinations in the population.

COADAPTATION

Segregation distortion has the effect of increasing certain gene combinations in the gametes of an individual and in the offspring that individual produces. In a sense it is similar to inbreeding, in that certain alleles increase and others are lost, in a manner not predicted by the Hardy-Weinberg theorem. However, whereas inbreeding affects all alleles and in a random fashion, and whereas it furthermore decreases the heterozygote level in the population, segregation distortion affects only specific gene combinations, and it can in certain cases (such as that of the *Drosophila pseudoobscura* inversions) increase the level of heterozygosity.

We may ask ourselves whether such phenomena as meiotic drive and linkage are beneficial to the individuals in the population, and whether they are consequently favored by selection. Many geneticists tend to think that they are beneficial because they maintain on a chromosome or part of a chromosome combinations of genes that work particularly well together. Such groups of nonallelic genes are termed "*coadapted gene complexes.*"

At present we lack incontrovertible proof of a biochemical nature for the concept of coadapted gene complexes, but there is a great deal of indirect evidence in favor of the concept.

The best-known evidence comes from the study of chromosomal inversions of *Drosophila pseudoobscura* (see also Chapter 9). When artificial populations of this species are established in the laboratory with varying mixtures of flies having either the Chiricahua (CH) arrangement or the standard arrangement (ST) taken from the same locality, after about a year, the population comes to equilibrium, with the ST chromosomes having a frequency of about 0.86 and CH a frequency of about 0.14 (Fig. 8.4). The relative fitness (W) of the three possible genotypes has been calculated by Wright and Dobzhansky to be ST/ST = 0.90, ST/CH = 1.00, CH/CH = 0.41, and selection coefficients of $s = 0.10$ for CH and $t = 0.59$ for ST (p. 144 and Chapter 9). From these values the expected chromosome frequency can be calculated (p. 192) as

$$p = \frac{t}{s+t} = \frac{0.58}{0.69} = 0.855$$

and

$$q = \frac{s}{s+t} = \frac{0.10}{0.69} = 0.145.$$

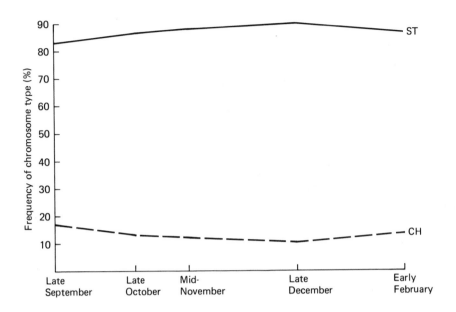

Figure 8.4

Frequency changes in chromosomal rearrangements under experimental conditions. A *Drosophila pseudoobscura* population containing the gene arrangements ST and CH was established and followed over time. The population was kept at 16°C. It is at equilibrium with about 86 percent ST and 14 percent CH. (Data from S. Wright and T. Dobzhansky, 1946. Genetics of natural populations. XII. Experimental reproduction of some of the changes caused by natural selection in certain populations of *Drosophila pseudoobscura. Genetics* 31:125–156)

However, when the flies in the artificial populations with the CH chromosome come from a different locality than the flies with the ST chromosome, the population becomes homozygous for ST/ST after about a year (Fig. 8.5). The fitnesses now are different. For example, if the ST chromosome comes from Piñon flats and the CH from either Piñon flats or Mather (both of these localities are in the Sierra Nevada of California) or from Mexico, the relative fitnesses shown in Table 8.4 are obtained. To make the comparison easier, the heterozygote has been assigned a fitness value of $W = 1.00$ in all cases. One can see that in crosses of flies from different localities, the ST/ST homozygote is superior. It is contended that the reason for the superiority of the heterozygote when the chromosomes come from the same locality is that the two chromosomes are "coadapted" to function together.

Another example is that of *Biston betularia*, mentioned briefly in Chapter 3, in which a new melanic form, f. *carbonaria*, has replaced the previously

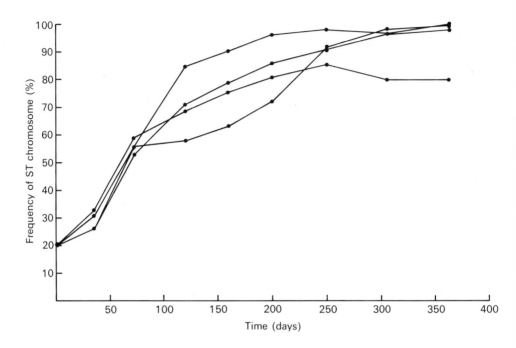

Figure 8.5

Frequency of ST (Standard) chromosomes in *Drosophila pseudoobscura* over time in four replicate experiments. Initially all four experimental populations had 20 percent ST chromosomes from Piñon flats, California, and 80 percent CH (Chiricahua) from Chihuahua, Mexico. Note that ST increases in frequency, and that it is about to become fixed in three of the four populations. Contrast with Fig. 8.4. (From data of T. Dobzhansky and O. Pavlovsky, 1953. *Evolution* 7:198–210)

dominant gray-spotted form in the last 150 years. This is the best studied case of the evolution of industrial melanism.

Evolution of industrial melanism

In the last hundred years or so, scientists have been able to witness and carefully record the replacement of light colors in many different species of moths by dark or black ones. *Industrial melanism*, as this phenomenon has come to be called, has been recognized as one of the most interesting of evolutionary phenomena. Through field observations and experiments, biologists have learned what the advantages and disadvantages of dark coloration are in the wild, and through laboratory experiments, they have demonstrated the various mechanisms that contribute to the inheritance of this character.

Dark coloration may confer on an animal three quite different attributes. First, it can make a dark animal harder to find on a dark background—in what

Table 8.4

Relative fitnesses in *Drosophila pseudoobscura* of ST × CH crosses with chromosomes of different geographical origins.

		Fitness (W)		
ST	*CH*	*ST/ST*	*ST/CH*	*CH/CH*
Piñon flats, Cal.	Piñon flats, Cal.	0.90	1.00	0.41
Piñon flats, Cal.	Mather, Cal.	1.38	1.00	0.43
Piñon flats, Cal.	Mexico	1.26	1.00	0.87

After T. Dobzhansky and O. Pavlovsky, 1953. Indeterminate outcome of certain experiments on *Drosophila* populations. *Evolution* 7:198–210.

is called *cryptic coloration*. Second, black absorbs heat more readily, so a dark insect will warm up quicker in the morning than a light one will. Finally, melanic forms are more protected against ultraviolet radiation.

From the middle of the nineteenth century onward, black forms of moths of widely different species and genera have been recorded in and around industrial areas in the northern hemisphere. In all cases, the melanic forms belong to cryptic species, that is, species that try to avoid predators by blending with the background in which they live. The most extensively studied are species of Great Britain, where the appearance of these melanic forms coincided with a great interest in the natural history of butterflies.

In Great Britain there are approximately 780 species of moths, and of these about 100 are undergoing the same process of substitution of their complicated and specialized pattern of coloration by a darker color. The basic story is similar in all cases: The industrial revolution in the nineteenth century brought about a change in the natural environment in and around the major industrial cities. The main effect was a general darkening of the environment, in part because of the deposition of soot, in part because of the disappearance of light-colored lichens. Light-colored insects that hid from predators by blending with the light background became conspicuous in a dark environment, but that was not so for dark-colored mutants. Consequently, birds that had previously found and eaten the dark animals now found and ate principally the light ones, so that the selection coefficients for dark and light became reversed.

For this explanation to be correct, two things are necessary: that moths do indeed choose a background that matches their coloration, and that the difference between light and dark forms is genetically simple. Kettlewell placed dark- and light-colored moths of *Biston betularia* in a barrel that was lined with black-and-white striped material and observed where they rested. As Table 8.5 shows, dark- and light-colored moths chose dark and light backgrounds, respectively, more often than would be predicted by chance. In the field, Kettlewell carefully recorded the resting places of the dark and light forms of the moth *Ectropis consonaria*. Again, their resting places were strongly correlated with their own color (Table 8.6).

In more than 90 percent of the cases of industrial melanism, the dark form is due to a single allele, and the melanic character is inherited as a Mendelian dominant. However, more complicated genetic controls are known. In the moth *Lymantria monacha*, the darkest forms are controlled by three genes. Other evidence indicates that although the basic genetic control is simple, the expression of the character can be and is modified by other genes in the genome.

Given that light and dark moths choose the background on which they rest during the day—a behavior that gives them some defense against insect-eating birds—the relative distribution of the background should be a good

Table 8.5

Resting position of *Biston betularia* in the laboratory.

| | Color Morph of Moth | | |
Background	Black (f. *carbonaria*)	White (f. *typica*)	Total
Black	38	20	58
White	21	39	60
Total	59	59	118
$\chi^2 = 10.9$	$P \simeq 0.001$		

From B. Kettlewell, 1973. *The Evolution of Melanism: The Study of a Recurring Necessity.* Oxford: Clarendon Press.

Table 8.6

Resting position of *Ectropis consonaria* in the wild.

| | Color Morph of Moth | | |
Background	Black (f. *nigra*)	White (f. *typica*)	Total
Dark	15	4	27
Light	5	22	19
Total	20	26	46
$\chi^2 = 14.202$	$P = 0.001$		

From B. Kettlewell, 1973. *The Evolution of Melanism: The Study of a Recurring Necessity.* Oxford: Clarendon Press.

predictor of the relative abundance of the melanic and nonmelanic forms in nature. Although in general this is true, the melanic forms are present in certain areas at a higher frequency than expected, and in others at a lower frequency. Consequently, additional explanations were needed. And indeed further study showed that melanic larvae sometimes have higher survival rates than light-colored ones; that heterozygotes have higher fitness than homozygotes; that moths from industrial areas are blown by wind currents into the country; and so on.

Over the years, crosses have been made between these two forms. Remember that the two forms differ by one gene, and that a 1 : 1 ratio is expected from the cross. Crosses made between 1900 and 1905 yielded approximately 47 percent melanics, but the same crosses performed in 1953–1956 yielded about 62 percent melanics! This shift from the expected 1 : 1 ratio can be interpreted as due to adaptation. When the melanic gene was introduced into the population, it presumably did not "fit" perfectly into the biochemical machinery of the animal. By 1950, when the melanic form had become dominant, it was the light forms that did not function so well. According to this interpretation, two kinds of selection were going on simultaneously. On one hand, the environmental change led to a lower mortality of the melanic forms because of predation pressure. On the other hand, small genetic changes were taking place that increased the viability of moths with the melanic gene (and presumably decreased that of moths with the normal gene).

In Chapter 4 we pointed out that each gene affects many characteristics of the organism, even though the effect on one characteristic may be most visible. In turn, each characteristic is affected by many genes. It should therefore not be surprising to find that the introduction of a new allele in the population by mutation and natural selection brings about a number of readjustments at other loci. Unfortunately, the evidence so far available is indirect. The study of adaptation at the biochemical level, which is a very promising field, may eventually provide some answers regarding the extent of coadaptation of genes within a population.

SUMMARY AND CONCLUSIONS

In this chapter we have explored recombination and have shown that it is a force that counters the action of those forces that reduce the genetic variability in populations. We showed, however, that there is a cost to maintaining genetic variability. This cost is the loss of homozygous genotypes from the population. The population can tolerate this cost only if it is counterbalanced by a benefit in the form of a higher survival probability of the heterozygous genotypes and their variable offspring. Since cost and benefits will differ from species to species, we expect different degrees of genetic variability in different species. In the next chapter, we explore what is presently known regarding genetic variability in natural populations of plants and animals.

SUGGESTED FURTHER READING

General references

Frankel, R., and E. Galun. 1977. *Pollination Mechanisms, Reproduction and Plant Breeding*. Berlin: Springer-Verlag.

Hawkes, J.G. (ed.). 1966. *Reproductive Biology and Taxonomy of Vascular Plants*. Oxford: Pergamon Press.

Jain, S.K. 1976. The evolution of inbreeding in plants. *Ann. Rev. Ecol. Syst.* 7:469–496.

Kettlewell, B. 1973. *The Evolution of Melanism: The Study of a Recurring Necessity*. Oxford: Clarendon Press.

Wallace, B. 1970. *Genetic Load: Its Biological and Conceptual Aspects*. Englewood Cliffs, N.J.: Prentice-Hall.

Zimmering, S., L. Sandler, and B. Nicoletti, 1970. Mechanisms of meiotic drive. *Ann. Rev. Genetics* 4:409–436.

Specialized articles and reviews

Carson, H.L. 1957. The species as a field for gene recombination. In E. Mayr (ed.), *The Species Problem*, Washington: American Association for the Advancement of Science.

Dobzhansky, T., and O. Pavlovsky, 1953. Indeterminate outcome of certain experiments on *Drosophila* populations. *Evolution* 7:198–210.

Grant, V. 1958. The regulation of recombination in plants. *Cold Spring Harbor Symposia on Quantitative Biology* 23:337–363.

Maguire, M.P. 1963. High transmission of a *Tripsacum* chromosome in corn. *Genetics* 48:1185–1194.

Müller, H.J. 1950. Our load of mutations. *Am. J. Hum. Gen.* 2:111–176.

Prakash, S., and R.C. Lewontin. 1968. A molecular approach to the study of genic heterozygosity in natural populations. III. Direct evidence of coadaptation in gene arrangements of *Drosophila*. *Proc. Nat. Acad. Sci.* 59:398–405.

Prakash, S., and R.C. Lewontin. 1971. A molecular approach to the study of genic heterozygosity in natural populations. V. Further direct evidence of coadaptation in inversions of *Drosophila*. *Genetics* 69:405–408.

Rick, C.M. 1959. Non-random gene distribution among tomato chromosomes. *Proc. Nat. Acad. Sci.* 45:1515–1519.

Sandler, L., and E. Novitski. 1957. Meiotic drive as an evolutionary force. *Am. Nat.* 91:105–110.

Solbrig, O.T. 1977. On the relative advantage of cross- and self-fertilization. *Ann. Mo. Bot. Garden* 63:262–276.

Williams, G.C. 1975. *Sex and Evolution*. Princeton, N.J.: Princeton University Press.

Wright, S., and T. Dobzhansky. 1946. Genetics of natural populations. XII. Experimental reproduction of some of the changes caused by natural selection in certain populations of *Drosophila pseudoobscura*. *Genetics* 31:125–156.

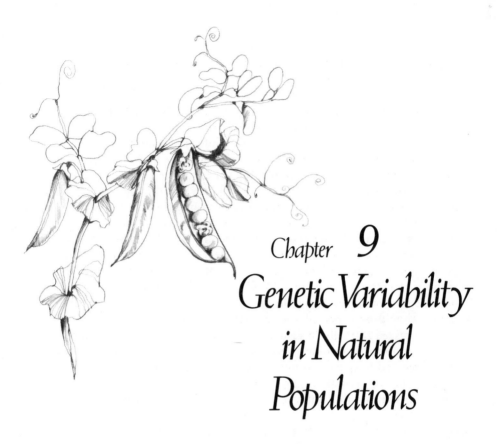

Chapter 9
Genetic Variability in Natural Populations

For evolution by natural selection to proceed, there must be a store of heritable variation in the population. In previous chapters we have shown that Mendelian theory provides a model of inheritance that accounts satisfactorily for the production and maintenance of variation. We also showed that the relationships between phenotype and genotype are not linear, and that at present it is usually not possible to precisely connect the two. However, we presented several examples (evolution of melanic forms in the peppered moth *Biston betularia*, of pesticide resistance in houseflies, and of influenza strains) in which the connection between genetic change and phenotypic selection is reasonably clear and unambiguous. But it is not always possible to account for the observed variation as neatly as in the examples presented so far. In this chapter we will review in slightly greater depth some of the patterns of genetic variability that are observed in natural populations, presenting and discussing some hypotheses that have been developed recently to try to explain the observed variability.

Individuals in natural populations differ from one another in both quantitative and qualitative characters. Some of this variability is genetic and some is environmental. Whenever a quantitative character is under strong selection, we expect the genetic component of the variance (Chapter 4) to be small and the variation to be largely environmental. For example, in Frisian cows, heritability $[h^2 = V_G/(V_G + V_E)]$ for amount of white spots is very high (0.95). This is a character that is *not* being actively selected by breeders. However, the heritability for milk yield is 0.3 and for conception rate in first service only 0.01. Any cow that does not conceive when serviced for the first time is normally eliminated from the herd, and that is as intense a regimen of selection as is possible. Milk yield is also under strong surveillance, but since selection cannot take place until *after* the cow has yielded milk for a year or two, selection tends to be exercised only through the males. That is, only bulls that are sons of high-yielding milk cows are chosen in breeding programs. This permits more genetic variation for milk yield than for conception rate to persist in herds. By a similar reasoning, in natural populations low amounts of genetic variation should be expected for characters under strong selection pressure. This is expressed formally in Fisher's Fundamental Theorem of Natural Selection (p. 154). Such reasoning presupposes a reasonably uniform environment both in time and in space and more or less constant selection over time. These conditions do not necessarily apply, and most organisms are now believed to live in environments that vary both in time and in space. An added complication is that when the heterozygote is the most fit genotype, there will always be genetic variation in the population. All this makes prediction of the expected level of genetic variation in populations impossible. It even makes the interpretation of the observed genetic variation very difficult.

PATTERNS OF GENETIC VARIATION IN POPULATIONS OF PLANTS AND ANIMALS

The degree of genetic variation in populations is not yet known for most characters and most organisms, although great strides have been made in the last ten years in developing appropriate techniques to study patterns of genetic variation in wild populations.

Since selection operates on the phenotype, at first glance one would think that in order to learn the degree of variation in the population, it would suffice to measure phenotypic variability. However, such an approach does not tell us the amount of this variability that is genetic, which is the desired information. By performing appropriate crosses and statistical manipulations, one can estimate the heritability of the character, as was explained in Chapter 4. This cumbersome and time-consuming method is used almost exclusively in programs of artificial breeding. In natural situations, the geneticist has tried instead to study characters with simple inheritance (preferably those controlled by a single gene) where alternative alleles have marked phenotypic

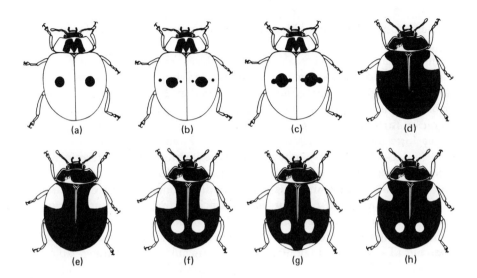

Figure 9.1

Morphs of *Adalia bipunctata*. (a–c) "Red" morphs (a, forma *typica*; b and c, f. *annulata*). (d–h) Black morphs (d, f. *sublunata*; e, f. 2-*maculata*; f, f. 4-*maculata*; g, f. 6-*postulata*; h, f. *lunigera*). (From N.W. Timofeff-Ressovsky, 1940. Zur Analyse des Polymorphismus bei *Adalia bipunctata*. *Biol. Zentral.* 60:130–137)

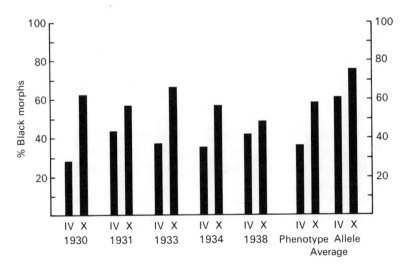

Figure 9.2

Percentage of black morphs of *Adalia bipunctata* in spring (April) and fall (October) censuses. The last two columns represent the frequency of the alleles for black. (From N.W. Timofeff-Ressovsky, 1940. Zur Analyse des Polymorphismus bei *Adalia bipunctata*. *Biol Zentral* 60:130–137)

effects, effects that can easily be measured or counted, such as different color forms.

The Russian geneticist N.W. Timofeff-Ressovsky in 1940 published an article on the color forms ("morphs") of the ladybug *Adalia bipunctata* in a population living in Berlin, Germany. Altogether he found eight different color patterns in this population (Fig. 9.1), although three of them were more common by far than the other five. Three of the morphs have black spots on a red background ("red" forms), and five have red spots on a black background ("black" forms). By appropriate experiments it was determined that the red forms are simple dominants over the black forms, and the exact pattern of spots is due to a multiallelic system. Timofeff-Ressovsky studied the population over a period of eight years, concentrating on the proportion of red and black forms. He was able to establish that the red forms were more abundant during the winter months, whereas the black forms increased during the summer (Table 9.1 and Fig. 9.2). He also was able to determine (Table 9.2)

Table 9.1

Spring and fall numbers of red and black morphs in populations of *Adalia bipunctata*.

Year	Morph	Spring		Fall	
		Number	Percent	Number	Percent
1930	Black	176	29.1	783	62.9
	Red	428	70.9	461	37.1
	Total	604	100.0	1244	100.0
1931	Black	334	43.0	622	57.7
	Red	443	57.0	494	44.3
	Total	777	100.0	1116	100.0
1933	Black	213	37.8	675	66.8
	Red	351	62.2	334	33.2
	Total	564	100.0	1011	100.0
1934	Black	149	34.5	708	57.2
	Red	283	65.5	529	42.8
	Total	432	100.0	1237	100.0
1938	Black	192	40.8	432	49.1
	Red	279	59.2	448	50.9
	Total	471	100.0	880	100.0
Total	Black	1064	37.4	3220	58.7
	Red	1784	62.6	2268	41.3
	Total	2448	100.0	5488	100.0

Source: N.W. Timofeff-Ressovsky, 1940. Zur Analyse des Polymorphismus bei *Adalia bipunctata*. Biol. Zentral. 60:130–137.

Table 9.2

Overwinter survivorship of red and black morphs of *Adalia bipunctata.*

Year	Morph	Original Number	Number Surviving	Percent Survivorship
1934	Black	739	23	3.11
	Red	334	21	6.29
	Total	1073	44	4.10
1937	Black	528	31	5.87
	Red	305	41	13.44
	Total	833	72	8.64
1938	Black	578	24	4.15
	Red	405	54	13.33
	Total	983	78	7.93
Total	Black	1845	78	4.23
	Red	1044	116	11.11
	Total	2889	194	6.72

Source: N.W. Timofeff-Ressovsky, 1940. Zur Analyse des Polymorphismus bei *Adalia bipunctata. Biol. Zentral.* 60:130–137.

that red forms show greater survivorship in the winter. Presumably black forms either reproduce faster or have greater survivorship during the summer months.

Another case of phenotypic variation, this time in a plant, is that of the habit of the grass *Agrostis stolonifera* in Britain, studied by J.L. Aston and A.D. Bradshaw. In inland habitats the plant is widely spreading through stolons (Fig. 9.3), whereas in coastal regions it is sometimes a small, compact tufted plant with many short stolons. The differences are retained when the plants are transplanted to a uniform garden, and they are also present in plants grown from seed, showing that the differences are inherited. Aston and Bradshaw studied plants from different localities with different stolon lengths. Their results are summarized in Table 9.3 and Fig. 9.4. Note that stolon length is correlated with the position that the plant occupies on a cliff, and that differences in this character between populations can be statistically significant. Even more interesting is the fact that where there is a sudden physical change, there is a corresponding sudden change in the length of the stolon, but in those cases where the physical change is gradual, the changes in plant morphology are also gradual. Presumably there is an advantage for this plant to have short stolons on cliffs and long ones in interior areas.

Figure 9.3

Schematic drawing of the inland form of *Agrostis stolonifera*. Note the connecting stolons that give the plant its name.

Table 9.3

General variation in mean stolon length of various maritime populations.

Population	Stolon Length, cm
1. Cliff base, exposed	14.5
15. Great Orme, exposed	15.9
8. Cliff top, exposed	16.5
3. Dry cliff, exposed	17.0
6. Dry exposed bluff	18.0
7. Cliff base, exposed and wet	24.1
16. Little Orme, cliff top pasture	32.2
2. Stream, exposed near sea	33.6
5. Pasture, exposed	35.5
18. Nash Valley (b)	41.9
11. Pasture, moderately exposed	42.3
14. Beaumaris cliff	44.0
17. Nash Valley (a)	45.4
10. Pasture, normal	46.7
4. Stream	47.5
9. Pasture, sheltered	47.9
12. Stream	48.7
20. Nash Valley (d)	49.8
19. Nash Valley (c)	51.2
13. Stream	57.0

From J.L. Aston and A.D. Bradshaw, 1966. Evolution in closely adjacent plant populations. II. *Agrostis stolonifera* in maritime habitats. *Heredity* 21:649–664.

COLOR POLYMORPHISM OF NATURAL POPULATIONS OF THE SNAIL CEPAEA NEMORALIS

One of the classical cases of variation is that of the color patterns of the shells of two closely related species of snails, *Cepaea hortensis* and *Cepaea nemoralis*. These snails have unusually variable shells, and consequently they have attracted the attention of European geneticists and evolutionists for more than half a century.

Individuals of *Cepaea nemoralis*, the better studied species of the two, can vary in the color of the shell (brown, pink, or yellow), the presence of dark longitudinal bands, and in the number of those bands (1–5); the bands can be dark colored or have no color at all; and finally, the lip, or peristome (Fig. 9.5), of the shell can be colored or white. Genetic studies have shown that shell color is determined by one gene with three alleles, with brown (Y^B) being dominant over pink (Y) and both of them dominant over yellow (y). An independent but tightly linked gene determines whether the snail will have bands or be smooth. This gene is epistatic over genes controlling the banding pattern. So, when the dominant (U) allele is present, the snail is unbanded; if the snail is double recessive (uu) it will have bands. The number of bands a uu snail will have is determined by still another gene, as is the presence or absence of pigment in the bands. The number of major genes controlling the pattern of shell color and banding in *Cepaea nemoralis* is therefore at least four, but probably more.

The species *Cepaea nemoralis* is widespread in western Europe from northern Spain and Italy to Great Britain. In recent times it has been introduced into eastern North America as well. It forms discrete populations of a few individuals to a couple of hundred, separated from one another by distances ranging from a few yards to several miles. The animals are found in hedges, tall grass, bushes, and the forest floor. Determining the evolutionary forces responsible for the observed variation in the shell phenotype of *Cepaea* snails has proved to be a formidable challenge, and it is thus a fine illustration of the complexity of evolution. The following information is derived in great extent from the work of the British evolutionists A. J. Cain and P. M. Sheppard and the French scientist Maxime Lamotte.

The population biologist studying the pattern of variation of a phenotypic character first tries to establish that the character is heritable. Once that fact is established, he or she carefully studies the pattern of variation in nature, searching for correlations between the character and some environmental factor that could possibly act as a selective agent. Concurrently with the field work, the population biologist conducts laboratory experiments to establish precisely how the population responds to specific physical or biological factors.

As we have already mentioned, genetic studies indicate that the shell color and banding pattern of *Cepaea nemoralis* are controlled by several major

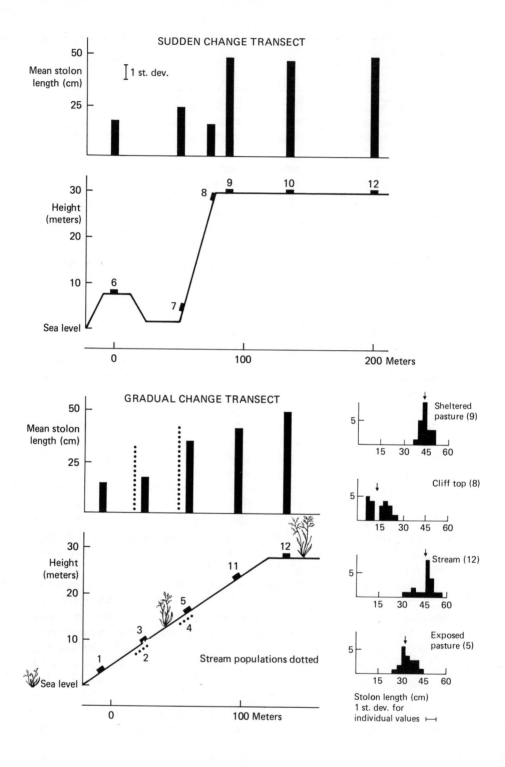

◀ *Figure 9.4*

Stolon length in populations of *Agrostis stolonifera*. Transects were taken along cliffs with gradual and abrupt changes in height. Populations were sampled along the transects. Mean stolon length is shown for the populations, and the values of individual plants within a population are shown for four populations. (From J.L. Aston and A.D. Bradshaw, 1966. Evolution in closely adjacent plant populations. II. *Agrostis stolonifera* in maritime habitats. *Heredity* 21:649–664)

genes and in addition are affected by other gene systems. As to the correlation with environmental factors, 25 years of careful and statistically sophisticated observations have revealed several correlations with environmental factors, but they do not suffice to explain entirely the observed patterns of variation. For example, snails with white lips are found only in some populations in the Pyrenees and Ireland, that is, at the borders of the distribution of the species, where they coexist with the brown-lipped individuals that prevail throughout. It is not clear what the Irish and Pyrenean environments have in common.

Figure 9.5

Shells of six individuals of *Cepaea nemoralis* from the grounds of the Palace of Versailles near Paris, France. Note the variation in banding pattern and peristome color. Shell at right is bandless.

The brown tint of the shell, on the other hand, shows a clearer pattern. Brown-shelled individuals are absent from nearly all populations in the south of France and the Mediterranean area; they are rare in central and northern France, although they are somewhat more frequent in beech woods in this area. They become fairly abundant in the Rhineland of Germany and in Great Britain, where they constitute 10 percent of all individuals. The frequency of brown shell color therefore appears to be related to climatic or microclimatic factors. This is further reinforced by the observation that snails with brown shells are never found in sunny places; they prefer to live in shady under-woods. On the other hand, the number of bands does not correlate so precisely with environmental factors. Sheppard has found a certain degree of correlation (Fig. 9.6) between the frequency of the various banding patterns

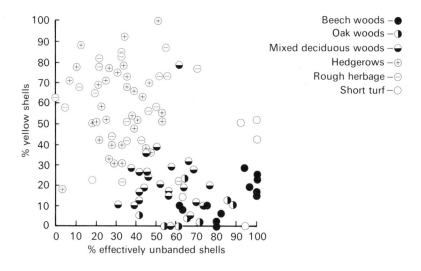

Figure 9.6

Correlation diagram for percent yellow shells, percent effectively unbanded shells (those with at least the upper two bands absent), and habitat of colonies of *Cepaea nemoralis*. The diagram shows the way in which natural selection has affected the frequency of the color and banding patterns. Thus in woods, which tend to have a uniform carpet of brown leaves (particularly beech woods), the proportion of the less conspicuous, effectively unbanded brown or pink shells is high. In contrast to this, habitats with rough herbage (including hedgerows) have the proportion of yellow banded shells high because these shells are less conspicuous in such places. The few exceptional colonies are also in accord with this view; thus the mixed deciduous wood population with 80 percent yellow shells lived in an exceptional wood in that the ground was carpeted with short green grass and not brown leaves, hence the high proportion of yellow, effectively unbanded shells. (From P.M. Sheppard, 1958. *Natural Selection and Heredity*. London: Hutchinson)

and microhabitat. He showed that in woods where the ground is brown with decaying leaves, and especially if it is fairly uniform in appearance, the un-banded brown or the unbanded or one-banded pink shells are particularly common, whereas in hedgerows and rough green herbage the yellow five-banded form tends to be more frequent. Moreover, he found that the less green the background, the lower the proportion of yellow shells, and the more uniform the background, the more common the unbanded or one-banded forms. The studies of Lamotte indicate that the same correlations are found in France but apparently with many exceptions. A possible reason for the difference in biotype distribution between France and Great Britain is that in France *Cepaea nemoralis* populations are often found along hedges where different species of plants coexist, or in modified woods, or in meadows dotted with trees and shrubs, whereas in Britain they are found both in dense woods and in meadows.

The selective agent behind this pattern appears to be predation by birds. In England the snail-eating song thrush, *Turdus ericetorum*, breaks open the shells of snails by smashing them against stones. By counting the banding and color pattern of the shells of snails eaten by the song thrush and comparing them with the population of live snails, one can show that when the habitat background is varied, as in hedgerows and rough herbage, banded individuals are more protected from predation, but that unbanded individuals have an edge in uniform habitats, such as woods. One can also show that individuals with yellow shells (which have a greenish tint when the animal is inside) are captured less often on green backgrounds, and that on brown backgrounds, such as open soil or dry leaves, brown- and pink-shelled snails are at an advantage. Similar results were obtained by Lamotte in France. Bird predation is consequently an important factor in maintaining the observed polymorphism. But it is not the only factor, as the far-from-absolute correlation of banding pattern and habitat type indicates. Another selective agent may be the physical conditions of the environment, since the different forms of the snail have different degrees of resistance to the various environmental factors. For example, bandless individuals, especially yellow ones, appear to be more resistant than banded individuals to heat (Table 9.4). Furthermore, yellow-shelled individuals are more resistant than pink ones to cold, and bandless snails more than banded ones. This fact may be more important than their susceptibility to heat because the snails have to survive the winter with little shelter. That heat and cold may indeed be a selective factor is shown by the observed correlation between the frequency of unbanded snails and July temperatures, and between the frequency of yellow and the January temperatures (Table 9.5).

In *Cepaea nemoralis*, then, we find several correlations between phenotypic variation and environmental factors that explain in part the patterns of variation found in nature. Nevertheless, when the cumulative action of these factors is taken into account, there always remains some unexplained residual diversity.

Table 9.4

Mortality in *Cepaea nemoralis* after 40 hours at 35°C.

	Yellow		Pink		Total	
	Number Tested	Percent Death	Number Tested	Percent Death	Number Tested	Percent Death
unbanded	66	13.6	15	33.3	81	17.2
one band, in position 3	45	26.6	26	42.3	71	32.3
five bands	39	38.4	22	36.3	61	37.7

Data from M. Lamotte, 1960. Polymorphism of natural populations of *Cepaea nemoralis*. *Cold Spring Harbor Symp. Quant. Biol.* 24:65–84.

Table 9.5

Frequency of "bandless" as a function of mean July temperature in *Cepaea nemoralis*.

Mean July temperature	16–18°	18–19°	19–20°	20–21°	21°
Average frequency of bandless (%)	22	24	26	29	30

b) Frequency of "yellow" as a function of mean January Temperature in *Cepaea nemoralis*.

Mean January temperature	4°	4–2°	2–0°	0°
Average frequency of yellow snails (%)	57.9	61.0	73.4	78.8

From M. Lamotte, 1960. Polymorphism of natural populations of *Cepaea nemoralis*. *Cold Spring Harbor Symp. Quant. Biol.* 24:65–84.

The number of known polymorphisms in phenotypic characters is very great, in both plants and animals. Few organisms have been studied as carefully as *Cepaea nemoralis*, but detailed inspection shows that the polymorphic character is usually correlated with varying physical and/or biotic factors of the environment, although in most cases the correlation is far from perfect. However, not only visible phenotypic characters are variable; cytological and molecular characteristics also vary.

CHROMOSOMAL INVERSION POLYMORPHISMS IN DROSOPHILA

A curious kind of genetic variability is found in species of the fruit fly *Drosophila*. With few exceptions, populations of all species of *Drosophila* have individuals with one or more chromosomal inversions (Chapter 5) in one, two, or three of their chromosomes. These inversions had already been detected in 1926 by Sturtevant, but they did not become an object of intensive study until the 1930s, after the giant polytenic chromosomes in the salivary glands were discovered (Fig. 9.7). This made investigations of inversions in nature an easy matter. It was the late Russian-American geneticist Theodosius Dobzhansky who pioneered in the study of this character.

Suppose that a fly has the gene arrangement ABCDEFGHI and AEDCBFGHI in two homologous chromosomes. For these chromosomes to pair during meiosis, one of them must form a loop (Fig. 9.8a). With time a second inversion may occur in the same chromosome. If it does not include the previously inverted segment, it is called *independent,* and two loops will be observed when it pairs with a normal chromosome (Fig. 9.8b). However, if the second inversion occurs within the first inversion, then during meiosis heterozygotes for these inversions form a loop within the first loop (Fig. 9.8c). Such inversions are called *included.* Finally, the second inversion may have one end inside and the other outside the limits of the first. Such inversions are *overlapping* (Fig. 9.8d) and form a figure eight during meiosis. Because the salivary gland chromosomes of *Drosophila* show a unique sequence of dark and light bands, it is possible to determine precisely what segment of a chromosome is the inverted one. By carefully establishing the order of the inversions, one can determine the probable *phylogeny,* or evolutionary origin, of the inversions, provided that all of them are connected by overlapping inversions, as follows (Fig. 9.9).

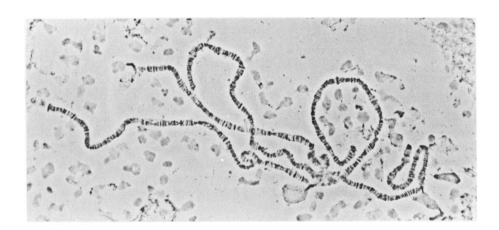

Figure 9.7

Polytene chromosome from the salivary gland of *Drosophila melanogaster.* Photograph courtesy of S. Elgin; previously published in L.M. Silver and S.C.R. Elgin, 1976. A method for determination of the *in situ* distribution of chromosomal proteins. (*Proc. Nat. Acad. Sci. USA* 73:423–427)

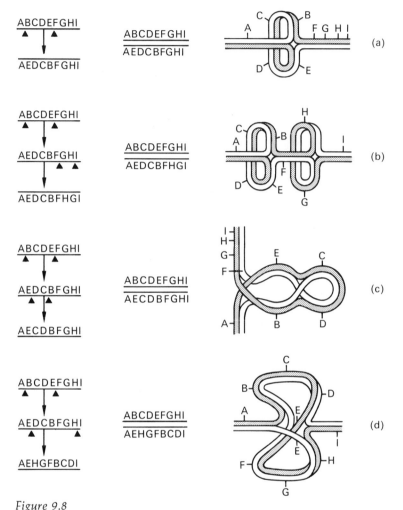

Figure 9.8

Chromosome pairing in *Drosophila* individuals heterozygous for inversions. (a) A single inversion. (b) Independent inversions. (c) Included inversions. (d) Overlapping inversions. (From T. Dobzhansky, 1970. *Genetics of the Evolutionary Process.* New York: Columbia University Press)

Suppose we observe three overlapping inversions with the following arrangements of bands: ABCDEFGHI, AEDCBFGHI, and AEHGFBCDI. We wish to determine the way they may have evolved, involving the fewest steps, the so-called most parsimonious way. The first inversion can arise from the second (or vice versa) in one step, and so can the second from the third, but for the third to originate from the first (or the first from the third), there must be at least two steps. Consequently, the most parsimonious phylogeny is 1 → 2 → 3 or 3 → 2 → 1 or 1 ← 2 → 3.

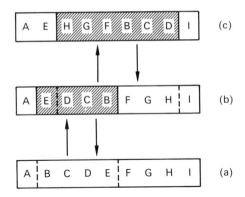

Figure 9.9

Hypothetical phylogenetic tree. Chromosome arrangement (b) can give rise to (a) and (c) by a simple inversion. It takes two inversions for (c) to be derived from (a), or vice versa.

By means of this procedure, all known inversions in a given chromosome of a species can be ordered in a phylogenetic sequence. Figure 9.10 shows such a phylogenetic tree for three very closely related species, *Drosophila pseudoobscura*, *D. persimilis*, and *D. miranda*. Note that two species (*D. pseudoobscura* and *D. persimilis*) share a chromosomal arrangement (*Standard*) and that one crucial arrangement, *Hypothetical*, has never been observed. The existence of predicted arrangements in previous cases was verified by subsequent findings in nature by Dobzhansky, thus lending credence to this phylogeny.

The variability of inversions in natural populations has been studied in greatest detail in the species *Drosophila pseudoobscura*. It has been determined that all populations are polymorphic for inversions in the third chromosome; that no specific arrangements occur over the entire distribution area of the species; that some arrangements (Standard, Chiricahua, Arrowhead) are quite frequent while others have been detected in only one population; and that consequently no population has a complete collection of the arrangements, although in some localities as many as eight different arrangements have been found.

In nature, both structural homozygotes (flies having two chromosomes with the same arrangement) and structural heterozygotes are found. Of even greater interest is the fact that the frequency of a given inversion varies from locality to locality, as well as within a locality throughout the year and over several years (Figs. 9.11 and 9.12). However, although these chromosomal inversions have now been studied for some 40 years in field and laboratory, we still do not know exactly what factor or factors account for their maintenance or for the frequency changes that are observed in natural populations. In the previous chapter we presented data that indicate some possible advantages of these inversions.

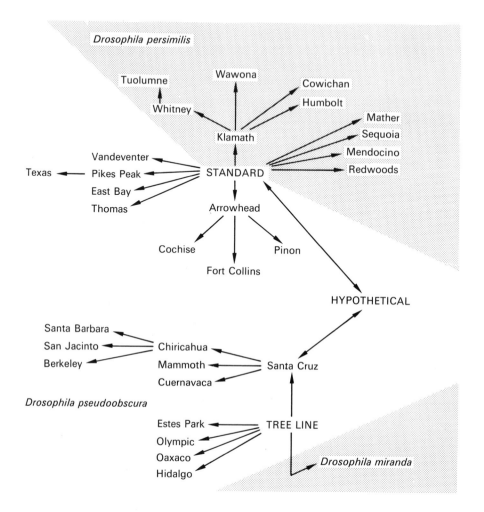

Figure 9.10

Phylogenetic relationships of the gene arrangements in the third chromosomes of *Drosophila pseudoobscura*, *D. persimilis*, and *D. miranda*. (From T. Dobzhansky, 1970. *Genetics of the Evolutionary Process*. New York: Columbia University Press)

GENOTYPIC VARIATION

The complex inheritance of phenotypic variation hinders understanding of the genetic and evolutionary forces at work. If statistical measures of phenotypic values could be replaced by the gene frequency of the underlying alleles responsible for the character, the evolutionary situation might possibly be better understood. Consequently, geneticists try to estimate gene frequencies directly

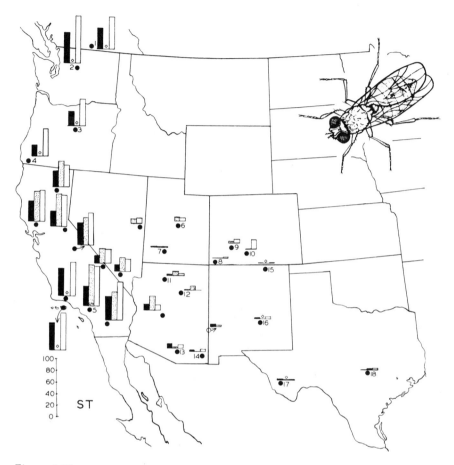

Figure 9.11

The frequencies of ST chromosomes in population samples of *Drosophila pseudo-obscura* taken in 1940 or thereabouts (black columns), in 1957 (stippled columns), and in 1963–1965 (white columns); a diamond means that no sample was taken during a given period. The scale in the lower left corner indicates the frequencies in percentages. (From T. Dobzhansky, W.W. Anderson, and O. Pavlovsky, 1966. Genetics of natural populations. XXXVIII. Continuity and change of populations of *Drosophila pseudoobscura* in western United States. *Evolution* 20:418–427)

in populations. A new technique, used for this purpose for the first time in 1966, allows easy and direct estimation of gene frequencies in a population.

The technique of gel electrophoresis

When a mixture of enzymes or other proteins supported by an appropriate medium, such as a strip of paper or of cellulose acetate or a slab of starch gel, is placed in an electrical field, the individual enzymes and other proteins mi-

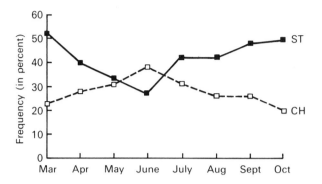

Figure 9.12

Seasonal changes in frequencies of chromosome inversions in *Drosophila pseudo-obscura*. Regular seasonal fluctuations in the ST and CH chromosome types were observed at a site on Mount San Jacinto in California during the years 1939–1946. (From T. Dobzhansky, 1970. *Genetics of the Evolutionary Process.* New York: Columbia University Press)

grate toward the cathode or anode if positively or negatively charged, but they don't migrate if uncharged (Fig. 9.13). The rate of migration toward the poles is a function of the charge of the molecule (those with the greatest charge migrate fastest) and of the resistance that the supporting medium offers to the movement of the enzyme. In broad terms, the resistance will be proportional to the size of the enzyme. If after a certain time the electricity is turned off and the medium is fixed and stained, a number of stained bands appear on the gel, each corresponding to one enzyme (or two or more with identical migration rates). Instead of fixing and staining, one can incubate the medium with an appropriate substrate and a soluble dye that will precipitate only when an enzyme is attacking the substrate. For example, an investigator who wishes to ascertain whether in the original mixture of proteins there is an *esterase* incubates the medium with α-naphthyl-acetate and a soluble dye called fast blue. Esterases will split α-naphthyl-acetate into acetic acid and α-naphthyl, using as a source of electrons the fast blue dye, which when oxidized becomes insoluble and precipitates. After a time the supporting medium is fixed, and the investigator can ascertain the rate of migration of proteins with *specific enzymatic activities.* A given band can be composed of one or more enzymes with equal rates of migration. This biochemical technique has had a tremendous impact on population genetics, as we will see. However, we must first review the conceptual basis for the use of this technique in estimating gene frequencies in populations.

In 1953, DNA was determined to be the genetic substance. Over the next ten years, molecular biologists ascertained that the DNA molecule acts as a code made up of three-letter words (Chapter 4). Each word is translated

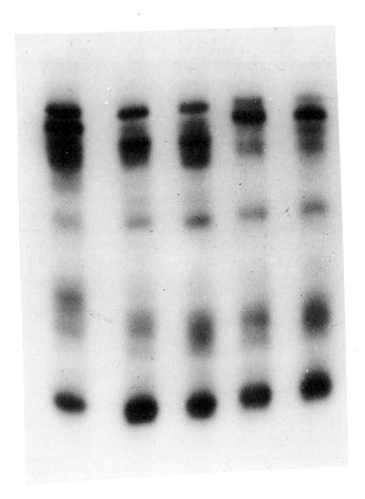

Figure 9.13

Acrylamide gel patterns of seed proteins subjected to electrophoresis. The seeds were collected from populations of *Agropyron scabriglume* and *A. tilcarense* in different parts of Argentina. (Photo courtesy J.H. Hunziker)

through a series of steps into an *amino acid* in a precise location within an *enzyme* or into a nonenzymatic protein in the cytoplasm. A *cistron*, which is a string of such words in the DNA molecule that together code for a polypeptide chain, is the closest molecular approximation to the gene of classical genetics. Enzymes can consist of one or more polypeptide chains. Sometimes the chains are identical and are coded by the same cistron, and sometimes they are different and are coded by more than one cistron (= gene). An example of the latter, which we discussed in detail in Chapter 5, was the hemoglobin molecule.

When a point mutation at one gene takes place, one or more amino acids are replaced by a different kind of amino acid. This can (but does not necessarily) change the charge of the protein, its conformation, or both, resulting in a protein with a different rate of migration but with the same basic enzymatic function, although possibly with modified *physiological* properties. An enzyme with the same biochemical function but different rate of migration is called an *allozyme* or *isozyme*. Since allozymes are primary gene products, they are inherited in Mendelian fashion. This was shown by Beckman, Scandalios, and Brewbaker in 1964 with the enzyme leucine-amino-peptidase (LAP) in corn (Table 9.6 and Fig. 9.14).

Armed with this insight, the population biologist can assay large numbers of individuals in a population for the electrophoretic mobility of a specific enzyme and, by determining the frequency of allozymes for that enzyme, can estimate the frequency of the various alleles at that locus that produce electrophoretically detectable modifications in the enzyme. To ascertain that two different bands in two individuals are coded by alleles of the same gene, the investigator needs to do only one cross and assay the offspring. If the two

Table 9.6

Genetic control of LAP A enzyme variants in corn.

| | | LAP A Zones in Offspring | | | |
| | | F | F + S | S | |
Female Parent	Male Parent	(F/F)	(F/S)	(S/S)	Total
AA4(F/F)	AA4(F/F)	30	0	0	30
P39(S/S)	P39(S/S)	0	0	30	30
AA4(F/F)	P39(S/S)	0	30	0	30
P39(S/S)	AA4(F/F)	0	30	0	30
(AA4 × P39)(F/S)	AA4(F/F)	16	14	0	30
AA4(F/F)	(AA4 × P39)(F/S)	17	13	0	30
(AA4 × P39)(F/S)	P39(S/S)	0	10	20	30
P39(S/S)	(AA4 × P39)(F/S)	0	12	18	30
(AA4 × P39)(F/S)	(AA4 × P39)(F/S)	20	22	18	60

The A enzyme for LAP has two variants, fast (F) and slow (S). The AA4 strain of corn has only the fast band, and the P39 strain has only the slow. The hybrid between them has both, and backcrosses to the parental strains give the 1:1 ratios appropriate to Mendelian inheritance at one locus. The genotypes in parentheses are deductions fitting the results.

From L. Beckman, J.G. Scandalios, and J.L. Brewbaker, 1964. Genetics of leucine amino peptidase enzymes in maize. *Genetics* 50:899–904.

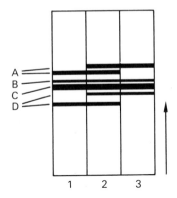

Figure 9.14

Electrophoretic patterns for leucine-amino-peptidase (LAP) in corn. Three strains
are shown: (1) P39, (3) AA4, and (2) F_1 hybrids between P39 and AA4. Bands
B and C are present in all strains with no genetic variation. Bands A and D,
however, have two forms, both of which show up in the hybrid. The arrow shows
the direction of migration in the gel. (From L. Beckman, J.G. Scandalios, and J.L.
Brewbaker, 1964. Genetics of leucine aminopeptidase isozymes in maize. *Genetics*
50:899–904)

bands segregate in a 1:2:1 pattern of homozygote fast:heterozygote:homo-
zygote slow, the allozymes in question are allelic.

Here, then, is a relatively easy technique for directly assaying allelic frequencies in populations. It was introduced in 1966 independently by the
British human geneticist Harris, and by the American biochemist Hubby and
the American geneticist Lewontin. Since then, all manner of species have
been assayed for their isozyme patterns, from deep sea invertebrates through
Drosophila species to the human species; from widespread to rare species;
from those that habitually outcross to those that self-fertilize; and so on. A
flood of data has been garnered whose interpretation has been very difficult.

One of the difficulties comes from the fact that not all the isozymes observed for a particular enzyme in the population are necessarily due to allelic
variation at the structural gene that codes for the enzyme. An alternative
hypothesis is that different isozymes may be the result of extensive modifications of a single enzyme by the products of other genes. This phenomenon is
called *post-translational modification* (Fig. 9.15). Such post-translational modification will have many of the characteristics expected from a single locus with
many alleles: (1) It is heritable; (2) if several modifying loci are active, very
many enzyme variants may be produced; and (3) if the modifying loci act
independently (i.e., vary different aspects of the enzyme), then their effect will
be multiplicative and a skewed frequency distribution will result, with some
common and many rare variants. This is the commonly observed pattern.

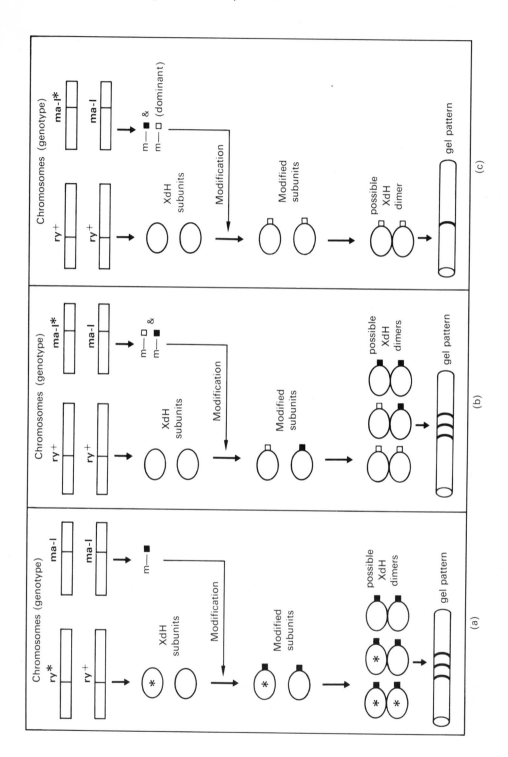

How to measure allozyme variation in populations

In order to compare populations of the same or different species, it is necessary to transform the allele frequency at a number of loci into a number that indicates the total or the average genic variation in the population.

Whenever a gene has more than one allele in the population, it is considered to be a *polymorphic* gene. The number of alleles of a polymorphic gene can be two or more, and each of them may be either rare (low-frequency) or abundant (high-frequency). If we square the frequency of each allele in the population and add these values, we obtain a number that is referred to as the *homozygosity* at that locus and symbolized by the letter j:

$$j = \Sigma x_i^2,$$

where x_i is the frequency of the ith allele. One minus the homozygosity is the *heterozygosity* for that locus, h:

$$h = 1 - j = 1 - \Sigma x_i^2.$$

If we average j or h over all loci, we obtain the *average homozygosity* (J) or *average heterozygosity* (H) for the population:

$$H = \frac{\Sigma h_i}{n}, \qquad J = \frac{\Sigma j_i}{n},$$

where h_i and j_i are the heterozygosity and homozygosity of the ith locus and n is the number of loci.

The concept of average homozygosity and average heterozygosity was developed with respect to a random mating population. For example, in a population where one locus is measured for allozyme variation and two alleles

◀ *Figure 9.15*

Sources of heritable variation in gel phenotypes. The example is from the xanthine dehydrogenase (XdH) system of *Drosophila melanogaster*. Alternative alleles are indicated by asterisks. The **ry** locus produces XdH subunits, which are altered by the modifier locus **ma-l**. The subunits combine in pairs to form dimers. Dimers with different subunits may have different mobilities in gel electrophoresis and give rise to different bands. The variation observed in the gels may be caused by variation at the modifier locus, as well as by variation at the structural locus. (a) Variation is entirely due to the structural gene (**ry**) polymorphism. (b) Variation is due to modifier (**ma-l**) polymorphism. (c) Although modifier is polymorphic, no variation is detected in gel because of dominance. (From V. Finnerty and G. Johnson, 1979. Post-translational modification as a potential explanation of high levels of enzyme polymorphism: xanthine dehydrogenase and aldehyde oxidase in *Drosophila melanogaster*. *Genetics*, in press)

are found, each with a frequency of $p = q = 0.5$, the homozygosity and heterozygosity at that locus are, respectively,

$$j = \Sigma x_i^2 = (0.5)^2 + (0.5)^2 = 0.25 + 0.25 = 0.5,$$
$$h = 1 - j = 0.5.$$

That is, half of the loci are heterozygous and half are homozygous. And indeed, that is the expectation in a random breeding population under Hardy-
Weinberg equilibrium (Chapter 4). However, when breeding is not random,
the formula does not apply as a measure of homo- and heterozygosity in the
population. Nevertheless it is a measure of *gene identity* (*J*) and of *gene diversity* (*H*) in the population, and as such it can be used. Since all populations
deviate to some extent from the random breeding model (Chapter 7), it is best
to think of *J* and *H* as measuring gene identity and gene diversity and to use
more direct measures to determine levels of heterozygosity.

GENETIC VARIATION IN DIFFERENT SPECIES OF PLANTS AND ANIMALS

Although the number of species that have been investigated for allozyme
variation numbers well over 100, the sample is still far too small when
matched against the two million or more species of plants and animals that
are known. However, the results in these species, though variable, allow us
to make some preliminary estimates.

The most surprising result to come out of these studies was the large
amount of variation that was found. For example, in many species 25 percent
to 50 percent of the loci in the population are polymorphic, and individuals
in the population may have 5 percent to 15 percent of their loci heterozygous.

Table 9.7 shows some estimates for heterozygosity and polymorphism
among different species belonging to different phyla of plants and animals.
Since the samples are small, any general conclusion would be premature. The
maximum variability among vertebrates appears to be less than in some other
groups such as plants and certain invertebrates, but much more data are
needed.

Another interesting finding is that the gene diversity (*J*) varies from gene
to gene, some being very monomorphic and some tremendously polymorphic.
The extreme of genetic variation so far known is in the enzyme *xanthine
dehydrogenase*. In the species *Drosophila persimilis*, Coyne found 23 alleles
among 60 females that had been made artificially homozygous in the laboratory. These flies came from three different populations. In the closely related
species *Drosophila pseudoobscura*, 37 alleles were found in an examination of
146 homozygous females, originally from 12 different populations. Because of
the laboratory manipulations it is not possible to state how common these
mutant alleles are, but the large number of alleles is surprising.

The enzyme xanthine dehydrogenase affects the metabolism of the normal

Table 9.7

Comparisons of genetic variation in animals and plants.

Group	Number of Species or Forms	Mean Number of Loci per Species	Mean Proportion of Loci Polymorphic per Population (P) (SD)	Heterozygous per Individual (H) (SD)
Insects				
Drosophila	28	24	0.529 (.030)	0.150 (.010)
Others	4	18	0.531	0.151
Haplodiploid wasps	6	15	0.243 (0.039)	0.062 (.007)
Marine invertebrates	9	26	0.587 (0.084)	0.147 (.019)
Snails				
Land	5	18	0.437	0.150
Marine	5	17	0.175	0.083
Total invertebrates	57	Mean = 21.8	Mean = 0.469	Mean = 0.135
Fish	14	21	0.306 (.047)	0.078 (.012)
Amphibians	11	22	0.336 (.034)	0.082 (.008)
Reptiles	9	21	0.231 (.032)	0.047 (.008)
Birds	4	19	0.145	0.042
Rodents	26	26	0.202 (.015)	0.043 (.005)
Large mammals	4	40	0.233	0.037
Total vertebrates	68	Mean = 24.1	Mean = 0.247	Mean = 0.061
Annual plants	36/33	11	0.446 (.296)	0.142 (.107)
Biennial plants	13/11	18	0.220 (.250)	0.079 (.081)
Herbaceous perennials	17/14	12	0.250 (.312)	0.116 (.118)
Woody trees and shrubs	10/7	12	0.652 (.314)	0.359 (.102)
Total plants	76/65	Mean = 12.6	Mean = 0.391	Mean = 0.149

From J.L. Hamrick, 1979. Genetic variation and longevity. In O.T. Solbrig *et al.* (eds.), *Topics in Plant Population Biology.* New York: Columbia University Press.

red pigment of the eyes of *Drosophila* flies. The rosy (**ry**) locus is the structural locus, but three other loci, maroon-like (**ma-l**), low xanthine dehydrogenase (**lxd**) and cinnamon (**cn**) (Fig. 9.16), also affect the levels of xanthine dehydrogenase and consequently the color of the eye. These other genes are known as modifiers, and it is generally assumed that their effect is through post-translational modification, although the precise nature of the modification is not known. Since the rosy gene (presumed to be the structural gene for xanthine dehydrogenase) is on the third chromosome of *Drosophila melanogaster*, the effect that post-translational modification has on isozymes can

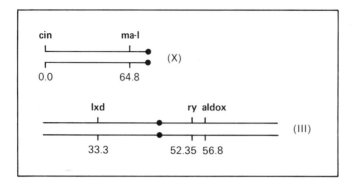

Figure 9.16

Genetic elements of the xanthine dehydrogenase (XdH) system in *Drosophila melanogaster*. The numbers are mapping units which indicate position on the chromosomes. Rosy (**ry**) is the structural gene for XdH, and **aldox** the structural gene for aldehyde oxidase (AO). Null mutants at **ry** lack XdH activity but are normal for AO, while **aldox**-negative mutants have no AO activity but are normal for XdH. **Ma-l** mutants are completely deficient for both XdH and AO. **Lxd** has a much reduced XdH and AO activity compared with wild type. Various **cin** mutants have XdH and AO activities far lower than are found in wild type. (From V. Finnerty and G. Johnson, 1979. Post-translational modification as a potential explanation of high levels of enzyme polymorphism: xanthine dehydrogenase and aldehyde oxidase in *Drosophila melanogaster*. *Genetics*, in press)

be tested by using flies that have identical third chromosomes (that is, chromosomes derived from exactly the same parent) but different X chromosomes, each X chromosome containing a different allele for the gene maroon-like. If maroon-like alleles affect the xanthine dehydrogenase enzyme, then in such an experiment different isoenzymes should be observed. When Finnerty and Johnson performed such an experiment, they found that in every instance there was a significant alteration in the protein shape resulting in different electrophoretic mobilities. Since Finnerty and Johnson worked with a different species than Coyne (*Drosophila melanogaster* rather than *D. persimilis*), it cannot be stated that the variation found by Coyne is partly or entirely due to post-translational modification. However, the experiment indicates that understanding genetic variation of characters under natural selection in natural populations remains an elusive task, made harder rather than easier by the discovery of allozyme variation.

MECHANISMS OF MAINTENANCE OF GENETIC VARIABILITY

The reasons for the high degree of genic variation in natural populations and the mechanisms by which they are maintained constitute at present central problems of population genetics.

A polymorphism at a given locus may be classified as *transient* or *stable*. A transient polymorphism is the situation that is created in a population when a once common allele is being replaced by an allele that imparts a higher fitness to its carrier. Transient polymorphisms are important, and they produce specific genetic effects already discussed in Chapter 8 (genetic load). Some of them will also be considered here. Stable polymorphisms are those that are maintained by some kind of balancing selection (Chapter 6). They can be artificially classified into cases of *single-gene heterosis, selection in favor of heterozygotes,* and *selection in a variable environment.*

Single-gene heterosis

In Chapter 6 we presented the derivation of the formula for gene change in a population differing at one locus and under directional selection (p. 142), and in Chapter 8 we derived the gene frequencies when the relative fitness (w) of the heterozygous genotype is higher than that of either homozygote and showed that heterozygote superiority entails a genetic load, as a result of the constant segregation of homozygous genotypes with lower fitness. The greater the advantage of the heterozygote, the greater the load.

The best case, and so far the only unmistakable one, of single-gene heterosis is that of sickle-cell disease in Africa.

Sickle-cell anemia is a genetic disease in humans that affects the oxygen transport ability of the hemoglobin β chain. It is due to a single mutation, which substitutes the amino acid valine at position 6 for the normal glutamic acid in the hemoglobin protein. Since the triplets on the DNA chain that code for glutamic acid are GAA and GAG (G = guanine A = adenine; see Chapter 4) and those for valine GUA* and GUG (U = uracil), one can see that a change in a single molecule—i.e., replacement of adenine by thymine in the DNA molecule—is so amplified by the process of growth and development that the individual that inherits this molecule from both parents dies of sickle-cell anemia at an early age.

A mutation that is lethal when homozygous should exist in a very low frequency in the population ($q = \sqrt{u/s}$, Chapter 6). However, the sickle-cell anemia gene is found in high frequency in certain regions of Africa (Fig. 9.17) and in higher than expected frequencies in New World black populations. The problem was studied by A.C. Allison, who established quite conclusively between 1955 and 1965 that the individual that is heterozygous for the sickle-cell trait is resistant to falciparum malaria. Consequently, here is a case in which the relative fitness of the heterozygote is 1 and that of the recessive homozygote zero or close to it, and the fitness of the homozygote dominant depends on the environment. In areas with a high frequency of malaria, the fitness of the homozygote dominant will be low because many such individuals will die of malaria infection (Table 9.8). In areas where malaria is not a

---

* The RNA equivalent is given here in accordance with usual practice.

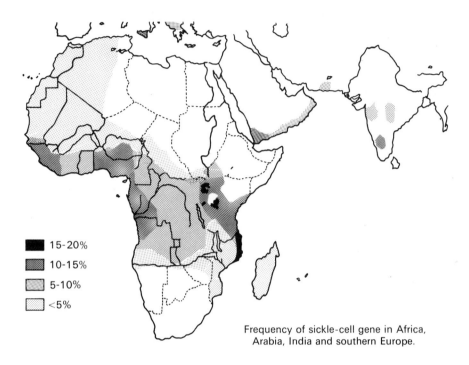

15-20%
10-15%
5-10%
<5%

Frequency of sickle-cell gene in Africa,
Arabia, India and southern Europe.

Figure 9.17
Frequency of the sickle-cell gene in various parts of the Old World. The key in the left-hand corner shows the percentages of the populations that bear the gene. (After A.C. Allison, 1961. Genetic factors in resistance to malaria. *Ann. N.Y. Acad. Sci.* 91(3):710–729)

Table 9.8

Deaths from malaria in relation to the sickle-cell trait in African children.

Locality	Number of Deaths	Number with Sickle-cell (Heterozygotes)	Incidence of Sickle-cell Trait in Population (in %)
Kampala, Uganda	16	0	0.20
Leopoldville, Zaire	23	0	0.235
Luluaborg, Zaire	23	1	0.25
Accra, Ghana	13	0	0.08
Ibadan, Nigeria	29	0	0.24

Modified from A.C. Allison, 1965. Polymorphism and selection in human populations. *Cold Spring Harbor Symp. Quant. Biol.* 29:137–149.

problem or does not exist, the fitness of the normal homozygote surpasses the heterozygote, since the oxygen transport system of such individuals functions better than that of heterozygous individuals. As a result of campaigns to·control malaria in different parts of the world, combined with emigration of African populations in the last century to both malaria-free and malaria-infected areas of the New World, this model can be tested in principle.

African blacks were transported as slaves from various parts of the African coast to the New World between 1520 and 1800. Historical evidence indicates that the largest number of people came from the neighborhood of Ghana, where the incidence of the sickle-cell trait is close to 20 percent, as it is on much of the west coast of Africa. However, there is no precise way of determining what the frequency of the gene in the emigrant population was. In all likelihood it varied, both with the geographical origin of the slaves and over time. In the New World, the black population mated with nonblacks, and it is not always easy to determine the extent of that mixing. Consequently, there are two sources of uncertainty regarding the expected decrease in the frequency of the sickle-cell gene: (1) The lowered frequency in New World populations could be due to an initial lower frequency in the emigrant population; (2) it could be due to mixing with populations that do not have the trait. However, there are research methods to partly circumvent these difficulties.

Table 9.9 shows the present frequency of the sickle-cell trait in various West Indian populations. Note that it varies from a high of 23 percent (equal to the African frequency) to a low of 7 percent. Interestingly, the high fre-

Table 9.9

Incidence of sickle-cell genes in West Indian populations.

Population	Number Tested	Percent with Sickle-cell Gene
Honduras, Black Caribs	705	23.3
Curaçao	1502	7.2
Surinam, Kabel	519	16.8
Surinam, Moengo	172	20.3
Surinam, Stoelman's Island	275	11.3
Surinam, Djukas	343	15.0
Jamaica	1018	10.9
St. Vincent	748	8.7
Dominiq	664	9.5
Barbados	912	7.0
St. Lucia	825	14.0

Modified from A.C. Allison, 1965. Polymorphism and selection in human populations. *Cold Spring Harbor Symp. Quant. Biol.* 29:137–149.

quencies are found in Honduras and Surinam, areas where the incidence of malaria is high, and the low frequencies are found in the islands of Curacao and Barbados, which have no malaria! In the United States investigators found, by studying 15 suitable genetic markers in the black population, that migration of white genes into the black population had been on the order of $m = 0.1$ to 0.2. However, for the sickle-cell trait, the calculated value of m was on the order of 0.46 to 0.69. Since that value is too high in relation to the other genes, the conclusion was that there had been significant selection against the sickle-cell trait in American populations of blacks.

In addition to the sickle-cell gene, there are a number of other polymorphic genes in the human population where single-gene heterosis is suspected, especially other hemoglobin variants. However, in no other case have the operating selective forces been elucidated as elegantly as they have been for sickle-cell anemia.

An interesting polymorphism in plants was reported by the Israeli botanists Apirion and Zohary in the common orchard grass, *Dactylis glomerata* ssp. *judaica*, a diploid ($2n = 14$) form of this species, which also has polyploid races. The polymorphism is interesting because it involves alleles for chlorophyll deficiency that are lethal when homozygous, and it appears to be explainable as a case of single-gene heterosis.

The subspecies *judaica* of orchard grass occurs in Israel mainly in three areas: the Judean hills around Jerusalem, the Nazareth hills, and around Sofad in Upper Galilee. In these areas, the orchard grass forms part of the common dwarf shrub and herbaceous plant communities common to the Mediterranean region. The plants are incapable of self-fertilization (see Chapter 8). One hundred seeds from a variety of plants at a number of localities were sown in uniform garden plots, and the percentage of albino seedlings (homozygotes for the chlorophyll deficiency) were estimated. The results are shown in Table 9.10. By the use of appropriate statistical techniques, the percentage of heterozygotes in the population can be estimated, as well as that of chlorophyll-deficient mutants. Note that the frequency of mutants is very high, ranging from a low of 2 percent in some populations to a high of 15 percent in others. Since such high frequencies cannot be the result of mutation alone, and since the homozygote is lethal, the only other possible explanation is strong selection for the heterozygote. A simple calculation shows that to be so. If the frequency of the recessive gene is $q = 0.15$ and that of the dominant $p = 1 - 0.15 = 0.85$ (population at Qiryat Anavim), the following gene frequencies can be predicted for the offspring generation:

for **AA**: $p^2 = 0.85^2 = 0.7225,$

for **Aa**: $2pq = 0.85^2 \times 0.15 \times 2 = 0.255,$

for **aa**: $q^2 = 0.15^2 = 0.225.$

But since the **aa** homozygotes are lethal, the predicted proportion of homozygotes will be $0.739 \simeq 74$ percent, and of heterozygotes, $0.26 = 26$ percent. But

Table 9.10

Dactylis glomerata populations containing chlorophyll-deficient mutants.

Locality and Soil Type	Total Number of Plants Tested	Number of Normal Homo- zygous Plants	Number of Hetero- zygous Plants	Percent of Hetero- zygotes in Sample	Estimated Gametic Frequency of Albino Gene (Percent)
Jerusalem hills:					
Matzleva Valley					
terra-rossa	47	35	12	26	13
Ora					
rendzina	79	65	14	18	9
Aminadav					
rendzina	94	81	13	14	7
Schneller Camp					
terra-rossa	78	55	23	29	14.5
University campus					
terra-rossa	64	58	6	9	4.5
Qiryat Anavim					
terra-rossa	80	66	24	30	15
Qiryat Anavim					
terra-rossa	94	66	28	30	15
Kastel					
terra-rossa	100	72	28	28	14
Nazareth hills:					
Kefar Hahoresh					
rendzina	82	79	3	4	2
Nazareth					
rendzina	95	90	5	5	2.5
Upper Galilee:					
Ein Zeitim					
rendzina	79	76	3	4	2
Safad					
rendzina	88	77	11	13	6.5
Biriya					
mixed soil	86	76	10	12	6
Miar					
rendzina	97	82	15	15	7.5

From D. Apirion and D. Zohary, 1961. Chlorophyll lethals in natural populations of the orchard grass (*Dactylis glomerata* L.). *Genetics* 46:393–399.

the actual proportion is (we assume that it stays constant from year to year) 66 percent and 30 percent, respectively. In other words, the heterozygote must produce more seed or have a lower rate of mortality than the homozygote, or both.

Although we do not know what advantage the heterozygotes have over the homozygotes, the results indicate that in all probability such an advantage exists. The situation is reminiscent of sickle-cell anemia.

Selection in favor of heterozygotes

In sickle-cell anemia there are strong balancing selection forces against both homozygotes: The homozygous dominant is susceptible to malaria, and the homozygous recessive is lethal. There are innumerable examples of hybrids that are more vigorous than either parent, the best known being hybrid corn and other crops. However, the fact that an artificial hybrid is more vigorous does not necessarily mean that in a natural situation it leaves more offspring, and that is the real test of hybrid superiority, or heterosis. Nevertheless, many geneticists and evolutionists feel that heterozygotes are usually more fit than homozygotes. Their reasoning is that with two different alleles there are more opportunities for the organism to adjust its biochemical machinery during development in a way as yet unknown. An example often cited in favor of this view is the bread mold *Neurospora crassa*. Strains that are heterozygous for the gene for para-aminobenzoic acid outgrow strains that are homozygous for either of the two alleles. Biochemical investigations revealed that the + homozygote produced a very high level of para-aminobenzoic acid, that the − homozygote produced too little, and that both the excess and the shortage of this substance inhibit growth.

There is a great deal of evidence indicating that perhaps heterozygotes (for one or more loci) have higher fitness than homozygotes. Dobzhansky and Spassky in 1963 published the result of a number of studies with *Drosophila pseudoobscura* flies caught in nature. By a special technique, these investigators were able to produce flies that were homozygous for one of their own chromosomes, and they then compared the viability of such animals with those that were heterozygous for the same chromosome (Fig. 9.18). The outcome of the experiment can be described quite simply: There is no correlation between the viability of a heterozygous fly carrying a pair of chromosomes from a natural population and the viability of the flies homozygous for the same chromosomes. Knowing that a chromosome is lethal or semilethal (because it carries lethal or semilethal genes) when homozygous does not enable us to predict its performance in the heterozygous state. The same is true of chromosomes that are of average or high fitness when homozygous. The heterozygote in either case could be normal or could have a fitness either less than or greater than normal. In other words, this experiment suggests that it is not necessarily the performance of genes in the homozygous state that determines heterosis; rather, it is the state of heterozygosity.

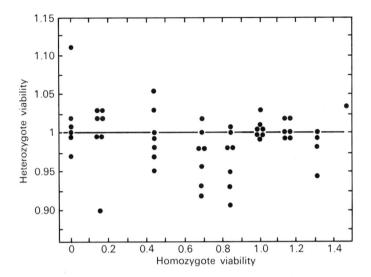

Figure 9.18

Viability of *Drosophila* heterozygous or homozygous for a wild-type chromosome. Flies homozygous for a chromosome found in nature were compared with flies heterozygous for that same chromosome. The viability is expressed in terms of the mean viability of the heterozygotes in the population, which is given a value of 1. The viability of each homozygote is graphed against the average viability of the heterozygotes in which the chromosome appears. (From T. Dobzhansky and B. Spassky, 1963. Genetics of natural populations. XXXIV. Adaptive norm, genetic load, and genetic elite in *Drosophila pseudoobscura*. *Genetics* 48:1467–1485)

Bruce Wallace of Cornell University has conducted a series of experiments involving the exposure of flies of *Drosophila melanogaster* to X-ray radiation at doses high enough to produce mutations. By special manipulations he was able to establish that some heterozygotes containing an irradiated chromosome and a homologous normal chromosome had a higher fitness than homozygotes containing two normal chromosomes (Fig. 9.19). In every instance the homozygote containing two identical irradiated chromosomes was lethal. Again, it appears to be heterozygosity *per se*, rather than the quality of the component genes, that is responsible for the higher fitness.

If heterozygotes have higher fitness than homozygotes but the segregating homozygotes are normal and vigorous, then the problem of the genetic load is not serious. Indeed, in such a population, only individuals homozygous for particularly deleterious alleles would be eliminated in each generation. This number would be relatively small. The rest of the population would consist of a mixture of heterozygotes and homozygotes, the former slightly more fit than the latter. Individuals would compete for resources and for mates, a

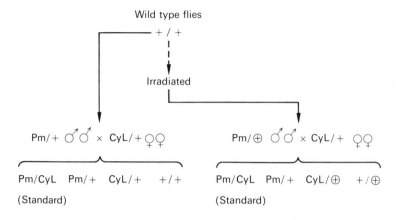

Figure 9.19

Bruce Wallace's irradiation experiment. Wild-type flies by appropriate manipulations are divided into stocks with identical chromosome containing the marker genes
Pm (plum eye color). One group is irradiated, the other not. They are then crossed
with homozygous tester stocks with **Cy** (curly wings) and **L** (lobe eyes). In this
manner, standards are produced that are identical and against which the homozygous wild type (+/+) and the irradiated heterozygous (+/⊕) can be compared.

combination of chance and genetic patrimony determining the outcome, with
the heterozygous genotypes having a slight competitive edge. Wallace has
called the elimination of individuals that carry deleterious alleles *hard selection,* and the elimination of normal individuals because they are outperformed
in the race for resources *soft selection* (Fig. 9.20).

Selection in a variable environment

Imagine again a population differing only at one locus with two alleles. We
have seen (Chapter 4) that most if not all genes show variable expressivity
and penetrance in different environments. Most alleles also affect the fitness
of their bearers in different ways under different environments. Consequently,
the fitness of a given genotype is not a constant number but a function of the
particular environment in which it lives. For example, we can imagine a situation where an organism lives in four slightly different environments that can
be ordered along an arbitrary axis. Let us further imagine that the phenotype
that is produced by the homozygous A_1A_1 genotype is most advantageous in
environment 1 and least in environment 4, and that the reverse is true for
the A_2A_2 genotype. The heterozygote is always as fit as the most fit homozygote but never more so. In such a situation we find the following fitnesses:

Environment	Genotype A_1A_1	A_1A_2	A_2A_2
1	1	1	$1 - 2t$
2	1	1	$1 - t$
3	$1 - s$	1	1
4	$1 - 2s$	1	1
Average	$1 - \frac{3}{4}s$	1	$1 - \frac{3}{4}t$

That is, over all four environments, the heterozygote is the most fit, even though in each environment one of the homozygotes would eventually become fixed, as a result of selection against the other homozygote. However, if there is gene flow from one environment into the other, a dynamic balance is established, a balance that results in a polymorphism. It does not matter whether we are dealing with different spatial environments or with a succession of temporal situations such as exists, for example, in a population throughout the year.

An example of a polymorphism that may be the result of changing selec-

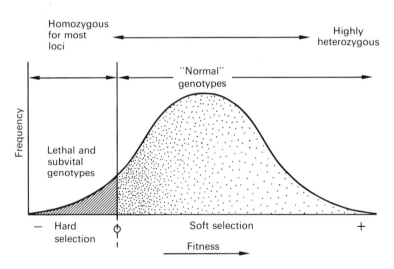

Figure 9.20

Depiction of hard and soft selection. Some of the genotypes in the population are eliminated because they are lethal (hatched area). This is *hard selection.* Among those genotypes that are not lethal, a certain number will not be able to survive to reproduction because of lack of resources, the number depending on resources and circumstances (dots). The lower the heterozygosity, the higher the probability of being eliminated.

tion pressures in time is provided by the inversion polymorphisms in *Drosophila pseudoobscura* that we discussed previously in this chapter. Dobzhansky carefully monitored a population of this species from Mount San Jacinto in California from 1939 to 1946 (Fig. 9.12). Three chromosomal arrangements prevail in this population. They are Standard (ST), Arrowhead (AR), and Chiricahua (CH). From March to June the frequency of ST decreases, and that of CH increases, whereas the opposite change takes place between June and August. Similar changes have been observed in other populations. In 1972 Dobzhansky and Ayala reported on a similar study at McDonald Ranch in Napa County, California. From April 1971 to October 1972 they took monthly samples confirming the seasonal changes in chromosomal polymorphisms, but also establishing significant changes in the frequency of enzyme polymorphisms (Fig. 9.21).

On the other hand, Wright and Dobzhansky showed that if a mixture of flies with the standard (ST) and Chiricahua (CH) chromosomal arrangement are reared in a population cage in the laboratory, the population at equilibrium will have 70 percent ST chromosomes and 30 percent CH chromosomes, from which one can deduce that the heterozygote ST/CH genotype has a relative

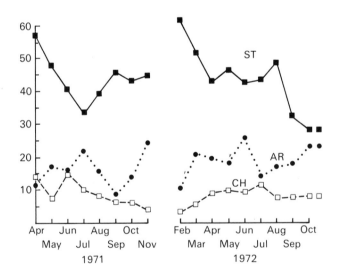

Figure 9.21

Changes in frequency of certain chromosome translocations in *Drosophila pseudoobscura*. The frequencies of Standard (ST), Chiricahua (CH), and Arrowhead (AR) in a population in California changed over the 1971 and 1972 seasons. (From data in T. Dobzhansky and F.J. Ayala, 1963. Temporal frequency changes of enzyme and chromosomal polymorphisms in natural populations of *Drosophila. Proc. Nat. Acad. Sci. USA* 70:680–683)

fitness under those conditions of 1, and the two homozygotes a fitness of 0.7 (ST/ST) and 0.3 (CH/CH). Consequently, in this case there is both selection in favor of heterozygotes and changing selection coefficients as a result of a varying environment.

In many cases there is an association between environmental parameters and genotypes, e.g., *Cepaea, Drosophila pseudoobscura, Adalia bipunctata, Biston betularia* already discussed. These and other examples constitute a substantial body of evidence suggesting that different environments cause different selective pressures, and that environmental variation in time and space can account for much of the genetic variation that is observed. However, these observations do not conclusively demonstrate that the selection pressures exerted by different environments are the sole cause of the variation patterns that are observed. What are needed are experimental demonstrations of what factor or factors in the environment control gene frequency. Of the known cases, only sickle-cell anemia falls in that category.

NEUTRAL VARIATION

Still another explanation for the presence of so much still unexplained variation in natural populations of plants and animals is that the variation is neutral. That is, if a new allele arises by mutation whose fitness is neither higher nor lower than the existing one in the population, it could establish itself by chance and persist (at least for a time) by chance alone.

This explanation, the so-called neutral, random-drift hypothesis as presented by the Japanese geneticists Kimura and Ohta in 1971, states that most new mutations will be harmful and therefore will be immediately eliminated by natural selection. A few will be beneficial and may become widespread. However, a small but significant number will be selectively neutral (neither beneficial nor harmful), and their fate is determined by random genetic drift.

Genes that are selectively neutral can arise in several ways. The genetic code (Chapter 4) is redundant, since there are 64 possible different triplets for only 26 amino acids. Consequently, certain gene mutations have no effect on the protein coded by the gene since the mutated triplet still codes for the same amino acid. In addition, it is believed that certain mutations (such as the change from GUU that codes for valine to GCU that codes for alanine) produce amino acids with very similar properties that do not affect the properties of the molecule and therefore may be selectively neutral. Still other mutations determine amino-acid changes with more of an effect on the enzyme, but some geneticists believe they are also selectively neutral.

Neutral alleles are not necessarily identical in their effect on population fitness. Instead, alleles are considered "neutral" when their effect on fitness is so small that their elimination or fixation in the population depends primarily on chance. Assume two alleles A_1 and A_2 in the population with a fitness for A_1 of 1 and for A_2 of $1 + s$, where s may be a positive or negative number;

that is, A_2 may be more or less fit than A_1. The two alleles are considered neutral when the absolute value of the product of s and the effective population number (p. 175) is much less than 1:

$$N_c \times s \ll 1$$

Consequently, whether an allele is considered neutral depends on the population in which it appears. The larger the effective population size, the smaller s has to be in order for the allele to be effectively neutral. This is so because the smaller the effective population size, the greater the probability of fixation of an allele by random drift. The question is whether such mutations exist and how frequent they are. Some investigators feel that they are very frequent, others that they are rare. The problem is that we have a technique (gel electrophoresis) that indicates that a species possesses many forms of proteins, but no direct way of ascertaining (with some notable exceptions, such as for hemoglobin) the effect of those enzymes and other proteins on fitness. In the absence of such information, two alternative arguments are brought forward: (1) Since no effect on fitness has been detected, the enzyme must be neutral. (2) Although a number of cases are known of enzymes that affect fitness, in no instance has neutrality been demonstrated; therefore all or most mutations must affect fitness. We have still a third alternative, namely, that most isozyme variants are due to post-translational modification by genes at other loci, not to single-locus mutations. In such a case, the necessary level of heterozygosity at a single locus might turn out to be a whole order of magnitude less than what we now believe it to be. If so, it could be explainable by the classical selectionist hypothesis.

SUMMARY AND CONCLUSIONS

Populations of plants and animals appear to have a great deal of genetic variation, a necessary condition for evolution by natural selection. The level of genetic variation, as revealed by the technique of gel electrophoresis, is much higher than was originally supposed. Exactly how much higher is not yet clear, because there are still technical and conceptual problems to be resolved (such as how much of the observed variation is due to allelic variation at the structural gene and how much is due to post-translational modification), and because the number of species and genes within species that have been studied is very small. We know already that there are substantial differences between species and even between groups of species (Table 9.7). Explaining case by case the mechanism by which that variation is maintained is a formidable task.

Although there are various general hypotheses to explain the existence of so much variation, there is at present no accepted single general explanation. Figure 9.22 is an attempt to depict the problem and the ways that various explanations account for the observed patterns. The expected distribution (dashed line) of genotypes in a large, random-breeding population is shown.

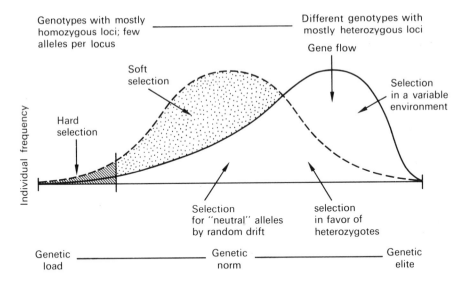

Genotypes with mostly homozygous loci; few alleles per locus _____ Different genotypes with mostly heterozygous loci

Gene flow

Soft selection

Selection in a variable environment

Hard selection

Individual frequency

Selection for "neutral" alleles by random drift

selection in favor of heterozygotes

Genetic load _____ Genetic norm _____ Genetic elite

Figure 9.22

Graph depicting some possible hypotheses that may explain the high frequency of heterozygote genotypes observed in natural populations. The dashed line shows the expected distribution with random breeding and no selection. The solid line approximates the observed distribution. Hard selection is the process of elimination of lethal and subvital homozygote genotypes. There may also be direct selection of heterozygotes because of heterozygote superiority or because of their greater ability to survive in a variable environment. Gene flow and random drift could also account for some of the observed excess of heterozygotes. Genotypes homozygous for lethal alleles constitute the "genetic load"; those with "superior" (in terms of fitness) genotypes have been called the "genetic elite." Intermediate genotypes are the "genetic norm."

A few individuals will be homozygous for most of their loci, and a few will be heterozygous for most of their loci. The majority should have about half of their loci homozygous and the other half heterozygous. Observations in natural populations, as described above, indicate that there is an excess of heterozygotes (solid line) and that the number of alleles per locus is much higher than expected. Some of the explanations brought forward are as follows: (1) *Hard selection*, the elimination of deleterious homozygous genotypes, would account for the elimination of the most homozygous genotypes. This concept is related to the mutation load due to unfavorable mutations, and the segregational load in cases of single-gene heterosis (e.g., sickle-cell anemia). (2) *Selection in favor* of heterozygotes at one or more loci. Dobzhansky called such heterozygous individuals at many loci the genetic elite. Although demonstrated for some cases, this hypothesis, as a general explanation, has the problem of accounting for the elimination of segregant homozygous genotypes. (3) To account for the problem of the elimination of segregants, Wallace proposed the concept of *soft selection*. Heterozygotes at many loci are favored (i.e., have a competitive and reproductive edge), but individuals with homozygous loci (what Dobzhansky called the genetic norm) are not necessarily eliminated. (4) An alternative explanation is that changing selection coefficients are due to a variable environment (in space and/or time). The heterozygote does not necessarily have a higher fitness than any homozygote, but it does better over many environments. (5) Some of the alleles in the population may be introduced by gene flow from neighboring populations, and (6) many of the alleles in the population are effectively "neutral" on their effect on fitness and are present because of random genetic drift. Since these various hypotheses are not necessarily mutually exclusive, the task today is to assess the relative importance of each of these factors in the genetic structure of the population.

Mutation, recombination, gene flow, and selection are the basic forces that determine evolution by natural selection. In Chapter 6 we described them and presented simple models of how they affect the gene frequency of a locus with two alleles in an infinitely large population living in a uniform environment. By now the reader has realized that those models are useful as learning tools but totally unrealistic in terms of real populations growing in nature. We showed that populations are finite in number, that they live in variable environments, that genes interact with each other in giving rise to the phenotype, and that they are packaged in chromosomes. Furthermore, we presented enough examples to convince the reader that there is a very large store of genetic variability in populations that is difficult or impossible to explain even when all known factors are taken into consideration. There is still much to be learned, much to be researched, and no doubt new theories will have to be proposed and proved before a really comprehensive theory of evolution by natural selection, as explained in Chapter 3 is available.

We now turn our attention to another dimension of evolution, namely, the origin and interactions of species.

SUGGESTED FURTHER READING

General references

Ayala, F.J. (ed.) 1976. *Molecular Evolution.* Sunderland, Mass.: Sinauer Associates.

Giblett, E.R. 1977. Genetic polymorphisms in human blood. *Ann. Rev. Genetics* 11:13–28.

Giesel, J.T. 1976. Reproductive strategies as adaptations to life in temporally heterogeneous environments. *Ann. Rev. Ecol. Syst.* 7:57–80.

Hedrick, P.W., M.E. Ginevan, and E.P. Ewing. 1976. Genetic polymorphism in heterogeneous environments. *Ann. Rev. Ecol. Syst.* 7:1–32.

Johnson, G.B. 1977. Assessing electrophoretic similarity. *Ann. Rev. Ecol. Syst.* 8:309–328.

Kimura, M., and T. Ohta. 1971. *Theoretical Aspects of Population Genetics.* Princeton, N.J.: Princeton University Press.

Powell, J.R. 1975. Protein variation in natural populations of animals. *Evol. Biol.* 8:79–120.

Sheppard, P.M. 1958. Natural selection and heredity. London: Hutchinson.

Selander, R.K. 1973. Genetic variation among vertebrate species. *Ann. Rev. Ecol. Syst.* 4:75–92.

Specialized articles and reviews

Allison, A.C. 1961. Genetic factors in resistance to malaria. *Ann. N.Y. Acad. Sci.* 91(3):710–729.

Allison, A.C. 1965. Polymorphism and natural selection in human populations. *Cold Spring Harbor Symp. Quant. Biol.* 29:137–149.

Anderson, W., T. Dobzhansky, O. Pavlovsky, J. Powell, and D. Yarkley. 1975. Genetics of natural populations. XLII. Three decades of genetic change in *Drosophila pseudoobscura. Evolution* 29:24–36.

Apirion, D., and D. Zohary. 1961. Chlorophyll lethals in natural populations of the orchard grass (*Dactylis glomerata* L.) *Genetics* 46:393–399.

Aston, J.L., and A.D. Bradshaw. 1966. Evolution in closely adjacent plant populations. II. *Agrostis stolonifera* in maritime habitats. *Heredity* 21:649–664.

Beckman, L., J.G. Scandalios, and J.L. Brewbaker. 1964. Genetics of leucine amino peptidase isozymes in maize. *Genetics* 50:899–904.

Brussard, P.F. 1975. Geographic variation in North American colonies of *Cepaea nemoralis. Evolution* 29:402–410.

Cain, A.J., and P.M. Sheppard. 1950. Selection in the polymorphic land snail *Cepaea nemoralis. Heredity* 4:275–294.

Coyne, J.A., and A.A. Felton. 1977. Genic heterogeneity at two alcohol dehydrogenase loci in *Drosophila pseudoobscura* and *Drosophila persimilis. Genetics* 87:285–304.

Dobzhansky, T., W. Anderson, and O. Pavlovsky. 1966. Genetics of natural populations. XXXVIII. Continuity and change of populations of *Drosophila pseudoobscura* in western United States. *Evolution* 20:418–427.

Dobzhansky, T., and F.J. Ayala. 1973. Temporal frequency changes of enzyme and chromosomal polymorphisms in natural populations of *Drosophila. Proc. Nat. Acad. Sci. USA* 70:680–683.

Dobzhansky, T., and B. Spassky. 1963. Genetics of natural populations. XXXIV. Adaptive norm, genetic load and genetic elite in *Drosophila pseudoobscura. Genetics* 48:1467–1485.

Finnerty, V., and G. Johnson. 1979. Post-translational modification as a potential explanation of high levels of enzyme polymorphism: xanthine dehydrogenase and aldehyde oxidase in *Drosophila melanogaster. Genetics* 91 (in press).

Harris, H. 1969. Genes and isozymes. *Proc. Roy. Soc. Ser. B* 174:1–31.

Hubby, J.L., and R.C. Lewontin. 1966. A molecular approach to the study of genic heterozygosity in natural populations. I. The number of alleles at different loci in *Drosophila pseudoobscura*. *Genetics* 54:577–594.

Jones, J.S., B.H. Leith, and P. Rawlings. 1977. Polymorphism in *Cepaea*. *Ann. Rev. Ecol. Syst.* 8:109–146.

Lamotte, M. 1960. Polymorphism of natural populations of *Cepaea nemoralis*. *Cold Spring Harbor Symp. Quant. Biol.* 24:65–84.

Livingstone, F.B. 1971. Malaria and human polymorphisms. *Ann. Rev. Genetics* 5:33–64.

Markert, C.L. 1968. The molecular basis for isozymes. *Ann. N.Y. Acad. Sci.* 151:14–40.

Markert, C.L., and F. Mollėr. 1959. Multiple forms of enzymes: tissue, ontogenetic, and species specific patterns. *Proc. Nat. Acad. Sci.* 45:753.

Ohta, T., and M. Kimura. 1975. Theoretical analysis of electrophoretically detectable polymorphisms: models of very slightly deleterious mutations. *Am. Nat.* 109:137–145.

Wallace, B. 1958. The average effect of radiation induced mutations on viability in *Drosophila melanogaster*. *Evolution* 12:532–552.

Wallace, B. 1975. Hard and soft selection revisited. *Evolution* 29:465–473.

Timofeff-Ressovsky, N.W. 1940. Zur Analyse des Polymorphismus bei *Adalia bipunctata*. *Biol. Zentral.* 60:130–137.

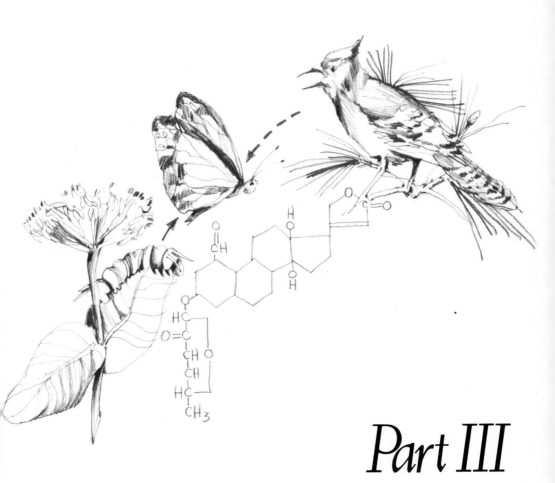

Part III

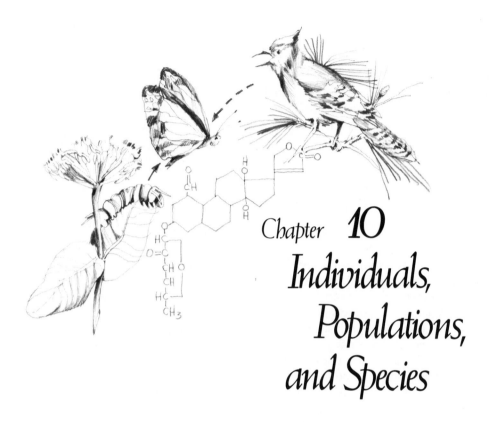

Chapter **10**

Individuals, Populations, and Species

In Chapter 3 we presented the basic axioms of the theory of evolution. For evolution to take place it is necessary that organisms be capable of reproducing their kind; that they produce an excess of offspring; that the survival of the offspring be related to their characteristics (biochemical, physiological, morphological, etc.); and that there be a mechanism by which these characteristics are transmitted to the offspring. In Chapters 4–9 we explored these premises in detail. We showed that organisms do not transmit their characters to the offspring, but only the *potential* of acquiring these characters if the appropriate internal and external environment is provided. We also showed that the advantages of possessing a given character are relative, not absolute, and depend on the properties of the environment and on the potential neighbors and competitors. We also indicated that chance plays an important role in evolution.

Mutation, recombination, and gene flow determine the genotype of the individuals of a population. These genotypes, acting in concert within the cells of an organism and in the context of the environment, will produce in each population a set of phenotypes. Natural selection is the differential survival of these phenotypes in the population, that is, the relative success of individual organisms in overcoming the rigors of the environment, in competing with other individual organisms for resources, and eventually in producing offspring. Consequently, the best way to study and describe mutation and recombination is in a context of genes and genotypes, i.e., in terms of changes in gene frequencies over time and the forces that change those frequencies. The best way to study natural selection is in a phenotypic context, i.e., in terms of changes of numbers of individuals over time.

GENOTYPE AND PHENOTYPE

It is generally accepted that for a given environment the genotypic composition uniquely determines the phenotypic composition of the population. However, experimental verification of this belief is not available. We know from a number of studies that some phenotypic differences have a genetic basis. For example, the differences in the color and shape of varieties of squash are determined by certain genes in the cells of the plants; the same is true for the kernel color of maize plants, or for eye color in *Drosophila* flies. But different genotypes can give rise to identical phenotypes (as in recessive mutations) and identical genotypes can give rise to different phenotypes if exposed to different environments. We also know that different genotypes can give rise to identical phenotypes in one environment, or different phenotypes in another. In Chapter 6 we presented two examples of this: human polydactyly and curly wings in *Drosophila*. Another example is the condition known as crossveinless in *Drosophila melanogaster*, which produces a defective pattern of veins in the wings (Fig. 10.1). When pupae of stocks of flies that produce normal wings are given a temperature shock of 40°C for four hours between the ages of 21 and 23 hours, approximately 40 percent of the flies are born with crossveinless wings. The differences between normal and crossveinless is a genetic one, and breeding together flies that respond to the heat treatment eventually produces a stock of flies that have crossveinless wings even in the absence of a heat shock.

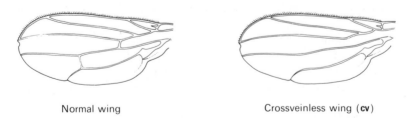

Normal wing   Crossveinless wing (**cv**)

Figure 10.1
Normal and crossveinless (**cv**) wings in *Drosophila melanogaster*.

Therefore, even if eventually it can be proven that each set of genotypes produces one and only one set of phenotypes in a given environment, the rules of transformation of genotypes into phenotypes will prove to be extremely complex. At present this area, known as developmental genetics, is largely uncharted territory.

The present impossibility of knowing exactly how genotypic changes affect the phenotypic composition of the population, and vice versa, hinders a full understanding of evolution. Two alternative ways of temporarily bypassing this troublesome problem are (1) to assume that natural selection changes gene frequencies and to study evolution at the genetic level, or (2) to assume that the interaction of mutation, recombination, gene flow, and recombinational distortion uniquely produces the different phenotypes that we observe in the population and to study evolution primarily at the phenotypic level. Both assumptions are reasonable. Changes in the frequency of phenotypes in the population brought about by natural selection produce changes in the population of gametes and of the alleles those gametes carry, and changes in gene and genotype frequency result in phenotypic changes. Nevertheless, when fitness values are assigned to genes and genotypes, as was done in Chapters 4–9, or when genotypic composition is assumed from phenotypic characteristics, two great simplifying assumptions are made, which usually cannot be verified experimentally.

In Chapters 4–9 we studied the major concepts of what we may call the "genotypic" form of the theory of evolution. That is, we presented salient aspects of the accepted theory regarding the processes that account for the generation, maintenance, and transformation of genotypic variability in populations of plants and animals. In the next six chapters we will study some aspects of what we may call the "phenotypic" form of the theory of evolution. That is, we will focus on the diversity of populations of organisms in space and time and explore their interactions. Remember, however, that this is only a didactical device and that evolutionary change affects all levels from genes to communities simultaneously.

THE LEVELS OF ORGANIZATION OF LIFE

Living organisms inhabit nearly every part of the earth's surface, from the tropics to the polar latitudes, and from deep in the ocean to high in the atmosphere. Their multitude is enormous; their forms, shapes, and ways of life are so varied that no limit is apparent. All this multiplicity expresses an advanced level of chemical organization that we call life. Nucleic acids, proteins, lipids, water, a certain number of minerals, and carbohydrate fuel, organized into little vesicles with semipermeable membranes, are common to all organisms, from the tiny blue-green algae and bacteria to the giant whales.

Wherever we look, whether up into the skies or down into the oceans, deep into matter or away out into the universe, we find a multiplicity of level upon level of systems of organization, all interrelated and all of them fantastically complex. When we observe and study this variety of forms, we instinctively think of the level of organization to which we as individuals belong, the level of the individual organism, as the unit of measure. Any system below that of the individual is thought of as "small" or "simple," anything above it as "large" and "complex." But the individual organism is a level of organization that is sometimes hard to define. The notion that an individual is not divisible without losing its properties of "life" is applicable to most higher animals. But most plants and many of the so-called lower animals can be broken up into smaller units, each of which is capable of regenerating the missing segments, a fact of which we take advantage when we obtain new plants through cuttings.

Furthermore, in some cases "individuals" cannot live by themselves but need to aggregate with other individuals. The lichens are the most striking example. A lichen consists of an alga and a fungus that live in close association. The algae that occur in lichens usually also occur alone, but nearly all the lichen fungi are restricted to their lichen occurrence. No other examples are as striking as the lichens, but a long list of parasites and symbionts are directly dependent on other individuals for their life, although such a relationship is not always mutual. Human beings, for example, depend not only on plants and animals for food but also on the colon bacterium, *Escherichia coli*, which manufactures vitamin K, without which we could not live. *Escherichia coli*, in turn, cannot live under natural conditions outside the colon.

Another type of organization is represented by colonial forms. Individuals of such diverse organisms as the green algae *Volvox* and *Pandorina*, honeybees, and ants, exist in close association with other members of their species, where each member performs roles that are vital to all the others. In these colonial forms individual members can no longer survive alone. The unit is the colony, not the individual member. But then, with the exception of hermaphroditic, self-fertile plants and a few parthenogenetic plants and animals, all organisms are incomplete, since they cannot reproduce by themselves. The minimum needed is two individuals of opposite sex. Furthermore, organisms do not occur in isolated pairs, but in groups of variable numbers of different sexes and ages at each locality. These groups are the *breeding populations*. A population is spatially confined, and a reality only in sexually reproducing species.

A further type of organization is the *community*. The population is a genetic unit, held together in time by reproduction and descent. The community is an ecological unit. It consists of one or more breeding populations of the same or different species that are adapted to the same environment. Communities include plants and animals, both microscopic and multicellular. To the extent that every organism modifies the environment in which it lives, it is a factor in the environment, and consequently there is a certain interrelationship between the members of the community. This relationship is very loose, as are the physical limits of the community. But communities are ecological units that conform to certain criteria and have a certain continuity in time, and as such they have an undeniable existence.

One of the major characteristics of matter, including living matter, is that it is organized. Individuals, symbionts, colonial forms, breeding populations are not the products of chance; they represent a highly integrated system of organization. But at the same time it is not always easy to define all these levels. The reasons are many and different for each particular instance. But this much can be generalized: No two situations in evolution are alike. The quality and rate of mutation, the recombination system, the ecological situation to which organisms have to adapt, are unique in each population, in each colony, in each community. The Douglas fir grows continuously over thousands of acres in North America, and no discrete breeding populations can be discovered. On the other hand, the swamp maple is restricted to wet areas, where it forms dense stands. Rabbits are found in loose pairs throughout the prairie and Great Basin, whereas prairie dogs, which also live in that area, form colonies of hundreds of individuals.

THE UNITS OF SELECTION

So far the discussion of natural selection has been concerned with what is usually called "individual selection." We have presented selection in terms of alternative alleles at a locus or alternative characters of individuals in a population; that is, two or more *individuals* differing in their genetic constitution have been considered. Since this is how Darwin primarily conceived of natural selection, this kind of selection is often called "Darwinian selection."

However, natural selection can take place among biological units other than individuals as long as these units meet the following criteria: (1) They have the ability to replicate; (2) they produce an excess of units in relation to what is needed to replace the parental generation; (3) their survival is correlated to their characteristics; and (4) they have a mechanism for the transmission of these characteristics (Chapter 3). We now explore four units of selection other than the individual and assess the evolutionary implications of selection at that level. These units are (1) gametes (gametic selection); (2) family groups (kin selection); (3) populations (group or deme selection); and (4) sexual selection.

Gametic selection

Gametes have a genetic constitution that is different from the diploid organism that produced them; there is an excess of them; they possess unique characters; and they transmit these characteristics (through the zygote) to other gametes. They consequently possess the necessary conditions for natural selection to act on a population of gametes produced by an individual diploid organism independent from the selection that operates on the diploid parent. The characters of gametes that should be under strong selection should be things such as the motility of sperm, the viability of different haploid gametic genotypes (e.g., ability to survive in different environments and longevity), and the ability to unite with another gamete to produce a zygote. In plants, for example, if one kind of pollen grain can produce a pollen tube that grows faster than the pollen tube of other pollen grains, then that kind will fertilize more eggs. Because the gametes are haploid and all genes are consequently exposed to selection, gametic selection is very effective. Therefore we expect any detrimental mutation to be eliminated very rapidly and any advantageous mutation to spread equally rapidly. However, if an allele is pleiotropic and confers an advantage to the gamete but reduces the fitness of the diploid organism, then the allele can be maintained at an intermediate frequency.

Kin selection

The effect of natural selection is to change the gene frequency in the population. The frequency of a given allele is increased when the individuals that carry it survive the rigors of the environment and produce a larger number of offspring than do members of the population without that gene. However, if an individual of a species is able to increase the survivorship of relatives with a common allele, there will also be an increase in the frequency of the genes that individual possesses even in cases where the individual's own survivorship or offspring size has not changed. For example, let us imagine a situation where three sibling prairie dogs, all three heterozygous for the same allele, are threatened by a predator while feeding in a meadow. Imagine further that the predator can silently catch each of the individuals without the others' noticing it. Now if one of them, on seeing the predator, gives an alarm call, it will draw attention to itself and thereby will (1) almost certainly be eaten by the predator and (2) give its siblings a chance to escape into a burrow. Clearly, giving an alarm call in this case is detrimental to the individual (increases the probability of death by attracting the predator to itself), but doing so nevertheless increases the survivorship probability of the allele, because only one out of three carriers dies. If the particular trait (giving an alarm call) is determined by the allele in question, one can imagine how it could spread in the population. We can see, then, that certain traits that may be detrimental to the owner may still be selected if they are sufficiently beneficial to related individuals. Such traits are called *altruistic*, and this mode of selection is called *kin selection*.

When should an altruistic gene be selected? It should be selected when the increase in fitness to closely related individuals (which are more apt to carry the gene than the general population) is great enough to compensate for the loss in fitness to the altruistic individual. Closely related individuals are more likely to carry the gene than members of the population at large. If the advantage to them is sufficiently great, the altruistic gene will increase in frequency. If it is not great enough, the gene will not be selected.

Thus, in order to predict the presence of altruism, it is important to know the degree of kinship among members of a group. The *coefficient of relationship* between two individuals is the probability for any locus that the two share an allele identical by descent. For a diploid parent and its offspring, the coefficient of relationship is 1/2, since the parent passes half its chromosomes to the offspring. Similarly, the coefficient of relationship between offspring of the same parents is 1/2. For any allele at any locus of one sibling, the chance that the parent from whom the allele was received also passed the same allele to the other sibling is 50 percent. For an altruistic act that favors a sibling, the altruistic genes will be selected if the fitness of the sibling is increased by twice the amount by which the fitness of the altruist is decreased.

The clearest examples of kin selection occur in the social insects. For example, in the honeybee (*Apis mellifera*) there is a caste of sterile female workers. How could natural selection, which is based on passing adaptive characteristics to offspring, produce a whole caste of *sterile* individuals? In the case of the honeybee, the male is haploid and the female is diploid. The coefficient of relationship between two workers is 3/4 since the set of genes each received from their father (half of their genome) is identical, and of the genes received from their mother, half on the average are identical. On the other hand, the coefficient of relationship between a mother and a daughter is only 1/2. Thus it can be more advantageous to rear sisters than to rear offspring.

To what extent does kin selection operate in our species? Here the situation is much more complex. Human behavioral adaptability makes it difficult to determine to what extent behavioral patterns are genetically, rather than culturally, determined. Acts that benefit others at some cost to the performer are widespread and encouraged in the cultures. The prevalence of behavior patterns whose immediate consequences benefit others more than the self suggests that there is a common base, a genetic predisposition toward some degree of altruism. There is a great deal of variability in kind and extent of altruism, and the exact form it takes seems to be shaped by each culture.

Through most of our early history, the basic social group is believed to have been the family. If this is so, then altruistic acts would benefit relatives, and kin selection would operate. With the increase in intelligence and the development of language as our species evolved, the capacity for more complex relationships increased. *Reciprocal altruism,* a concept introduced by R. Trivers, is one possible development. Reciprocal altruism means that an altruistic act, which could benefit an unrelated individual, is performed in the expectation of some reward. The reward could be the expectation of future aid. It could also be favorable recognition by the rest of the community. Thus unselfish acts are not necessarily truly unselfish. Because there is some benefit attached, the capacity for altruism can, in the right circumstances, be selected.

Group selection

Imagine a population of animals that required a renewable environmental resource of limited supply at any one time, such as grass or some prey. Individuals in the population would grow and reproduce, and natural selection, as always, would favor those phenotypes that produce the largest number of surviving offspring. Soon the population might have used up all the available resources before they were renewed. Such a population would then become extinct. Imagine another population in which, for lack of the appropriate mutations, no phenotypes had evolved that can outeat the resource. That population would not only survive until the resource was renewed, but it could colonize the areas formerly occupied by the fast-growing but now extinct populations. Such selection, which operates through groups and which opposes individual selection, is called *group* or *interdeme* selection.

For group selection to operate, there must be differential extinction of groups made up of "selfish" members, followed by recolonization of the site by groups of altruists. At the same time, gene flow cannot be too widespread; otherwise the "selfish" genes will invade the colonies of "altruists," where they will be favored by individual natural selection.

Group selection is a controversial subject. There is no direct evidence for it, and the evolution of most if not all characteristics of organisms that are favorable to groups can be explained by individual and/or kin selection. However, the lack of direct evidence does not necessarily deny its existence. As a matter of fact, certain traits, especially behavioral traits, are easier to explain by invoking group selection. Whether group selection can act in opposition to individual natural selection is a different matter, one that is unresolved but hotly debated at present.

Sexual selection

If for each female in the population there are several available males, the female may choose the male of her liking. On the other hand, if one male is capable of driving away other males, either by threats or by actual combat, and as a result manages to fertilize one or more females, the successful male will produce more offspring than the males he drove away. (In extreme cases the other males may be excluded from breeding altogether.) Those characteristics that make the female or the male more attractive to the other sex or more threatening to rivals of the same sex and that contribute to her or his success will be of adaptive significance and will be selected. This is the essence of the theory of *sexual selection*, presented for the first time by Charles Darwin in 1871 in his book *The Descent of Man* to account for the difference between males and females found in many species of animals—features such as fancy feathers or huge antlers that do not appear to be of general adaptive significance.

In his original presentation, Darwin stressed combat as a very important selective force, which can produce armament such as antlers in deer, wing and leg spurs in many birds, horns in stag beetles. This kind of sexual selection, which arises from competition among individuals of one sex, is called *intrasexual selection*. Actual combat occurs in many instances, but display of aggressive structures such as horns or tusks, or behavior such as raising of fur or feathers to give the appearance of greater size as a threat without actual combat, may play a more important role than actual fighting. Over-aggressive behavior may reduce rather than increase fitness, as in many species of elk, where actual combat may eliminate both aggressors or may allow other males to run away with the does while the fight goes on. Structures of use in combat may consequently evolve with little actual combat, at least between members of the same species.

Intrasexual selection will advance with the greatest momentum in species where the competition between members of one sex is keenest. Darwin perceived this and therefore attached great significance to the mating system of a species. Strict monogamy with numerical equality between the sexes is the one system in which intrasexual selection will be ineffective, provided that the mating takes place for at least the length of the breeding season. All other mating types—polygamy, promiscuity, and numerical inequality of the sexes —will allow for at least some intrasexual selection. Since the most common mating systems are monogamy with an excess of males and polygyny (one male fertilizing many females), intramale selection should be more common than intrafemale selection, and it is. In these systems, males that ward off other males by threat or combat or possess characters that will attract more females leave a greater progeny to inherit their features.

The theory of sexual selection as proposed by Darwin has not been universally accepted. Some authors feel that procuring mates is just one more aspect of natural selection. For example, if a sterile male is very successful at warding off other males and in mating with all the females, he still will be without offspring, and his genes will not be perpetuated. On the other hand, a brightly colored male (among birds or insects) can be more successful in mating, but his offspring may suffer more predation because of their conspicuousness, so that a generation later there may be fewer rather than more brightly colored animals in the population. Clearly, sexual selection is a component of total fitness, and the evolution of sexual dimorphism cannot proceed if it is detrimental to the individual. The question remains, however, whether characters can evolve that are of advantage only or primarily in procuring mates.

In *Drosophila*, mutant flies are known that have a yellow body instead of the normal gray-yellow. If "yellow" and gray-yellow males are introduced into a population of normal gray-yellow females, the gray-yellow males leave many more offspring, since the yellow males are seldom accepted by the females. On the other hand, yellow females do not discriminate between yellow and gray-yellow males. This observation indicates that mating discrimination can indeed influence the fate of a character otherwise neutral.

Sexual selection, which is very important in the evolution of behavior and sociality, will be taken up again in Chapter 16.

Cultural evolution

So far we have discussed natural selection in relation to genetically transmitted characteristics. However, some characteristics are spread within a population by nongenetic means. We will call this phenomenon *culture*, although that term is often used in a less general sense to refer only to certain activities of human societies, or even as the characteristic that distinguishes us from other animals.

Following our definition, there are many species of animals with cultural traits. Species of migrating birds follow the same routes, year after year and generation after generation. Young prairie dogs learn from older members of their colony to recognize the boundaries of the colony. Some species of birds learn dialects of their species' basic song from other birds in their vicinity.

An interesting example of the spread of a cultural characteristic occurred in a troop of Japanese macaques (*Macaca fuscata*). The troop had become accustomed to a new food, sweet potatoes. A 1½-year-old female began to wash her sweet potato in the sea before eating it, rather than merely to brush off the sand. This behavior was picked up by other juveniles. Mothers learned it from their juvenile children and taught it in turn to younger offspring. Eventually the whole troop, except the older males, who did not interact much with the rest of the troop, washed their sweet potatoes before eating.

The same female, at age 4, initiated another technique. Wheat was being thrown on the sand as feed for the monkeys. They picked it up, grain by grain. Our young female invented the placer technique, which involved throwing a handful of grain and sand on the water and collecting the grain that floated on the surface after the sand sank. This characteristic also spread through the troop.

Cultural evolution is obviously influenced by genetic characters. The ability to learn is essential to cultural evolution, and this is at least in part genetic. There can also be feedback between physical and cultural evolution. Such feedback is particularly important in human evolution. Many people now believe that the development of culturally transmitted activities, such as the use of tools and language, after being made possible by a certain brain structure, created selection pressure to develop the brain further, permitting in turn new kinds of cultural advances. Human behavior patterns were once believed to be entirely cultural, with no genetic input. Now, however, it seems more likely that we have inherited not only the capability for culture but also the tendency or even need to develop certain cultural traits, such as marriage, dance, or language. All known human societies have these traits, though the precise form varies greatly.

Cultural evolution has certain features different from genetic evolution. It is Lamarckian, in that acquired characters can be inherited. It is also possible to pass on characters to individuals other than offspring. In addition, cultural characteristics can spread through a population much more quickly than genetic characters.

There are also basic similarities between cultural and genetic evolution. There are differences between populations in their cultural characteristics. These differences can be transmitted from one to another with results similar to gene flow. There can also be cultural drift, in which characters change as a result of chance. And of course, natural selection can operate on cultural as well as genetic characters.

THE SPECIES

So far in this chapter we have seen that living matter is highly organized and that several of these units—gametes, individuals, family groups, and perhaps even populations—are or can be independent selection units because they possess the necessary prerequisites (Chapter 3). However, the most important biological unit is probably the species. The species is the basic unit of classification. Even to the nonscientist different members of the cat family—lions, tigers, mountain lions—are easily recognizable. However, a closer look reveals that defining and interpreting the nature of species is a difficult matter.

The concept of the species

Groups of organisms called *species* have been recognized for a long time, but there has been and still is much confusion about what a species really is.

There are three main species concepts. The first, that of the philosopher and the logician, goes all the way back to Aristotle. Species is a category of thought, more specifically a logical class, to which belong all the objects (living or not) that have certain common properties. For example, fork is a species, all forks having certain basic common properties, such as the possession of prongs or tines and a handle arranged in such a way that objects can be both lifted and pierced. Within these limitations, forks can be of different shapes, sizes, and materials.

The second concept, the so-called morphological criterion of biologists, is an application of the logical definition of species to organisms. Species are defined strictly on the possession by their members of certain characteristics—morphological, physiological, behavioral, chemical—not possessed by members of other species. This was the concept of species held by Linnaeus and most eighteenth- and nineteenth-century taxonomists, and even in our day, some adhere to it strictly.

Finally, the third concept is the so-called biological species concept. The individuals of populations that under natural conditions are potentially capable of interbreeding constitute the species. The morphological species concept stresses likeness and similarity, and the biological species concept stresses the breeding relationships. In general there is a high degree of coincidence between the species defined by those two concepts. Individuals that look alike tend to interbreed; and individuals that interbreed tend to look alike, since they have large numbers of common genes. Nevertheless, this is not always so. Sexual dimorphism is a striking illustration. Linnaeus, applying a strict morphological criterion, described the male and female of the mallard duck as different species. Other cases of dimorphism are not tied to sex. Many species of flowering plants have individuals with different flower colors. In the species *Leavenworthia stylosa* of the mustard family, for example, some individuals have white flowers and other individuals have yellow flowers. Both white and pink wild radishes belong to the same breeding population. Snails of the species *Cepaea nemoralis,* as we saw in Chapter 9, have different banding patterns in the shell, and so forth.

The species can be considered the largest populational unit. Individuals in a population usually mate with individuals of the same population. Occasionally an animal will wander away from its breeding population, or a seed from a plant will be blown away by the wind. In such instances individuals from different populations will exchange genes, but with members of their own species (with some exceptions). There are usually, especially among the higher vertebrates, definite barriers to the exchange of genes between different species. These can be spatial, temporal, ecological, behavioral, cytological, or genetical.

Higher categories

Species are grouped into larger categories called genera, which in turn are grouped into families, orders, classes, phyla, and kingdoms (see Table 10.1).

Table 10.1

The major categories used in classification.

Kingdom
  Phylum (Animal); Division (Plant)
    Class
      Order
        Family
          Tribe
            Genus
              Species

Intermediate categories can be erected using the prefixes *sub* and (in animal taxonomy) *super,* such as subclass, superfamily, and the like.

The categories discussed so far—individuals, breeding populations, communities, and species—are natural, nonarbitrary categories, but the higher ones are ideal, logical categories (Fig. 10.2). Each is a class that includes all the categories of the next lower level having certain characters in common. For example, the genus *Felis* comprises all the species of mammals with claws and certain types of dentition, such as cats, lions, tigers. Note, however, that although the genus *Felis* is a useful concept that helps us express in one word a certain relationship, and although probably all the species of *Felis* have had a common origin, there is no tangible reality to the genus *Felis*, as there is to the species *Felis leo*, the lion. The different species of the genus usually cannot breed with one another, they are not subjected to the same selection pressures, they do not live in the same environment, and although they may have had a common origin, they do not have a common present or future.

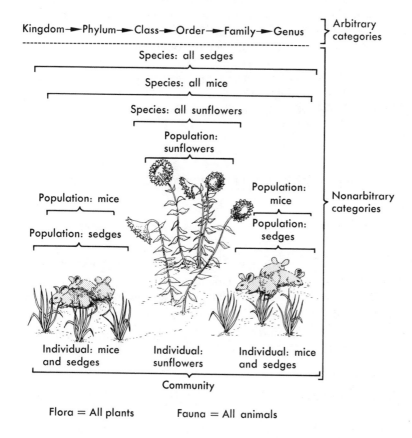

Figure 10.2
Relationships among different categories.

A hierarchical system of categories in which the groups on each level are formed by placing together those of the level below has become a fundamental aspect of the taxonomic method. It was essentially established by Linnaeus in the eighteenth century, and it still persists today. This system has provided a means for revealing relationships, but it has often proved inadequate as a way of expressing certain conclusions about the evolutionary relationship of organisms. This inadequacy has led some scientists, especially cytologists, geneticists, and ecologists, to reject the classical system and to construct their own system of categories. Another result has been that some of the more orthodox taxonomists have disregarded or minimized genetic and breeding information in systematic studies. However, the apparent conflict is a false one. The objective of the taxonomist is to order and to provide a way to classify and name organisms. The objective of the evolutionist is to explain the diversity of organisms and their past history, as well as to develop general predictive theory. Evolutionists could not study the diversity of the living world if it was not somehow ordered by the taxonomist, and the taxonomist depends on others for the understanding of the properties of organisms. We will return to this point in Chapter 18 when discussing the operations involved in classification.

In the next chapter we explore some of the ways by which new species originate, and in so doing we will discuss in greater detail the difficulties involved in applying a morphologicial concept of the species as well as the so-called biological species concept.

SUGGESTED FURTHER READING

General references

Darwin, C. 1871. *The Descent of Man and Selection in Relation to Sex.* London: Murray.

Fisher, R.A. 1958. *The Genetical Theory of Natural Selection.* New York: Dover.

Fox, R. 1971. The cultural animal. In J.F. Eisenberg and W.S. Dillon (eds.), *Man and Beast: Comparative Social Behavior* (Smithsonian Annual III), pp. 273–311.

Grant, V. 1971. *Plant Speciation.* New York: Columbia University Press.

Maynard-Smith, J. 1978. *The Evolution of Sex.* Cambridge: Cambridge University Press.

Articles and reviews

Hamilton, W.D. 1964. The genetical evolution of social behavior. I, II. *J. Theor. Biol.* 7:1–16; 17–52.

Kawai, M. 1965. Newly acquired pre-cultural behavior of the natural troop of Japanese monkeys on Koshima inlet. *Primates* 6(1):1–30.

King, J.L., and T.H. Jukes, 1969. Non-Darwinian evolution. *Science* 164:788–798.

Kawamura, S. 1963. The process of subculture propagation among Japanese macaques. In C.H. Southwick (ed.), *Primate Social Behavior: An Enduring Problem.* Princeton, N.J.: Van Nostrand.

Lewontin, R.C. 1970. The units of selection. *Ann. Rev. Evol. Syst.* 1:1–18.

Maynard-Smith, J. 1964. Kin selection and group selection. *Nature* 201:1145–1147.

Sammeta, K.P.V., and R. Levins. 1970. Genetics and ecology. *Ann. Rev. Genetics* 4:469–488.

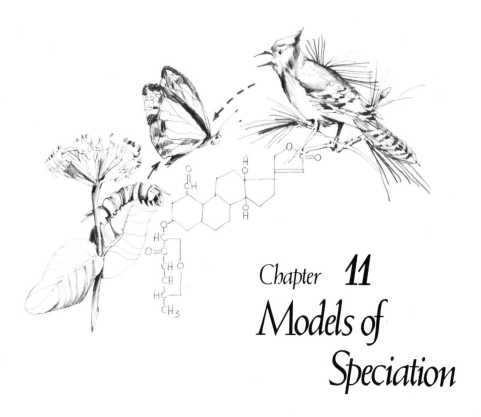

Chapter **11**

Models of
Speciation

One of the central problems in population biology is the way species originate. Darwin chose to entitle his famous book *On the Origin of Species* (although he did not solve the problem), and the way species originate has been the subject of many scholarly discussions and some heated arguments ever since. In modern times, the Harvard biologist Ernst Mayr has made great progress toward the elucidation of this question, but there are still a number of problems left. The reason for so much interest in this subject is that the formation of new species is one of the crucial stages in evolution.

When we speak about the origin of species, we are talking about the coming into existence of a group of individuals that we call a species and recognize as distinct from other such groups. However, it is usually believed that a species arises by selection of *individuals* with certain new characteristics, which by multiplication give rise to the species. Group selection is not necessarily excluded, but its operation is difficult to prove.

In the previous chapter we defined a species as a group of potentially interbreeding populations that are reproductively isolated from other such groups. This is the so-called *biological species concept*. It is not a good definition in an absolute sense, since the phrase "reproductively isolated" can assume different meanings and is therefore ambiguous. Furthermore, it is in practice very difficult and time-consuming to establish whether two populations are potentially capable of interbreeding or not. However, alternative definitions pose still more problems (Chapter 12). Ultimately speciation is the establishment of barriers to gene flow between closely related populations. Presumably the two populations or groups of populations that become reproductively isolated have adapted uniquely to their local environments, and introduction of genes from one population into the other is detrimental to the fitness of the individuals involved.

MODES OF SPECIATION

There are two major modes by which new species can originate (Table 11.1). The first is by transformation in time of one species into another. This mode is called *phyletic speciation*. Hybridization and fusion with other species to various degrees provides other means of transformation of species in time. Neither of these two systems involves the development of isolating mechanisms nor the multiplication of the number of species.

The second major mode of speciation is the process by which one ancestral species gives rise to one or more species without necessarily losing its identity. This process is called *multiplication of species, or true speciation*. We will be concerned in this chapter with the latter process. The speciation pro-

Table 11.1

Potential modes of origin of species.

I. Transformation of species (phyletic speciation)

   1. Autogenous transformation (owing to mutation, selection, etc.)

   2. Allogenous transformation (owing to introgression from other species)

II. Multiplication of species (true speciation)

   1. Abrupt speciation (through individuals)

      a. Chromosomal rearrangements

      b. Polyploidy

   2. Gradual speciation (through populations)

      a. Sympatric speciation

      b. Parapatric speciation

      c. Allopatric speciation (geographic speciation)

After E. Mayr, 1963. *Animal Species and Evolution.* Cambridge, Mass.: Belknap Press of Harvard University Press.

cess can be divided into instantaneous speciation and gradual speciation, and the latter can be further divided into three major types: (1) allopatric or geographical speciation, (2) parapatric speciation, and (3) sympatric speciation.

The most widely accepted hypothesis explaining the process of multiplication of species, or true speciation, is what has been called the geographical theory of speciation, first enunciated by the German biologist Moritz Wagner in 1868. It states succinctly that the first step in speciation is reproductive isolation brought about by physical (geographical) separation. A second step is the independent evolution of these reproductively isolated populations. If the ranges of the populations eventually merge again, and if the reproductive isolation persists because some other isolation mechanism, such as genetic incompatibility, has evolved, the speciation process is considered completed. We now consider the three main types of gradual speciation separately, starting with allopatric or geographic speciation.

Allopatric speciation

Four steps can be identified in the allopatric speciation process (Fig. 11.1): (1) separation of the original gene pool into two or more isolated groups (isolation); (2) independent evolution of the two gene pools (differentiation); (3) secondary merger; and (4) competition between the new gene pools. Let us analyze them separately.

Isolation. A group of interbreeding populations may become separated into two or more isolated groups for a number of reasons. Foremost is change in the physical environment that results in a shrinkage of the range of the species. Destruction of the environment by humans or other types of disturbance can have the same effect. Actually, physical separation of populations so that they become partially or completely isolated is a common phenomenon, so much so that it is the expected pattern (Fig. 11.2). Isolated or semi-isolated populations can also be the result of long-range dispersal, when a new population becomes established by a small number of founders.

Isolation is a necessary but not a sufficient condition for allopatric speciation.

Differentiation. By differentiation is meant the process by which two populations in isolation evolve in different directions so that they become morphologically and/or physiologically different. Differentiation does not necessarily follow upon isolation, though apparently it often does. The factors controlling the process of differentiation still need to be investigated in more detail. However, there is ample evidence for the concept that populations adapt closely to their local environment. In Chapters 4–9 we analyzed the dynamics of breeding populations and the forces that tend to produce change, as well as the forces that oppose change. These forces—mutation, recombination, and

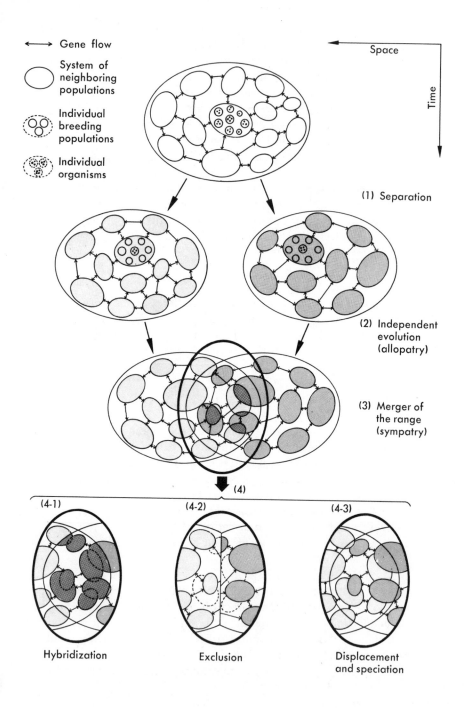

Gene flow

System of
neighboring
populations

Individual
breeding
populations

Individual
organisms

Space

Time

(1) Separation

(2) Independent
evolution
(allopatry)

(3) Merger of
the range
(sympatry)

(4)

(4-1)   (4-2)   (4-3)

Hybridization   Exclusion   Displacement
and speciation

Figure 11.1

◄ The stages of the speciation process. (1) The original population gives rise to two isolated population systems that (2) evolve independently under different selection pressures. When (and if) they (3) merge their ranges partially or entirely, they may (4-1) hybridize, restoring the original situation of one species; or (4-2) they may exclude each other, restoring the case of two geographically isolated population systems; or (4-3) they may displace and form the new situation of two sympatric species. The last can also occur if the species have differentiated before they merged their range. (From O.T. Solbrig, 1970. *Principles and Methods of Plant Biosystematics.* New York: Macmillan)

[A] DISCONTINUOUS SPECIES RANGE DUE TO EXPANSION

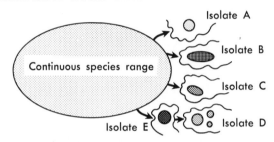

[B] DISCONTINUOUS SPECIES RANGE DUE TO SHRINKAGE

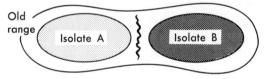

Figure 11.2

Two modes of allopatric speciation. (A) By means of geographic isolates. (B) By splitting of a former continuous range. (From O.T. Solbrig, 1966. *Evolution and Systematics.* New York: Macmillan)

selection—acting independently on the individuals of different breeding populations, tend to produce slightly different adjustments to the environment. How different these adjustments are will depend on how much the environments differ between different breeding populations, on the amount of gene flow, and to a lesser extent on random phenomena. The result is a tendency for different breeding populations of a species to be slightly different genetically, physiologically, and morphologically. The greater the gene flow between two populations, the more closely related they will be, and vice versa. However, gene flow is usually quite low. Patterns of variation have been studied most intensively in plants.

One of the most thoroughly studied plants is a species of cinquefoil, *Potentilla glandulosa*, which was investigated by the team of Jens Clausen, David Keck, and William Hiesey for more than 30 years, starting in the early 1930s. *Potentilla glandulosa* is an herb that is found scattered throughout western North America forming a series of local populations. The team of Clausen, Keck, and Hiesey sampled a large number of California populations. The species can be divided on the basis of its appearance into four subspecies (Fig. 11.3): subspecies *glandulosa*, which occupies the coast ranges of California and isolated localities in the foothills of the Sierra Nevada; subspecies *reflexa*, which occurs on warm, sunny slopes of the foothills and mid-altitudes of the Sierra Nevada from 300 to 2000 m elevation; subspecies *hanseni*, which is a meadow form occurring between 1300 and 3000 m altitude; and finally, subspecies *nevadensis*, a high-altitude form found in meadows or on slopes between 3000 and 3600 m. The subspecies of *Potentilla glandulosa* have no barriers to interbreeding except those provided by their spatial ecological separation.

Samples from several populations from each of these subspecies were transplanted to three experimental gardens in California at approximately 38° north latitude. The first station was located at Stanford near San Francisco Bay. It has a mild coastal climate, and the altitude of the station is only 30 meters. Although mild winter frosts do occur, conditions in general are favorable for a considerable amount of growth, even during the winter, because the minimum temperature remains relatively high. Winter conditions are much less likely to hinder growth at Stanford than are the heat and drought of the summer. The second station is located at Mather, at 1400 m elevation on the west slope of the Sierra Nevada. In Mather the climate favors active growth of plants only during the period between May and October. Spring weather ordinarily begins early in April, although frost may occur until the beginning of June. The summers are mild with warm days and cool nights and little rain. The fall is cool and dry. Snow can be expected between October and April, rarely exceeding two or three feet in depth. The third station is located at Timberline in the Sierra Nevada at 3050 m. Here the climate is quite severe, with only three to six weeks of frost-free weather during the year, and frost may occur at any time. The accumulated snow may reach ten to twenty feet in the winter.

The following results were obtained. The coast range form of subspecies *glandulosa* grew tallest at Stanford, grew less well at Mather, and failed to survive the winter at Timberline (Fig. 11.4). Subspecies *reflexa* grew best at Mather, where it is native, grew less vigorously at Stanford, and also died at Timberline. Subspecies *hanseni* was strongly winter-dormant at Stanford, in spite of the mild winters, grew best at Mather, and survived at Timberline, although it was seldom able to set ripe seed at the high altitude. Finally, sub-

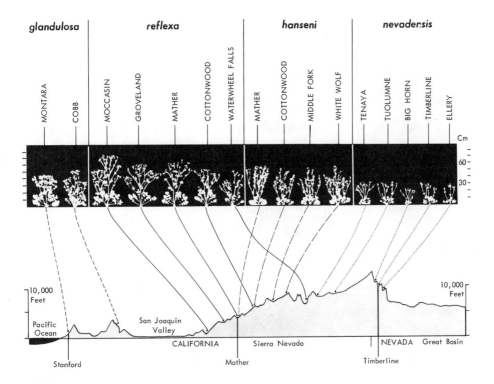

Figure 11.3

Herbarium specimen silhouettes of sample plants of *Potentilla glandulosa* of the four subspecies grown at Stanford under uniform conditions, with a profile across central California at approximately 38° north latitude, indicating the original localities of the populations. Note how the morphology of the plants changes from east to west, and also note that ssp. *reflexa* and ssp. *hanseni* overlap in part of their range. The profile map also indicates the position of the three transplant stations at Stanford, Mather, and Timberline. (After J. Clausen, D.D. Keck, and W. Hiesey, 1940. Experimental studies on the nature of species. I. Effects of varied environments on western North American plants. *Carnegie Inst. Wash. Publ. 520*)

species *nevadensis* barely survived at Stanford, where it was winter-dormant; it was most vigorous at Mather, and it was vigorous and consistently set seed at Timberline. We can see, then, that the four morphological subspecies also represent four distinct physiological types. Furthermore, each one performed best in terms of growth and seed set at the station that was closest to its native environment, with the exception of subspecies *nevadensis*, which did better at Mather than at Timberline. However, subspecies *nevadensis* was the only one of the subspecies capable of functioning normally in the extremely harsh conditions of the Timberline station.

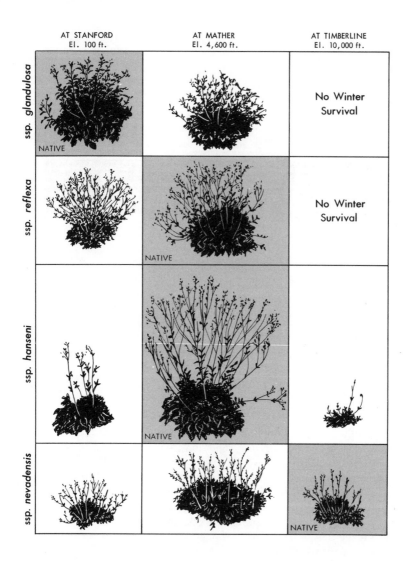

Figure 11.4

Responses of four ecotypic subspecies of *Potentilla glandulosa* to the three contrasting environments in the Carnegie Institution gardens at (left to right) Stanford, Mather, and Timberline. The four subspecies (bottom to top) are: *nevadensis* from 10,000 feet elevation; *hanseni* from 4,600 feet; *reflexa* from 2,500 feet in the Sierra Nevada foothills; and *glandulosa* from 900 feet in the outer coast ranges of California. (After J. Clausen, D.D. Keck, and W. Hiesey, 1940. Experimental studies on the nature of species. I. Effects of varied environments on western North American plants. *Carnegie Inst. Wash. Publ. 520*)

Differentiation can also occur over very short distances, as the British ecologist Anthony Bradshaw and his collaborators showed with the species *Agrostis tenuis.* This common, perennial, self-incompatible species of grass grows in a variety of habitats in temperate regions. In Britain, populations of *Agrostis tenuis* are found growing on mine workings in the vicinity of abandoned mines and in surrounding pastures. Some of these mine workings have a high concentration of lead, copper, or zinc and are toxic to most vegetation. Consequently, few plants are able to grow on them, including plants of *Agrostis tenuis* from normal soil. However, populations have evolved that are tolerant to these metals. This tolerance does not appear to be induced by conditioning processes, is not lost in cultivation in the absence of the metal, and is heritable. In a detailed study of the populations of the Goginan lead mine in Wales, Bradshaw was able to demonstrate that the transition zone between lead-tolerant and intolerant populations is less than 20 meters wide. He and his collaborators also demonstrated that the plants on each side of the zone have the same level of tolerance and intolerance as the plants considerably farther away from the boundary (Fig. 11.5). Analysis of the ecological and historical factors indicates that the lead-tolerant forms must have developed recently and that they evolved in place, that is, within a few meters from the nontolerant forms and within pollination range. However, the selection pressures are apparently so strong that the hybrids have nowhere to become established. They are not able to tolerate the high lead content on one side of the line or the competition from the other plants on the other side. Lack of hybrid survival constitutes in this case an absolute or almost absolute barrier to gene exchange.

Many more examples of local adaptation of populations can be cited, both in plants and in animals. We expect the formation of different gene combinations, which determine different adaptive combinations, to have the incidental effect of reducing the possibilities of interbreeding between populations. That is, we expect that populations evolving in isolation will eventually develop some kind of reproductive isolation. Figures 11.6 and 11.7 illustrate crossing relationships in two species in the sunflower family studied by Jens Clausen, indicating the appearance of some sterility between geographically separated populations.

Populations from opposite extremes of the range of a species (both in the geographical and ecological sense) will tend to be the most different. Gene flow through intermediate populations may keep a certain cohesiveness to the species. However, the process of differentiation is accelerated on the fringes of the distribution of a species. In effect, suitable environments are often scarce there, and they are isolated. For example, where a forest merges with grassland, we will usually have an area of small pockets of woods in the midst of the grassland. The same is true, for example, at the upper limits of the forest in the mountains. A population of a forest species living in one of the pockets of forest in the prairie becomes isolated from the other breeding populations.

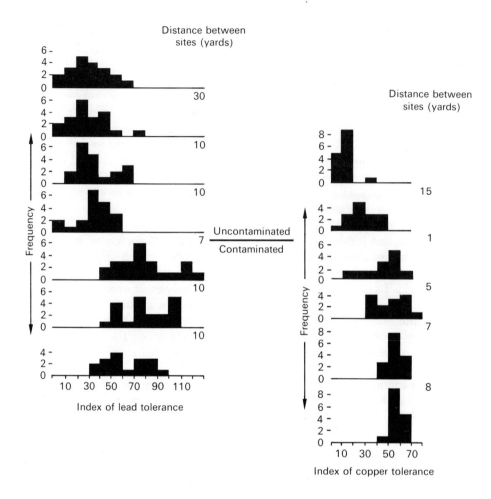

Figure 11.5

Histograms showing the patterns of distribution of tolerance of two linear series of populations of *Agrostis tenuis* across the boundary of (a) Goginan lead and (b) Drws y coed copper contaminated areas. (From A.D. Bradshaw, T.S. McNeilly, and R.P. Gregory, 1965. Industrialization, evolution, and the development of heavy metal tolerance in plants. *Brit. Ecol. Symp.* 6:327–343)

Such populations are called geographic or peripheral isolates. In peripheral isolates, gene flow is no longer acting as a cohesive force, and the population can evolve toward a balance with its environment at a faster rate. Since the environment of isolates is marginal to the species, as we have seen, natural selection will tend to modify peripheral populations considerably. In due time some of the many forms of reproductive isolation may build up between the isolate and the original species. When that happens, the isolate becomes a

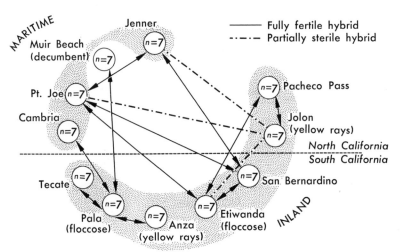

Figure 11.6

Intraspecific crosses within the species *Layia platiglossa*. Note also how certain populations have acquired novel morphological characters. Nevertheless, there are only incipient sterility barriers. (Reprinted from Jens Clausen, *Stages in the Evolution of Plant Species*. Copyright © 1951 by Cornell University. Used by permission of the publisher, Cornell University Press)

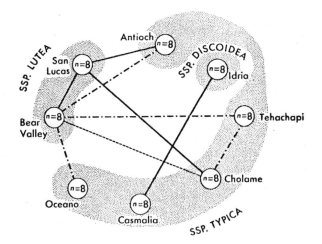

Figure 11.7

Crossing relationships between populations of *Layia glandulosa*. In contrast to the case of *L. platiglossa*, in this species there are already some sterility barriers between populations. (Reprinted from Jens Clausen, *Stages in the Evolution of Plant Species.* Copyright © 1951 by Cornell University. Used by permission of the publisher, Cornell University Press)

new species adapted to the intermediate habitat, in this case the pockets of woodland in the midst of the grassland (Fig. 11.8).

Actually, it is fairly rare for peripheral isolates to become new species for many reasons. Species colonize peripheral situations to which they are not too well adapted in years when the environmental conditions are extremely favorable. When conditions worsen, the invaders usually cannot compete, and

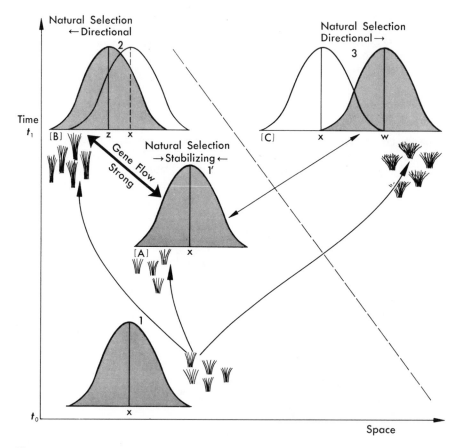

Figure 11.8

Some of the evolutionary forces at work in a situation involving geographical variation in space and change in time. At time t_0, population 1 was adapted and growing at site A with a mean x (to represent a mean of all character states). At time t_1, population 1 has given rise to populations 1′, 2, and 3 at sites A, B, and C. At B, natural selection favors a mean z, at A it still favors a mean x, and at C it favors a mean w. Since A and B are very close and environmentally similar, gene flow is strong and directional selection not very different; the opposite is true for C. (From O.T. Solbrig, 1970. *Principles and Methods of Plant Biosystematics.* New York: Macmillan)

they are eliminated. Furthermore, the processes leading to speciation require a series of conditions, such as favorable mutations and/or recombinations, evolution of an isolating mechanism, and conditions that permit survival of the isolated population—a combination of requirements seldom found together.

But speciation can take place also during times of shrinkage of the range of the species. If for reasons of unfavorable environmental conditions the species decreases its range, the continuity of the area occupied may be split into two or more isolated areas. We have already mentioned that within a continuous area, populations at the extremes tend to be quite different, sometimes to the point that there is some reduction of fertility when individuals of these populations are bred with each other (see Fig. 11.6). Since the gene flow can take place through the intermediate populations, the integrity of the species is not affected. But if suddenly these intermediate populations disappear, the differences between the extremes become paramount, and if they survive, they are often on their way toward becoming new species (see Fig. 11.7).

Merger of the Range. The forces that lead to an overlapping of the ranges of two hitherto isolated populations are similar to those that lead to their separation. Basically these forces are environmental changes that lead to expansion of the range or migration of the population or population systems into an area not previously occupied. It is also possible that the evolution of the populations produces physiological or morphological changes (such as the development of a better seed dispersal apparatus) that are directly or indirectly responsible for expansion. This step is not a necessary condition in the process of speciation. When it takes place, however, it leads to important interactions between the two hitherto separated units.

The superposition of the range of the populations so that their members are close enough to cross and the resulting interaction constitute the last step in the speciation process. The two populations that come together may have differentiated physiologically and genetically so that they will have different ecological requirements and so that their gametes will no longer fuse to form viable zygotes. In such cases the speciation process has been completed without further interaction.

However, if the populations that come into contact have the same ecological requirements and are also capable of hybridizing, they will interact with each other. The result of this interaction can have one of three outcomes:

1. One species outcompetes the other, and the previous nonoverlap of the range is restored.
2. As a result of selection, the two species divide the environment in such a way that they no longer compete, and at the same time, some kind of breeding barrier is established so that few or no hybrids are formed.
3. The two populations hybridize and form one interfertile population.

Competitive Exclusion.   When two hitherto separated populations come into contact with each other, individuals of different populations will interact. Even when behaviorally and/or genetically isolated, individuals may be in need of identical or similar resources, such as food, space, nesting sites, etc. If not yet reproductively isolated, they may also attempt to interbreed. Competition may favor members of one population so strongly that coexistence is not possible.

Every breeding population has a specific set of environmental requirements, such as habitat, food, water tolerance, and so on. Whenever two populations share the requirement for some or all of the environmental components in a way that causes mutual interference, the two populations are competing for these resources. When two populations are competing, they tend to exclude each other unless the exact requirements of one or both populations change in such a way that the competition is lessened. This is the essence of the so-called *exclusion principle* enunciated by the Russian biologist G.F. Gause in 1934: "No two forms can share exactly the same environmental requirements for an indefinite period of time; eventually one form will replace the other."

We believe that most of the species that make up the native biota are in a state of dynamic equilibrium. That is, they have divided the environment in such a way that interference is minimal. Of course, there will always be some competition for environmental factors because their supply is limited. For example, to take an extreme case, all plants require space to grow, and where one plant is growing, obviously no other can grow. Light is also a requirement of all plants. In a forest the components of the various layers of vegetation, such as the canopy trees, the shrubs, and the herbs, have adjusted their photosynthetic light requirements so that they can coexist. Most trees cannot live in the shade whereas shrubs and herbs can, and herbs can often exist with less light than shrubs. And all plants need water. However, some can exist where water is scarce and others cannot, and often adjustments are made for relatively small differences. For example, within the northern oak-beech forest, swamp maples are always found where the soil moisture is highest.

However, competition can be best appreciated where equilibrium is lacking. For example, if any cultivated field or garden is not attended, it will very soon change its floristic composition. The cultivated plants are usually unable to reproduce and are replaced by annual plants adapted to the disturbed conditions that cultivation has created. After a few years, however, as the soil becames more stabilized, the annual vegetation is gradually replaced by the vegetation typical of the region. Plants introduced involuntarily from other continents by humans are, in general, agricultural weeds that cannot compete with the native vegetation under undisturbed conditions. They are restricted in their distribution to fields and gardens, roadsides and railroad yards, and other such habitats that we have created. The native vegetation adapted to another set of conditions cannot grow in these disturbed habitats.

Exclusion, that is, the elimination of one population, is one outcome of range overlap. The excluding population may be the ancestral one or it may be the descendant one. Another outcome of range overlap is displacement.

Character Displacement.   Only where the cruising ranges of two similar animal species or the areas of effective pollination of two like plant species overlap can there be absolute certainty that two distinct but closely related breeding populations are indeed two species. In such cases there cannot be any doubt: If they have developed breeding isolation, they will not mix; if they mix, they have not yet developed breeding isolation.

When two such closely related animal or plant species formed allopatrically come together in their ranges again, occasional individuals might confuse members of other species and breed with them; the progeny, if any, is sterile or largely so. Since such individuals produce few or no progeny, they and their genotypes are eliminated by natural selection. Within each population there is phenotypic variability. The cases of mistaken identity will most likely occur among the individuals that most resemble the other species. The result is that those individuals in each species that look less similar to the members of the other species are more likely to produce viable offspring. Consequently, the two species will tend to look more different as time goes by. Such a phenomenon has been called *character displacement*. Since character displacement is possible only where two species overlap, the populations living separately will be less distinct than those living together. Furthermore, since animals actively choose their mates, but plants do not, the phenomenon will be more obvious among the former, although structures that impede fertilization by the wrong species may develop in plants as a result of character displacement.

Character displacement is illustrated by two species of rock nuthatches, *Sitta neumayer* and *Sitta tephronota*. The former species ranges from the Balkans eastward through the western half of Iran, and the latter ranges from the Turkestan to Armenia. The two species overlap in several sectors of Iran. They are very similar to each other in their separate zones, but they diverge in the zones of overlap, where *S. neumayer* shows a distinctive reduction in overall size, beak length, and the intensity of a prominent facial stripe, characters that are unchanged in *S. tephronota*.

Hybridization.   The third outcome of a range merger is hybridization, that is, the union of the gene pools of the two populations. We take up this topic in detail in the next chapter.

Parapatric speciation

In organisms with low vagility (that is, organisms that move very little or not at all), such as plants, fossorial (i.e., burrow-inhabiting) rodents such as moles, or flightless insects, speciation can occur when genetically unique organisms arise that are (1) capable of exploiting a slightly different but physically contiguous habitat, and (2) reproductively isolated from the rest of the species. Parapatric speciation is probably not as common as allopatric speciation, but in plants particularly it may be more common than is usually recognized. Parapatric speciation is illustrated by two examples, one from the animal kingdom, the other from the plant kingdom.

Spalax ehrenbergi is an Old World mole rat. It lives during the day in burrows that it digs. Males defend their burrows and surrounding areas. Consequently, once the animal has established a burrow, it remains there for most if not all of its life. When the young reach a certain age, they move out of the nest and establish their own burrows. The Israeli biologist Eviatar Nevo and his collaborators have studied *Spalax ehrenbergi* in great detail and have found that it is a complex of several closely related species, each characterized by a different chromosome number.

In Israel four chromosomal forms are encountered (Fig. 11.9). (1) A 52-chromosome type lives in the Upper Galilee Mountains in a humid to subhumid climate. (2) A sibling species with 54 chromosome prefers the subhumid to humid but colder region around Mount Hermon and the Golan Heights. (3) The 58-chromosome form is found in the Lower Galilee Mountains and in Central Israel, a region with a subhumid to semiarid climate. (4) A sibling species with 60 chromosomes is found in the semiarid to arid region of the Samaria and Judea Mountains. These four sibling species have been intensively studied by Nevo and collaborators, who have investigated a number of morphological, behavioral, and biochemical characteristics. The results indicate that the four chromosomal types are indeed very similar. For example, the four groups not only are very similar morphologically and have identical adaptations to a subterranean life-style but also appear to be almost identical in their genes, as evidenced by information obtained through protein electrophoresis (96 percent of tested proteins, representing 25 loci, were identical). On the other hand, the four groups exhibit physiological differences in their metabolic rate, which decreases from the humid north to the arid south. How did these forms arise, and are they reproductively isolated? Fortunately both questions can be answered. *Spalax ehrenbergi* probably arose from a common ancestral form, *Spalax mimtus*, that lived in the area during the ice ages and whose fossilized remains have been found in deposits dated at a half-million years before the present. Fossils indistinguishable from present *S. ehrenbergi* have been found in deposits near Jerusalem and dated at 250,000 years before the present. Unfortunately the chromosome number of fossils cannot be determined. Gradual speciation accounts for the morphological change observed between ancestral *S. mimtus* and derived *S. ehrenbergi*. On the other hand, the chromosomal races are thought to have arisen in the more recent past through parapatric speciation. Which of them was the ancestral one cannot be stated.

Although the chromosomal races differ in their physiological preferences, they live within sight of one another in the zone of contact. There is therefore

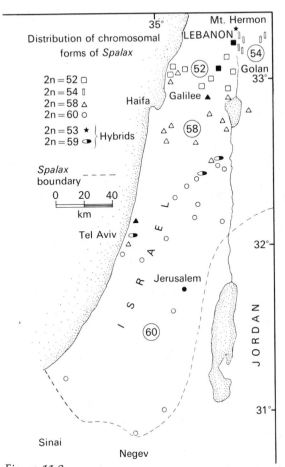

Figure 11.9
Distribution of chromosome forms of *Spalax ehrenbergi* in Israel. (From E. Nevo and A. Shkolnik, 1974. Adaptive metabolic variation of chromosome forms in mole rats, *Spalax. Experientia* 30:724–726)

the possibility of crosses between members of different chromosomal races. Such crosses can be made in the laboratory. The hybrids, which have an intermediate chromosome number, are partially fertile and can cross with either parent. They occur in nature in the areas of contact, but only rarely and only in a very narrow band in the zone of contact. Nevo was able to show that females can discriminate between a male with a chromosome number equal to hers and one with a different chromosome number. The combination of reduced fertility of the hybrid and behavioral discrimination keeps gene flow close to zero. All this points to speciation having taken place parapatrically. That is, the species expanded into new territories that were ecologically slightly different, and then chromosomal changes took place that probably also brought about small physiological changes. The reproductive isolation resulting from the chromosomal and ecological differences was subsequently reinforced by the evolution of behavioral discrimination.

An example of parapatric speciation in the plant kingdom involves a pair of very closely related species in the sunflower family: *Stephanomeria exigua* ssp. *coronaria* and *S. malheurensis*, intensively studied by the American evolutionist Leslie Gottlieb.

Stephanomeria malheurensis is found in a single locality in eastern Oregon, which in 1973 comprised fewer than 250 individuals. It differs—but very slightly—from its ancestor, *S. exigua* ssp. *coronaria*, in five morphological characters (Fig. 11.10). In addition, approximately 98 percent of the individuals of *S. exigua* ssp. *coronaria* have pink flowers and 2 percent have white flowers, whereas 80 percent of *S. malheurensis* have pink flowers and 20 percent have white. Studies of the genic relationships using protein electrophoresis indicate that all or most alleles present in *S. malheurensis* are also found in *S. exigua* ssp. *coronaria*, but not vice versa. Some physiological differences exist between the two species in growth rates and germination requirements, but it is not clear whether these differences have any significance. The two species are reproductively isolated, however. *Stephanomeria malheurensis* is very highly self-pollinating, and that means that the movement of pollen between it and the genetically self-incompatible and outcrossing *S. exigua* ssp. *coronaria* is drastically curtailed. Furthermore, when a plant of one species is pollinated with pollen from the other, it produces only half the normal number of seeds. Finally, when these F_1 hybrids are formed, only ¼ of their pollen grains are viable, because the two species differ in chromosomal structure in ways that include a reciprocal translocation.

The evidence indicates that the process of speciation has been very similar in both these cases. At the edge of the distribution of the species but not in isolation, individuals arose that possessed both new physiological characteristics, which subsequently allowed them to occupy a new ecological zone (a new "niche"), and some degree of reproductive isolation. At least for the mole rats reproductive isolation was subsequently reinforced by selection. Whereas in allopatric speciation differentiation *precedes* the development of reproductive isolation, in parapatric speciation both processes are simultaneous or nearly so.

Sympatric speciation

When reproductive isolation precedes differentiation and when it takes place within a population or within the dispersal range of a population, we have sympatric speciation.

Sympatric speciation is a rare phenomenon. Furthermore, one cannot always separate it from parapatric speciation, since one can rarely verify

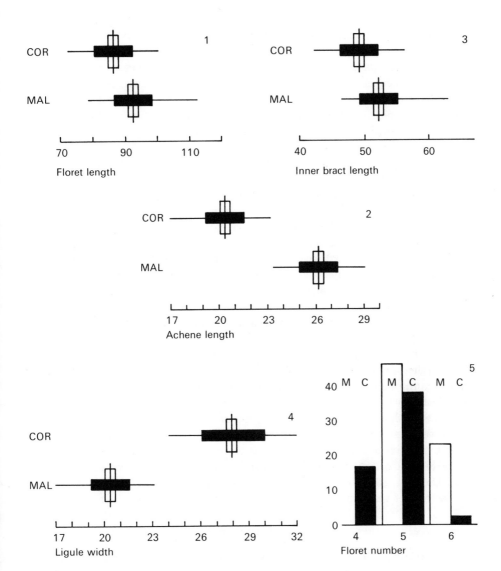

Figure 11.10

Differences between *Stephanomeria exigua* ssp. *coronaria* and *S. malheurensis*. The range, mean, and standard deviation are shown for floret length, inner bract length, achene length, and ligule width (the units shown are 1/6 mm). The distribution of floret number is also shown. (From L.D. Gottlieb, 1973. Genetic differentiation, sympatric speciation, and the origin of a diploid species of *Stephanomeria. Am. J. Bot.* 60:545–553) whether isolation preceded differentiation or whether they occurred simultaneously.

The best-known examples are in insects that are parasites of plants. The Texas biologist Guy Bush, who has investigated this matter in great detail, thinks that sympatric speciation may be restricted to parasites and parasitoid animals because parasites usually mate on the host. However, the number of species involved is more than 500,000. The simple process of moving to a new host species can result in reproductive isolation. The population on the new host will now have to adapt to the new host, and that adaptation will normally entail morphological and physiological changes.

The best-known examples are new host races of the hawthorne fly, *Rhagoletis pomonella*. This species, which parasitizes the fruits of North American species of hawthorne (*Crataegus*), developed in 1864 or thereabouts a race that parasitizes certain late varieties of apples that had been introduced from the Old World. The race appeared first in New York and subsequently spread to other regions. In 1960, a new race of the hawthorne fly appeared in a cherry orchard in Wisconsin.

Change to a new host involves minor genetic changes, according to Bush. Two types of genes appear to play a major role: (1) those involved in the recognition and selection of the host and (2) those involved with survival on the host. The number of changes involved in recognition and selection of a new host may be few in number and, at least in insects, may be involved with chemoreception, since discrimination of hosts appears to be based mostly on chemical rather than physical cues. To move to a new host, the first step is the establishment of the ability to transmit to the offspring the ability to recognize the new host. Of course, the animal must be able to survive on the new host. To the degree that the new host differs chemically and physically from the old host, a small or large degree of differentiation will follow.

In summary, allopatric, parapatric, and sympatric speciation are different ways by which a population or group of interbreeding populations adapted to a given set of environmental and biological conditions gives rise to a new population or group of populations adapted to a different set of conditions and reproductively isolated from its ancestors.

Abrupt speciation

To complicate matters further, sometimes speciation occurs abruptly rather than in a more or less gradual way. This usually occurs as a result of the doubling of the chromosome number, a process called polyploidy, or as a result of special evolutionary mechanisms, such as what is known as catastrophic selection.

Polyploidy. The subject of polyploidy was covered extensively in Chapter 5 (page 107). Polyploidy can be defined as the phenomenon that results in doubling of the number of chromosomes of an organism. The resultant organism in most cases is incapable of forming fertile offspring with members of the ancestral diploid population. It has therefore acquired instant reproductive isolation, and if it is capable of reproducing and of finding an ecological situation that it can exploit, a new species has been formed. Polyploidy is common in plants but rare in the animal kingdom. Among the flowering plants, close to 40 percent of all species appear to be polyploid, so that polyploidy is a very important phenomenon indeed.

Polyploids are basically of two kinds (see Fig. 5.20): (1) autopolyploids, which are the result of the doubling of the chromosomes of a normal individual, and (2) allopolyploids, which are the result of the doubling of the chromosomes of a hybrid individual. The latter kind is apparently more frequent. The probability of establishment of an allopolyploid is higher for cytological reasons (meiosis is more likely to be normal) and because the hybrid may have novel characteristics that will allow it to exploit habitats not open to either of the parents. Polyploidy can therefore be considered a case of parapatric or sympatric speciation.

Catastrophic Speciation. Apparently in some instances very rapid speciation can take place, leading to genetic isolation with little or no morphological differentiation but without polyploidy. This phenomenon is known as catastrophic selection. The best example is furnished by the plant genus *Clarkia* of the family Onagraceae.

Clarkia is a group of plants found in the western United States, especially in California. They have been the subject of intense biosystematic studies over the last 20 years by the California evolutionist Harlan Lewis and his students. In the process they have attempted interspecific crosses between most of the species. The cytological analysis of the hybrids so obtained showed that many species differ drastically in the linear order of the genes. The difference is illustrated particularly well by *Clarkia franciscana*. This species is known from only one locality, a serpentine-covered slope overlooking the Golden Gate at the east end of the Presidio in San Francisco (Fig. 11.11). The species is morphologically similar to two other species, *C. rubricunda* and *C. amoena*, that occur close by. The first of these species is found along the coast and in the valleys of the California coast ranges from northern San Luis Obispo County to just north of San Francisco Bay in Marin County. *Clarkia amoena* replaces *C. rubricunda* in central Marin County and continues northward along the coast to British Columbia and Vancouver Island. Interspecific hybrids among *Clarkia franciscana*, *C. amoena*, and *C. rubricunda* were made by Lewis and Raven in all combinations to determine the fertility of the hybrids and to study the differences in chromosome architecture, as indicated by chromosome pairing at meiosis. The fertility of the pollen in every case was not more than 2 percent. The pairing of the chromosomes in the hybrids showed that the chromosomes of these three species differed drastically in the arrangement of homologous segments. *Clarkia franciscana*

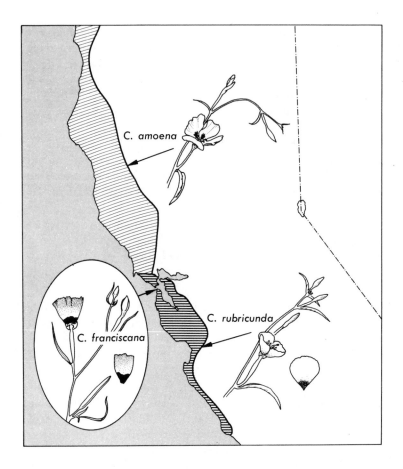

Figure 11.11

Distribution of *Clarkia amoena, C. franciscana,* and *C. rubricunda* in California. Note how morphologically similar the three species are. The best distinguishing feature is the petal, particularly its spot. The spot is bright red and near the center in *C. amoena;* it is well defined, conspicuous, reddish purple, and basal in *C. franciscana;* and it is not so well defined and merges somewhat in *C. rubricunda.* (From H. Lewis, 1958. Rapid evolution in *Clarkia. Evolution* 12:319–336)

differs from *C. rubricunda* by at least three large translocations and four paracentric inversions, and from *C. amoena* by at least two translocations and two or more paracentric inversions, as well as possibly by a series of smaller rearrangements. *Clarkia amoena,* in turn, differs from *C. rubricunda* by three translocations and two inversions. The chromosomal changes are of an entirely different magnitude than in most species, such as *Glandularia,* mentioned in Chapter 5 (see also Fig. 5.9).

When two closely related species or populations differ in the order of the genes in their chromosomes, it is obvious that in one or both of the differing species or populations, changes have taken place in the architecture of the chromosomes, because the common ancestor of these species or populations must have had one arrangement only (at least at one time). How does the change in chromosomal architecture come about? The usual assumption is that a small chromosomal change takes place in an individual of a population that gives that individual a higher fitness. In time all individuals of the population will have the new chromosomal arrangement. By the repetition of this process, the populations can become quite distinct cytologically. These changes presumably are accompanied by morphological and physiological changes as the populations become adapted to different ecological conditions.

Does the great dissimilarity in the chromosomal arrangement of the genes between *Clarkia franciscana, C. amoena,* and *C. rubricunda* and their great morphological similarity suggest that they have originated in ways other than by the gradual accumulation of small changes? Suppose we have an individual with a different order of the genes in one of its chromosomes. If the individual is heterozygous, crossing over within the altered segment will produce two sterile gametes out of the four gametes that are the normal product of meiosis, because of gene deletions. If the altered segment is small, the probability that a crossover will occur within the segment is very small; consequently, a crossover will not occur in most cells undergoing meiosis, and the new arrangement can become established in the population. However, if the segment is large, the converse is true. Therefore, the probability that the chromosomal arrangements of *Clarkia franciscana* are the result of accumulation of small changes is indeed very low, because the repeated establishment of small chromosomal changes without corresponding changes in morphology is an improbable event. Harlan Lewis has therefore hypothesized a mechanism that he terms "catastrophic selection."

According to this hypothesis, under certain unknown conditions (which might include environmental stress, such as drought, or the presence of "mutator genotypes," i.e., genotypes that produce a high frequency of mutations) drastic chromosomal rearrangements take place. If an individual with such a drastic chromosomal rearrangement becomes detached from the population, it can establish a new population with a new chromosomal arrangement that is genetically isolated from the ancestral population from its very inception.

The forces that bring about such drastic changes in some cases and so little change in others remain a mystery, particularly because some species that have been isolated for millions of years have not developed any genetic or chromosomal differences. One example of long isolation without genetic change is that of the two species of sycamores, *Platanus orientalis* and *Platanus occidentalis.* The first is a native of the Middle East, the second of eastern North America. They are morphologically very distinct and also physiologically so, because they grow in different climatic regions. It is impossible to say with any certainty how long these two species have been isolated, but it is likely to have been at least from the end of the Tertiary Period, a million or more years. The artificial hybrid between these species, known as *P. acerifolia* and widely cultivated as an ornamental tree, is vigorous and highly fertile and has a normal meiosis, indicating that no drastic changes in the order of the genes has taken place.

Catastrophic speciation is a special case of parapatric speciation, and the same applies for most cases of polyploidy. Thus allopatric speciation is always gradual, whereas parapatric and sympatric speciation are more likely to take place abruptly.

SUGGESTED FURTHER READING

General references

Clausen, J. 1951. *Stages in the Evolution of Plant Species.* Ithaca, N.Y.: Cornell University Press.

Gause, G.F. 1935. *The Struggle for Existence.* Baltimore: Williams and Wilkins.

Grant, V. 1971. *Plant Speciation.* New York: Columbia University Press.

Mayr, E. 1963. *Animal Species and Evolution.* Cambridge, Mass.: Belknap Press of Harvard University Press.

Solbrig, O.T. 1970. *Principles and Methods of Plant Biosystematics.* New York: Macmillan.

Stebbins, G.L. 1974. *Flowering Plants: Evolution Above the Species Level.* Cambridge, Mass.: Belknap Press of Harvard University Press.

Articles and reviews

Ayala, F.J. 1975. Genetic differentiation during the speciation process. *Evol. Biol.* 8:1–78.

Ayala, F.J., M.L. Tracy, D. Hedgecock, and R.C. Richmond, 1974. Genetic differentiation during the speciation process in *Drosophila. Evolution* 28:576–592.

Bradshaw, A.D. 1959. Population differentiation in *Agrostis tenuis* Sibth.: I. Morphological differences. *New Phytol.* 58:208–227.

Bradshaw, A.D. 1960. Population differentiation in *Agrostis tenuis* Sibth.: III. Populations in varied environments. *New Phytol.* 59:92–103.

Bradshaw, A.D., T.S. McNeilly, and R.P. Gregory, 1965. Industrialization, evolution and the development of heavy metal tolerance in plants. *Brit. Ecol. Symp.* 6:327–343.

Bush, G.L. 1975. Modes of animal speciation. *Ann. Rev. Ecol. Syst.* 6:339–364.

Clausen, J., D.D. Keck, and W. Hiesey. 1940. Experimental studies on the nature of species. I. Effects of varied environments on western North American plants. *Carnegie Inst. Wash. Publ. 520.*

Clausen, J., and W. Hiesey. 1958. Experimental studies on the nature of species. IV. Genetic structure of ecological races. *Carnegie Inst. Wash. Publ. 615.*

Ehrlich, P.R., and P.H. Raven. 1969. Differentiation of populations. *Science* 165: 1228–1232.

Gottlieb, L.D. 1973. Genetic differentiation, sympatric speciation, and the origin of a diploid species of *Stephanomeria*. *Am. J. Bot.* 60:545–553.

Jain, S.K., and A.D. Bradshaw. 1966. Evolutionary divergence among adjacent plant populations: I. The evidence and its theoretical analysis. *Heredity* 21:407–442.

Lewis, H. 1958. Rapid evolution in *Clarkia*. *Evolution* 12:319–336.

Nevo, E., and H. Bar-El. 1976. Hybridization and speciation in fossorial mole rats. *Evolution* 30:831–840.

Nevo, E., and A. Shkolnik. 1974. Adaptive metabolic variation of chromosome forms in mole rats, *Spalax. Experientia* 30:724–726.

Wilson, E.O. 1965. The challenge from related species. In H.G. Baker and G.L. Stebbins (eds.), *The Genetics of Colonizing Species*. New York: Academic Press.

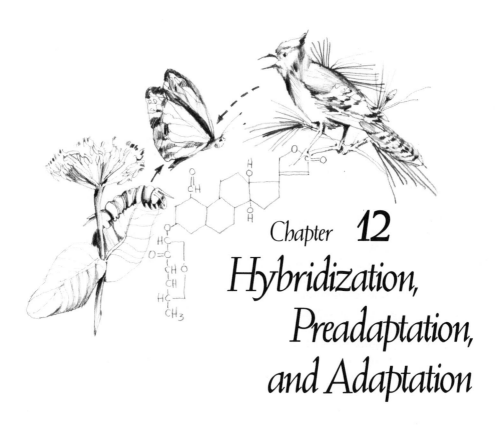

Chapter **12**

Hybridization, Preadaptation, and Adaptation

In the previous chapter we discussed the three major modes by which species originate: allopatric or geographical speciation, believed by many to be the usual way by which species originate; parapatric speciation, and sympatric speciation. Speciation is a process of divergence: One population or group of populations sharing common genes and interbreeding freely (at least within the population) become two separate groups of populations, no longer sharing their genes. In this chapter we look at species hybridization, the process by which two distinct but closely related species occasionally share their genes. We also look at the processes by which species acquire entirely new characteristics.

Speciation divides the gene pool, but in itself the process does not produce new characteristics. The variability that exists within a species becomes a variability between the new species. But the origin of species over time is associated with the emergence of new characteristics—of what are sometimes called "evolutionary novelties"—which is verified by the fossil record. We end this chapter with another look at the meaning of adaptation in the context of what has been learned so far about the working of natural selection.

HYBRIDIZATION

Hybridization is the third possible outcome when two partially differentiated and hitherto geographically isolated populations come together (exclusion and displacement, discussed in the previous chapter, are the other two). Strictly speaking, hybridization is synonymous with outcrossing. Thus, whenever a zygote is formed by the union of an egg with a sperm from another individual, a hybrid is formed. The sperm may come from an individual in the same population, from an individual in another population of the same species, or from a different species. To distinguish between these different mating possibilities, we will restrict the use of the term hybridization to crosses between individuals presumed to belong to different species. For crosses within the same species, the term recombination will be used.

We usually assume that if two populations interbreed, eventually the two gene pools will completely mix so that, in the last analysis, only one integrated breeding population will survive. But we know today that there are mechanisms that allow populations of different species to hybridize occasionally, but that prevent them from becoming swamped by foreign genes. Anyone who wonders why we consider those two populations as distinct species is in good company, since one of the knottiest problems of biology is deciding how much interbreeding can take place before two populations are considered one rather than two species.

Hybridization is rather frequent among the higher plants but rare among the vertebrates. Following Ernst Mayr, we can distinguish five main types of hybrids. We have to remember, though, that these are not distinct categories and that there are intermediate situations.

1. The occasional crossing of sympatric species resulting in the production of hybrid individuals that are inviable or sterile and therefore do not backcross with the parental species.

2. The production of more or less fertile hybrids between sympatric species, some of which backcross to one or both parental species.

3. The formation of a secondary zone of contact and of partial interbreeding between two formerly isolated populations that failed to acquire complete reproductive isolation during the preceding period of geographic isolation.

4. The complete local breakdown of reproductive isolation between two sympatric species resulting in the formation of hybrid swarms that may include the total range of variability of the parental species.

5. The production of a new species as a result of hybridization and subsequent doubling of the chromosomes (allopolyploidy).

The first category is of no consequence, and natural selection will take care of the hybrids and their parents. We have already discussed allopolyploidy, and the fourth category produces total swamping between the parental types. Hybridization of the second and third categories, known as *introgression*, will be considered now.

Interspecific hybrids will occasionally be formed that are almost totally sterile. In plants they will often not produce any viable seed, but they will produce some fertile pollen. This pollen can fertilize individuals of the parental species. In other cases the hybrids are fertile, but they are not well adapted. Two species of sage studied by Epling, *Salvia apiana* and *S. mellifera*, occur together over thousands of square miles in the coastal regions of southern California. The species are reproductively isolated, but in areas where the habitat has been disturbed by humans, hybrid swarms are formed, which then backcross to the parents (Fig. 12.1). Anderson has investigated two species of spiderwort, *Tradescantia caniculata* and *T. subaspera*, that grow together in the Ozark region, the former on rocky slopes in full sun and the latter on rich

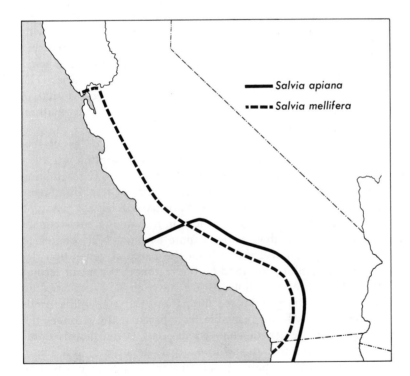

Figure 12.1

Map of California showing the distribution of *Salvia mellifera* and *Salvia apiana* and the area of overlap of the two species. (From C. Epling, 1947. Natural hybridization of *Salvia apiana* and *S. mellifera*. *Evolution* 1:69–78)

soil in deep shade. But in ravines and on gradually intergrading cliffs, intermediate habitats are found. In these areas grow hybrids between the two species that also produce backcrosses to the parental species. To understand why the two species maintain their identity in the face of hybridization, we have to take natural selection into account. The hybrids are obviously not well adapted to either of the two parental habitats, and they are short-lived. The backcrosses will not be as well adapted as the pure species and will be selected against. The integrity of the species is therefore preserved by selection. But if the number of hybrids were to become very large—for example, as a result of human activities in disturbing the habitat—the chances that a plant of either species will not get pollen from hybrids decreases, and eventually most if not all plants may be hybrids or backcrosses. Thus the fate of two species that hybridize will depend in part on the amount of interbreeding and the degree of genetic and ecological sterility of the hybrids and the backcrosses. However, the evolutionary role of hybridization depends on the effect that hybridization has on the genetic composition of the parental populations and not necessarily on the frequency with which they are formed. That is, in most cases of hybridization it is not the hybrids but the backcrosses of the hybrids with the parents that play an evolutionary role. This process of backcrossing is known as *introgressive hybridization*. By introgressive hybridization, genes from one species can be introduced into populations of another species. The late brilliant evolutionist Edgar Anderson has repeatedly suggested that in cases of introgressive hybridization, the backcrosses are favored by natural selection because of the new genes they carry.

The individual organisms of a breeding population are more or less closely adapted to the environment in which they grow. In each generation mutation and recombination will produce individuals with genotypes closely resembling the parents and also individuals differing from them more or less. In a stable environment most of the young that survive to maturity will have a genotype and phenotype similar to the parental generation; in a changing environment the offspring will not resemble their parents quite as much. The degree to which the offspring generation will differ from the parental generation depends on the degree of recombination (and all the factors that control recombination) and the intensity of selection, as we have already explained in Chapters 4–9. The greater the genetic recombination, the higher the probability is that the offspring will not resemble their parents; the stronger the intensity of selection, the more genetic change will occur. If selection is stabilizing, the changes will be minimal and will be largely genotypic; if selection is directional, the changes will probably be of greater magnitude and will be reflected both in the genotype and in the phenotype (p. 143).

Hybridization introduces new genes into a population and consequently increases the probability of genetic change. It acts as a factor that increases variability.

Effect of hybridization and introgression: the sunflower example

An excellent example of the role of hybridization in evolution is that of the sunflowers, studied by Charles Heiser of Indiana University, particularly the common sunflower, *Helianthus annuus*. This species is found in central and western North America from southern Canada to northern Mexico. Morphologically the common sunflower is quite variable, not only within but especially between populations. Heiser subdivided the species into three subspecies and one variety: subspecies *lenticularis*, subspecies *texanus*, subspecies *annuus* (which includes the cultivated sunflower, variety *macrocarpus*). Subspecies *annuus* is the ruderal or weedy sunflower common in the middle western United States, particularly in railroad yards and vacant lots in cities and towns. Subspecies *lenticularis* is the wild sunflower of roadsides and abandoned fields in the west and southwest of the United States. Finally, subspecies *texanus* is restricted to eastern Texas. The story of the probable origin of these three forms is a fine example of the role of hybridization in evolution.

The ancestral form of the sunflower is not known with certainty. However, on the basis of extensive studies, Heiser postulates that subspecies *lenticularis* of western North America is more like the original form of the species than any other living subspecies. This subspecies, according to Heiser, gave rise to subspecies *annuus* under indirect human influence. The seeds of subspecies *lenticularis* were gathered for food by various tribes of Indians, and the sunflower may have become an early Indian camp weed, adapted to disturbed areas around the camps and villages. In time, through either conscious or unconscious human selection, subspecies *lenticularis* could have given rise to the slightly larger forms that developed into subspecies *annuus*. Subspecies *annuus* variety *macrocarpa*, the cultivated sunflower, differs basically from variety *annuus* by a single mutation that restricts branching and leads to the production of a single or very few large heads. In addition, of course, we have selected forms for higher yield, especially in recent times.

The history of the third subspecies, *H. annuus* subspecies *texanus,* is more involved. It appears to have arisen after natural or human introduction of either subspecies *lenticularis* or *annuus* into Texas. Today *H. annuus* subspecies *texanus* is very common at roadsides and in abandoned fields, especially in central Texas. In eastern Texas another annual species of sunflower, *Helianthus debilis* variety *cucumerifolius*, grows as a weed both at roadsides and in undisturbed sites. Where *H. debilis* variety *cucumerifolius* and *H. annuus* subspecies *texanus* grow together, occasionally a few hybrid plants are found. These are more stunted in growth than the parental species that grow in the vicinity, and they are mostly sterile. However, they occasionally produce a small proportion of viable pollen.

The variation of *Helianthus annuus* in eastern Texas (Table 12.1) is clearly in the direction of *H. debilis* as indicated by the large proportion of leaves

Table 12.1

Comparison of the number of ray flowers and disk diameter of the common sunflower, *Helianthus annuus*, and the Texas sunflower, *H. debilis* var. *cucumerifolius*.

Species	Locality	Number of Ray Flowers	Disk Diameter (cm)
H. debilis, var.	Nacogdoches Co., Texas	13.7	1.62
cucumerifolius	Frio Co., Texas	15.7	1.84
	Houston Co., Texas	19.0	2.24
H. annuus	Galveston Co., Texas	17.3	2.30
	Frio Co., Texas	20.4	2.58
	Galveston Co., Texas	22.0	3.28
	Collin Co., Texas*	23.2	3.60
	Pittsburgh Co., Okla.*	25.9	3.54

Data from C. Heiser, 1951. Hybridization in the annual sunflowers: *Helianthus annuus H. debilis* var. *cucumerifolius. Evolution* 5:42–51.

Note how populations of *H. debilis* acquire more rays and a larger disk as they move north and west (*H. annuus* territory), whereas *H. annuus* populations lose rays and acquire a smaller disk as they move south and east (*H. debilis* territory).

* These populations are outside the territory of *H. debilis.*

with jagged serration, speckled stems, small head diameters, small achenes, low number of ray flowers, small bract width, and the tendency to branch from the base. Consequently, Heiser has postulated that subspecies *texanus* arose as a result of hybridization and introgression of *H. debilis* variety *cucumerifolius* with *H. annuus* subspecies *lenticularis* or subspecies *annuus* when either of the latter subspecies was introduced into Texas. In addition to the morphological characters that *H. annuus* acquired from *H. debilis*, it presumably also acquired physiological characteristics that adapted it to withstand better the climate and environment of Texas, characteristics acquired by *H. debilis* over many years of existence in Texas.

At the other end of the distribution of *Helianthus annuus* in California, a similar phenomenon of hybridization can be observed. The sunflower is a relatively recent resident of California. It was probably introduced by the Indians, who used the seeds as food and who also used the flower in their religious rites. In the beginning it was probably restricted in its distribution to disturbed sites around Indian camps. In the last hundred years or so, as a result of the activities of western settlers, disturbed sites have multiplied enormously in California, and so has the sunflower, which is now very widespread along roadsides and railroad tracks and as a weed in cultivated fields. *Helianthus bolanderi* is a sunflower native to California and Oregon. It grows in two kinds of habitats. One race is found in the serpentine outcrops of the coastal ranges. These serpentine outcrops are soils with an extremely low calcium level and a high content of magnesium and heavy metals. They sustain a highly specialized flora. In addition, *H. bolanderi* occurs as a ruderal weed in the central and northern valleys of California at the same sites as *H. annuus*. In a detailed study, Heiser has shown that these two species hybridize occasionally. Although the hybrid is largely sterile, it can produce some good pollen. According to Heiser, when *H. annuus* was introduced into California, it hybridized with *H. bolanderi*, which probably was then restricted to the serpentine outcrops. From repeated crosses and backcrosses of these two species emerged a form of *H. annuus* that resembled *H. bolanderi* (Table 12.2; Fig. 12.2) and adapted well to the California environment, presumably by borrowing genes from *H. bolanderi*. From introgression of genes of *H. annuus* into *H. bolanderi* emerged the form of *H. bolanderi* that is capable of growing in disturbed habitats.

In addition to *H. bolanderi* and *H. debilis* variety *cucumerifolius*, other annual sunflower species have ranges that overlap with *Helianthus annuus*. One of them is *H. petiolaris*, a species that is widespread in the western United States and occasionally eastward, with a distribution similar to that of *H. annuus*, although the last species is more widespread. Although both species grow together in many places, they seem to have some slightly different ecological preferences. In general, *H. annuus* seems to be more restricted to heavy soils and *H. petiolaris* to sandy soils. In addition, *H. petiolaris* comes into bloom in June and blooms through July and into August, but *H. annuus* generally does not bloom until July and continues to bloom into September, restricting somewhat the opportunities for gene exchange. Hybrid swarms

Table 12.2

Comparison of the number of ray flowers and disk diameter of the common sunflower, *Helianthus annuus*, and the California sunflower, *H. bolanderi*.

Species	Mean Number of Ray Flowers	Disk Diameter
H. bolanderi—serpentine	12.7*	1.70*
H. bolanderi—valley	14.8*	2.07*
H. bolanderi × H. annuus hybrid	14–20†	2–3†
H. annuus, Calif.	18.2*	2.41*
H. annuus, non-Calif.	20.7*	2.98*

Data from C. Heiser, 1949. Study in the evolution of the sunflower species *Helianthus annuus* and *H. bolanderi*. *Univ. Calif. Publ. Bot.* 23:157–208.

* Mean value for garden-grown plants (29 to 61 plants measured according to population).
† Range of values for 6 plants.

Involucral
bracts

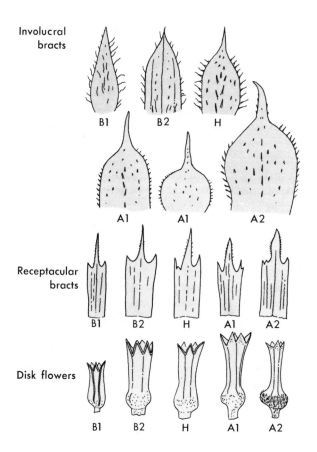

Receptacular
bracts

Disk flowers

Figure 12.2

Effect of hybridization on morphological characteristics of *Helianthus annuus* and *H. bolanderi*. B1, *H. bolanderi*, serpentine race; B2, *H. bolanderi*, nonserpentine race, presumed to have introgressed with *H. annuus*; H, natural hybrid between *H. annuus* and *H. bolanderi*; A1, *H. annuus*, from California, presumed to have introgressed with *H. bolanderi*; A2, *H. annuus* from St. Louis, Mo., outside the range of *H. bolanderi* and consequently free of any contamination by *H. bolanderi*. Note how the presumed introgressed types look more like the hybrid than either of the "pure" species. (From C. Heiser, 1949. Study in the evolution of the sunflower species *Helianthus annuus* and *H. bolanderi*. *Univ. Calif. Pub. Bot.* 23:157–208)

between these two species are not frequent, but they have been observed several times. Again, the hybrid is quite sterile but produces some viable pollen, and consequently introgression has occurred. Introgressed populations of *H. petiolaris* have broader and more serrated leaves, hispid involucral bracts, and a general increase in the size of all the parts of the plant. Populations of *H. annuus* subjected to introgression from *H. petiolaris*, on the other hand, have narrower leaves, narrower involucral bracts, smaller heads, a more densely pubescent chaff, and a reduced number of ray flowers (Table 12.3, Fig. 12.3).

From these three examples of hybridization and introgression a pattern emerges that has been observed in several other species. The four species of sunflowers we have discussed are likely to have developed in isolation from one another. Under these circumstances they differentiated and became adapted to specific ecological conditions. Nevertheless, they did not acquire total genetic isolation. When *H. annuus* increased its range (probably as a result of human activities), it came into contact with the three other species. Hybrids were formed, and they in turn backcrossed with the parents, transmitting genes from one species to the other. Some of the genes so transmitted conferred on the recipient species characteristics that made it better adapted to the new environment. Because *H. annuus* was the invading species, it probably acquired physiological characteristics that allowed it to adapt better to the conditions of Texas, California, and the Midwest. It in turn probably conferred on the other three species characteristics that permitted them to grow on disturbed soils. Harlan and De Wet have termed species such as *H. annuus* compilospecies, "robber species," because through hybridization they "borrow" genes from native species and in this way become more widely adapted.

We can see, then, that the probable role of introgressive hybridization in evolution is to enable some species to acquire genes from other species and in this way increase their fitness and adaptation to new habitats.

ISOLATION MECHANISMS

By now it must be clear that the key property of a species is the fact that it is *reproductively isolated* from all other species in nature. Reproductively iso-

Table 12.3

Comparison of the number of ray flowers and disk diameter of the common sunflower, *Helianthus annuus*, and *H. petiolaris*.

Species	Number of Ray Flowers*	Disk Diameter[†] (cm)
H. petiolaris	13–16	2.3
H. petiolaris × H. annuus	15–21	2.9
H. annuus	21–30	3.3

Data from C. Heiser, 1947. Hybridization between the sunflower species *Helianthus annuus* and *H. petiolaris*. *Evolution* 1:249–262.

* Range for several plants grown in the garden.
† Mean for populations from East St. Louis.

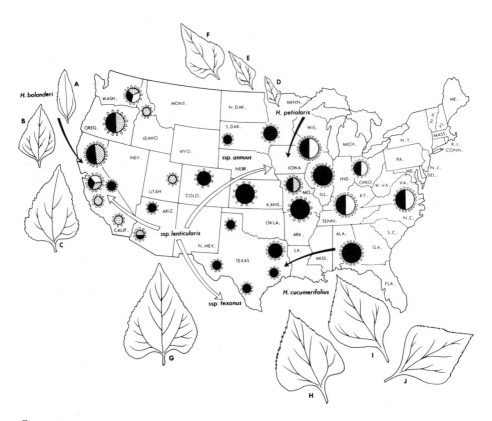

Figure 12.3

Probable evolution of *Helianthus annuus* in the United States, according to Heiser. Subspecies *lenticularis* (or the derived subspecies *annuus*) spread from the southwestern United States to California, where it hybridized with *H. bolanderi*. Subspecies *lenticularis* (or ssp. *annuus*) spread also to the Midwest, where it hybridized with *H. petiolaris*, and to Texas, where it hybridized with *H. debilis* var. *cucumerifolius*, giving rise to subspecies *texanus*. Differentiation as a result of selection was also taking place, as can be seen by the different head sizes, anther color, and number of ray flowers. Diagrams of heads are drawn proportional to values of natural populations. "Petals" each represent two ray flowers. Shading represents anther color: black, purple anthers; hatching, red anthers; white, black anthers; combinations, mixtures of colors in the population. Outlines of leaves: A, *H. bolanderi*; B, *H. annuus* × *H. bolanderi*; C, *H. annuus* ssp. *annuus*; D, *H. petiolaris*; E, *H. petiolaris* × *H. annuus*; F, *H. annuus* ssp. *annuus*; G, *H. annuus* ssp. *lenticularis*; H, *H. annuus* ssp. *texanus*; I, *H. annuus* × *H. debilis* var. *cucumerifolius*; J, *H. debilis* var. *cucumerifolius*. (From O.T. Solbrig, 1970. *Principles and Methods in Plant Biosystematics.* New York: Macmillan)

lated does not mean that it is absolutely prevented from crossing with other species, or that such a cross, if it occurs, is sterile. Many species can and occasionally do cross in nature, and the hybrids are fertile. Hybridization may even result in the acquisition of advantageous genes, such as in the sunflower example. Reproductive isolation means that habitually the majority of the members of the species do not cross with other species. Two further examples, one from the animal, the other from the plant kingdom, illustrate this.

The mallard duck, *Anas platyrhynchos*, and the pintail duck, *Anas acuta*, are perhaps the two most common fresh-water ducks in the Northern Hemisphere. According to Ernst Mayr, the world population of these two species is well over 100,000,000 ducks, which nest side by side on millions of lakes, ponds, and creeks in Asia, Europe, and North America. In captivity the two species will hybridize, producing normal and totally fertile hybrids in F_1, F_2, and following generations. But in nature less than one in a thousand ducks is a hybrid, and there is no evidence of backcrossing or introgression.

Salvia apiana and *S. mellifera* are two species of sage that grow in the coastal sage communities of southern and central California (Fig. 12.1). Their ranges overlap in the area between Santa Barbara and the Mexican border. *Salvia apiana* shows a slight preference for drier soils, although the two species often grow intermixed. *Salvia mellifera* blooms earlier, approximately from April to the end of May, whereas *S. apiana* blooms from the middle of May to the end of June. Consequently, there is little overlap in their blooming times. *Salvia mellifera* has a small flower, and *S. apiana* has a much larger flower. Finally, the first species is typically pollinated by small bees, flies, and butterflies, whereas *S. apiana* is pollinated by larger carpenter bees (Fig. 12.4). These ecological and mechanical differences combine to keep the two species reproductively isolated. Nevertheless, hybrids are formed occasionally, most frequently in disturbed habitats. When these hybrids grow to maturity, they are only partially fertile. Although in such instances some introgression occurs, it does not seem to be very widespread.

The two species of ducks and the two species of sage illustrate how species with no absolute genetic isolating barriers can live side by side with little or no hybridization. The sage example also shows that there is no absolute reproductive isolating barrier, but a large number of partial barriers that reinforce one another to produce almost total reproductive isolation.

Table 12.4 tabulates the principal kinds of isolation mechanisms. Of these the only truly absolute (and probably irreversible) isolating barriers are (1) gametic incompatibility, that is, the inability of the sperm of one species to form a viable zygote with the egg of a second species, and (2) absolute hybrid sterility, that is, the incapacity of the hybrid to form viable gametes. All other barriers are likely to break down occasionally and can be termed "leaky" isolating barriers. Species that are geographically separated may come together in the course of time; some pollen grains may in some instances germinate on the style of the supposedly incompatible species; inbreeders and apomicts

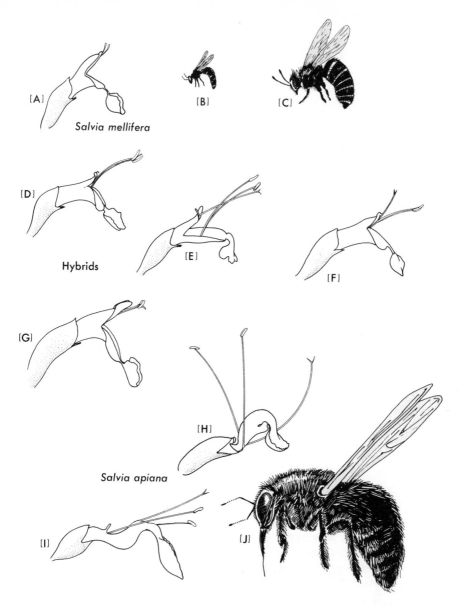

Figure 12.4

Flowers of *Salvia mellifera*, *S. apiana*, and their hybrids and backcrosses, and some specific pollinators. (A) *Salvia mellifera*; (B) *Chloralictus* sp., a small, solitary bee, pollinator of *S. mellifera*; (C) *Anthophora* sp., a medium-sized bee that pollinates *S. mellifera*; (D) F₁ hybrid between *S. mellifera* × *S. apiana*; (E–G) backcrosses of (D) to both parents; (H–I) *S. apiana*, untripped and tripped position of flower; (J) *Xylocopa brasilianorum*, a pollinator of *S. apiana*. One can readily see that the different sizes of the flowers of these two species and their pollinators act as an isolating mechanism. (Flowers redrawn from Epling, 1947. Natural hybridization of *Salvia apiana* and *S. mellifera*. *Evolution* 1:69–78. Bees redrawn from K.A. Grant and V. Grant, 1964. Mechanical isolation of *Salvia apiana* and *S. mellifera* (Labiatae). *Evolution* 18:196–212)

Table 12.4

Isolation mechanisms.

Prefertilization	Reduction of contact	1. Geographical separation*
		2. Ecological separation
		3. Gametic incompatibility† (gametic isolation)
	Reduction of mating frequency	4. Inbreeding and asexual reproduction
		5. Ethological isolation
		6. Different breeding times (allochronic isolation)
Postfertilization	Reduction of zygote formation (prezygotic)	7. Gametic incompatibility†
	Reduction of gene flow through hybrids (postzygotic)	8. Hybrid sterility†
		9. Hybrid weakness or breakdown†
		10. Lack of hybrid establishment (environmental isolation)

* Not always considered as a true isolation mechanism.
† May be absolute and irreversible.

occasionally outbreed and hybridize; the tail end of the breeding period of one species may overlap with the beginning of the breeding period of a second species; and hybrids can live in areas where the environment is intermediate between that of the two parental species, even in cases where the hybrids are not very vigorous. As we have seen, once hybrids are established, they are capable of crossing with the parental species, leading to the formation of backcrosses and gene exchange between populations.

Because so many species are separated by "leaky" isolating barriers and consequently may hybridize, is there not a fallacy in our species concept? In effect, a strict application of the biological species concept will lead to a circumscription of only those taxa that possess absolute barriers to hybridization. This means that many species of oaks would have to be considered as one species, as well as practically all birches and willows, many insects, fishes, ferns, orchids, and grasses, and so on. Obviously they are not the same but different species. Consequently, even in cases where no absolute isolating mechanisms are present, new species can be formed and maintained.

When one is considering species formation, one must take into account selection and gene flow in addition to reproductive isolation. Biological phenomena are never unidimensional. For ease of understanding, we study them one at a time, but often this can be misleading.

THE PROCESS OF SPECIATION AND THE FORCES THAT CONTROL IT

Evolutionarily, the basic unit is the gene pool, which is equated with the local breeding population. But even the concept of the gene pool is an abstraction to a certain extent because it cannot be defined in strict terms. Also, as we have already seen in Chapter 8, the breeding mechanism has a great effect on the size and variability of the gene pool. In some outbreeding species, the gene pool may be rather large; at the other extreme, some populations of asexual organisms may exchange hardly any genes at all, generation after generation.

Any mechanism that restricts gene flow may favor speciation. The converse, namely, that speciation will proceed whenever gene flow is restricted, is not true. When two populations are subjected to different selection forces, they will tend to diverge genetically and thus physiologically and morphologically. Gene exchange will slow down the process of divergence, because intermediate forms will be constantly formed. If selection is not sufficiently great, the formation of two sharply distinct populations is not possible. However, if selection is intense and the intermediate zone very narrow, the formation of two distinct populations can occur and has occurred. Good examples are the evolution of populations adapted to extreme edaphic conditions, such as *Agrostis tenuis*, studied by Bradshaw and already mentioned (p. 273).

On the other hand, a number of cases are now known where absolute barriers to gene exchange exist with little or no morphological differentiation. One such case is the *Gilia inconspicua* species group studied by the American evolutionist Verne Grant. This group of annual plants is found throughout the arid regions of western North America and southern South America, where the plants grow in open sandy places in a variety of habitats, such as desert washes or sandy river floodplains, in openings in the desert scrub vegetation, and on the slopes of mountains. The individuals of these species are self-compatible, and the pollen is deposited on the stigma lobes automatically. The products of self-pollination are fully vigorous generation after generation as Grant showed in experimental progenies. At maturity the plants belonging to the *Gilia inconspicua* complex bear small, dull-colored flowers on the upper branches. Although alike in their general characteristics, they differ in various fine details of external morphology. On the basis of detailed morphological, cytological, and crossing experiments, it was possible to demonstrate that what had been considered one species was actually a group of six independently derived groups of species, comprising no less than 25 species altogether (Fig. 12.5). With very few exceptions, all 25 of these species are completely genetically isolated from each other, no hybrid seed having been formed in spite of repeated attempts at crossing them. In the few cases where hybrids were produced, they were sterile or had low fertility (highest fertility, 28%).

Another example is *Lasthenia*, belonging to the family Compositae. In a study of this group of small winter annual herbs of vernal pools in California,

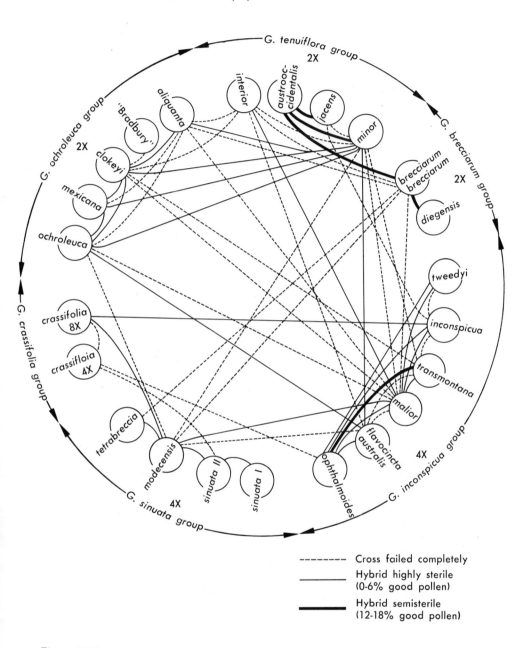

Figure 12.5

Crossability and sterility barriers between 23 sibling species in the *Gilia inconspicua* species complex. (From V. Grant, 1964. The biological composition of a taxonomic species in *Gilia*. *Adv. in Genetics* 12:281–327. Copyright © 1964 Academic Press.)

Ornduff found a composite picture. In section *Baeria*, for example, *Lasthenia chrysosthoma* was found to have diploid and tetraploid races, whereas the other species, *L. macrantha*, had diploid and hexaploid races. Crosses between populations of *L. chrysosthoma* with different chromosome numbers yielded hybrids that had reduced pollen fertility, although occasionally the fertility was as high as 50 percent. However, hybrids between diploid populations also often had reduced fertility, in one instance less than 20 percent (Fig. 12.6).

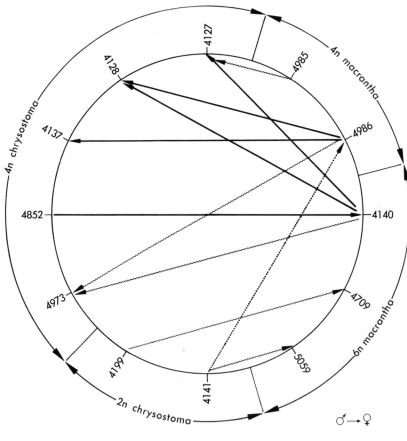

Per Cent Pollen Viability: ·················· 0-19 ⌇⌇⌇⌇⌇⌇ 60-79 ————

Figure 12.6

Diagram summarizing average pollen viabilities of progenies obtained from artificial interspecific hybridizations in *Lasthenia*, section *Baeria*. (From R. Ornduff, 1966. A biosystematic survey of the goldfield genus *Lasthenia*. *Univ. Calif. Publ. Botany* 40:1–92)

Crosses within *L. macrantha,* on the other hand, tended to be highly fertile, even where the chromosome number differed. Finally, interspecific hybrids were relatively fertile or were sterile, according to the populations crossed and the chromosome number of these populations, but they usually had better than 50 percent fertility (Fig. 12.7).

A very fine example in animals is provided by the intensive studies conducted by the California geneticist Francisco Ayala and his collaborators with

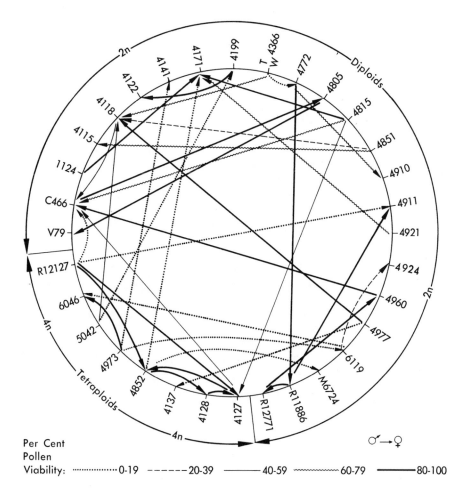

Figure 12.7

Diagram summarizing average pollen viabilities of progenies obtained from intraspecific hybridizations within and between diploid and tetraploid races of *Lasthenia chrysostoma.* (From R. Ornduff, 1966. A biosystematic survey of the goldfield genus *Lasthenia. Univ. Calif. Publ. Botany* 40:1–92)

six species of the so-called *Drosophila willistoni* group in the Western Hemisphere (Fig. 12.8). Four of these species are quite widespread (*D. willistoni*, *D. paulistorum*, *D. equinoxalis*, and *D. tropicalis*) in Central and South America; one, *D. insularis*, is found only in some islands of the Lesser Antilles and *D. pavlovskiana* is found only in Guiana. At any one locality within the range of the *D. willistoni* group, more than one species may exist sympatrically. The six species are practically identical in morphological appearance; the only difference lies in the male genitalia. When reared in the laboratory, the species show strong ethological isolation, which presumably also operates in nature.

Figure 12.8

Distribution of four sibling species of *Drosophila* belonging to the *D. willistoni* group. (From F.J. Ayala, 1975. Genetic differentiation during the speciation process. *Evol. Biol.* 8:1–78)

When an occasional hybrid appears in the laboratory, males are always sterile; hybrid females are occasionally fertile. No hybrids have been found in nature. In addition, Ayala and collaborators were able to distinguish within the species *D. paulistorum* six groups that manifested strong preferences for mating within their own group, even though the groups were completely undistinguishable morphologically. The ethological isolation between these groups, however, is not as strong as that which exists between species. When hybrids between these groups are formed, the males are once more sterile, but the females are always fertile. Each of these groups has a distinct geographical distribution (Fig. 12.9), but there is some overlap between them. No hy-

▲ Centro-American
▼ Amazonian
■ Orinocan
● Andean-Brazilian
× Transitional
◆ Interior

Figure 12.9

Distribution of the semispecies of *Drosophila paulistorum*. There are no morphological differences among the semispecies, but reproductive barriers exist among some of them. Most hybrids between semispecies yield fertile females but sterile males. (From F.J. Ayala, 1975. Genetic differentiation during the speciation process. *Evol. Biol.* 8:1–78)

brids have been found in nature. Because the groups within *D. paulistorum* show many characteristics of species, they have been called *semispecies*.

We see, then, that there are species that are not separated by absolute isolating barriers and, equally important, that there are species whose populations are not entirely interfertile. In addition, remember that most populations of a species are isolated from other populations because of space and time. Gene exchange between a population of field mice from Nashville, Tennessee, and one from Gainesville, Florida, must be very small indeed. Because the probability of direct exchange between the populations is so small as to be almost nil, any exchange is through geographically intermediate populations. Such a process would take at best tens or hundreds of generations. However, experiments show that the Nashville, Tennessee, and the Gainesville, Florida, populations have not acquired reproductive isolation.

SPECIES DEFINITION

How can species be defined? According to the biological species concept, a species is "a group of interbreeding populations reproductively isolated from any other such group of populations." In too many instances this concept is inapplicable in an absolute sense. However, it still remains the only attempt to define a species in nonarbitrary terms—nonarbitrary because a well-defined biological property, interbreeding, is used to define the species. The fact remains, nevertheless, that there is no way of defining species in a nonarbitrary way that will reflect all aspects of the evolutionary situation. If breeding behavior is chosen as the yardstick, many distinct morphological groups can be shown to be interfertile; if morphological similarity is chosen instead, it will soon be apparent that these groups often comprise an assemblage of isolated or partly isolated populations. If physiological and ecological factors are brought to bear, a similar lack of correlation can be observed.

The enunciation of the biological species concept more than a quarter of a century ago was a great advance in the history of biology. It pointed out the role of isolating mechanisms and the importance of defining species in a nonarbitrary way. However, in the intervening years much has been learned about population structure and genetics and the mechanisms of speciation. Science, it has been said, is a series of increasingly accurate approximations to the truth. The species concept reflects this aphorism very well.

PREADAPTATION AND THE ORIGIN OF EVOLUTIONARY NOVELTIES

In the example of the evolution of the sunflower, we saw that this species acquired new characteristics by occasional hybridization with other species. In Chapter 11, we saw that character displacement during species differentiation leads to the modification of a character or, by accentuation of a given feature, to the development of what may be considered a new character (e.g., a different mating call in frogs derived from emphasizing a part of the original call). However, occasionally during evolution, species acquire characters that differ more than quantitatively from the character that gave rise to them. Particularly important are the acquisition by organisms during evolutionary time of structures or properties that permit them to assume a new function, as, for example, when animals invaded land and acquired lungs and legs, or when they learned to fly and acquired wings. Unless the early land animals could somehow breathe air, they could not survive on land. Likewise, the first animals to take to air must have had some way of stabilizing themselves. But how can an animal acquire the ability to breathe air through the accumulation of favorable mutations *before* it invaded land?

Clearly it is not possible for an organism to perform a new function, such as breathing air, until the new structure has been acquired. Consequently, the structure must be present before. That is, the organism must possess a structure that has been selected in one environment and for one function to the point where it is able to perform a new function for which it was not selected. We call such structures *preadapted*. The concept is well illustrated by the history of the early amphibians.

Aquatic animals are kept moist, supplied with oxygen, and kept buoyant by the medium (water) in which they live. If they leave water for land, they face the necessity of holding themselves up and moving in a less buoyant medium, of extracting oxygen from air, and of keeping their cells and tissues moist. Primitive lungs were acquired first by fishes as an outgrowth from the pharyngeal region of the gut. These lungs originally enabled such fishes to live in waters of low oxygen content, and primitive lungs have persisted to this day in certain lines of bony fishes (lung fishes). These same lines of fishes also acquired pedunculated fins, presumably used for moving along rocks and over the bottom. Animals with some kind of lung and the possibility of moving over the bottom first evolved among fishes in response to pressures within that environment, i.e., the environment of shallow coastal areas, of stagnant waters, low in oxygen. Once such characteristics had been acquired, these fishes could leave the aquatic medium temporarily, either to escape from a predator or to escape from a drying pool. This would have been possible only during humid nights or rainy days, in the same manner that certain fishes (such as the Bali catfish that has been introduced into Florida) do today. Once the animals acquired this new behavior from which they derived immediate advantages, they set up a powerful array of new selective pressures, which led to rapid evolution and specialization of the preadapted structure, in this case lungs and limbs. The selection pressures that originally brought about the lung to breathe and the limbs to stay erect in air and to move were related to life in shallow, oxygen-poor waters, so that when new functions were required (breathing air, moving on land), the structures were there. The new selective force reinforced and perfected the lungs and the forelimbs and also had a great impact on the rib cage and the head, but it would not have taken place unless the primitive lung and fins had not been *preadapted* to life on land.

Preadaptation is an important concept, and it explains how we can have evolution and adaptation at the same time in a changing environment. The fate of nonadapted individuals, populations, and species is extinction. As a result of changed environmental conditions, many specialized and thriving organisms became unable to cope with their surroundings and died. Thousands of once-dominant species, sometimes after flourishing for long periods, have become extinct. Examples are the dinosaurs and the notoungulates of South America. Among plants, there were the giant calamites and seed ferns, which once dominated the vegetation of the earth. In more recent times we have witnessed and are still witnessing the extinction of many animals and plants, such as passenger pigeons, American bison, the pretty *Franklinia* shrub (known today only in cultivation), and many others, which are incapable of coping with an environment changed drastically by hostile human beings. That this is not a necessary fate of all organisms is demonstrated by such diverse organisms as wheat rusts, cockroaches, Japanese beetles, rats, and starlings, which are thriving in spite of and, in fact, largely because of human intervention.

ADAPTATION

The word *adaptation* has been used by evolutionists and ecologists for a very long time, but with different meanings. It is used primarily in three different ways: (1) in conjunction with a character ("long fur is adaptive to cold"); (2) in relation to the state of being adapted ("cacti are adapted to deserts"); and (3) in relation to the process ("plants adapt to drought by closing their stomata"). These different usages of the word have created communication and semantic problems. We will use adaptation principally to mean the *state of being adapted*.

Scientists have been interested for a long time in the very elaborate and marvelous ways by which organisms acquire characteristics to perform certain tasks with great efficiency. Race horses have thin, long legs; thin, streamlined bodies; long necks and heads—all artificially selected to increase the horse's speed. Draft horses have stocky, thick, strong legs; big chests and short necks —all selected for strength. Likewise cheetahs, which catch their prey by pursuit, have thin, long bodies and long necks, and lions and tigers, which ambush their prey, have strong, wide bodies. It is therefore reasonable to conclude that natural selection molds characteristics of the organisms so as to increase their efficiency in performing a given task. We call such structures adapted.

However, when the matter is pursued further and in a more rigorous way, many questions arise. Are all characteristics of organisms adaptive, or just some, and then which? Are only morphological characteristics adaptive, or also behavioral, chemical, and reproductive characteristics? Exactly how does an organism that is adapted to one environment become adapted to a new environment? And is the accumulation of small mutations by natural selection sufficient to explain the evolution of elaborate structures such as the mammalian eye?

Adaptation can be viewed as an all-or-none phenomenon. In this context existence is the sole criterion for adaptation. A population is adapted to an environment if it can live and reproduce there. But if adaptation is viewed only as an all-or-none phenomenon, the concept is of very limited use. Can we determine degrees of adaptation?

In common usage, when we refer to an organism (or a human being) as having "adapted," we imply that the organism acts "appropriately" for the task at hand. We can follow this up by saying that a structure is adaptive when it is appropriate for a particular task. But some structures are more "appropriate" than others. Consequently, it is meaningful to speak of degrees of adaptation. The problem is to establish criteria that can help us decide how appropriate or advantageous a characteristic is to an organism.

The process of natural selection, as explained in Chapter 3, provides us with a criterion to determine the degree of "adaptiveness" of a character. In effect, natural selection results in the increase of certain variations in the population and the decrease of others. Those that increase are more "appropriate" in that they enhance the survival of the organisms that possess them; i.e., they are more adaptive. In other words, an adaptation is a characteristic of an organism selected in preference to its absence, and better adaptations are selected in preference to poorer ones. For example, dark coloration in *Biston betularia* is adaptive in relation to light coloration in polluted environments because it is favored by natural selection. Note that we make no judgment regarding whether dark is better or more efficient, or possessed of any other value. Note also that degree of adaptiveness of a character state can be judged only in relation to other states of a character, such as dark versus light coloration in moths or palatable versus unpalatable butterflies (Chapter 15) or DDT-resistant versus DDT-susceptible flies (Chapter 6). For characters that are fixed and invariable in the population, any statement regarding degree of adaptation is conjectural.

Are adaptive characters optimal?

It is often asserted that evolution by natural selection leads to the development of optimal phenotypes, i.e., phenotypes that are theoretically the most efficient in aiding organisms to survive and reproduce.

The basic problem is, how is it possible to tell what is the most efficient genotype? In their efforts to formulate optimality models, scientists cannot be sure that they have considered every alternative. Something completely different may work better than what they consider to be optimal. Similarly, natural selection can work only on characters or combinations of characters that actually appear. A drastically different character (or combination) may allow an organism to take advantage of a new environmental situation in a way its ancestors could never have done, even though it may be very inefficient there in comparison with a theoretical optimum. The earliest land vertebrates are an example. They were better adapted to land than fishes but were very inefficient in an absolute sense.

The fact that evolution over the course of time keeps producing organisms differing from their ancestors, even when the environment remains the same, is an indication that the optimum has not been achieved.

Natural selection is a process that favors the least detrimental among available phenotypes. But natural selection is not a process that necessarily creates optimal phenotypes. Selected phenotypes satisfy only two basic requirements: (1) They are adapted in an absolute sense, and (2) they maintain themselves from generation to generation.

There are additional considerations, mostly of a genetic nature, that can prevent evolution of optimal phenotypes. First, there is the random nature of mutation. Most mutations are deleterious, and since each individual in the population carries some detrimental alleles, no phenotype will ever be truly optimal. Second, most genes are pleiotropic (i.e., have multiple effects), and although some of these effects will be beneficial, others will not. Third, there is linkage. Slightly deleterious alleles can increase in the population if they are tightly linked to beneficial genes. Given sufficient time, such linkage disequilibrium (Chapter 8) can be broken, but it can be a slow process. Fourth, in heterozygotes there is segregation of less-fit homozygous genotypes. Although each of these genetic factors in itself may be trivial, their additive effects can be considerable.

In conclusion, natural selection eliminates the worst (in terms of survivorship and fecundity) phenotypes in the population, but surviving phenotypes, though better fit than those that are eliminated, are by no means optimal; that is, they are not necessarily the best possible ones.

We now proceed to investigate in greater detail how individuals, populations, and species interact and how these processes affect evolution by natural selection.

SUGGESTED FURTHER READING

General references

Anderson, E. 1949. *Introgressive Hybridization.* New York: Wiley.

Dawson, G.W.P. 1962. *An Introduction to the Cytogenetics of Polyploids.* Oxford: Blackwell Scientific Publications.

Solbrig, O.T. 1970. *Principles and Methods of Plant Biosystematics.* New York: Macmillan.

Stebbins, G.L. 1971. *Chromosomal Evolution in Higher Plants.* London: Edward Arnold.

Articles and reviews

Ayala, F.J. 1975. Genetic differentiation during the speciation process. *Evol. Biol.* 8:1–78.

Epling, C. 1947. Natural hybridization of *Salvia apiana* and *S. mellifera*. *Evolution* 1:69–78.

Grant, V. 1964. The biological composition of a taxonomic species in *Gilia*. *Adv. in Genetics* 12:281–327.

Heiser, C. 1951. Hybridization in the annual sunflowers: *Helianthus annuus* × *H. debilis* var. *cucumerifolius*. *Evolution* 5:42–51.

Heiser, C. 1961. Natural hybridization and introgression with particular reference to *Helianthus*. *Recent Adv. in Botany* 1:874–877.

Jackson, R.C. 1976. Evolution and systematic significance of polyploidy. *Ann. Rev. Ecol. Syst.* 7:209–234.

Lewis, H. 1962. Catastrophic selection as a factor in speciation. *Evolution* 16:257–271.

Ornduff, R. 1966. A biosystematic survey of the goldfield genus *Lasthenia*. *Univ. Calif. Publ. Botany* 40:1–92.

Packard, G.C. 1974. The evolution of air-breathing in paleozoic gnathostome fishes. *Evolution* 28:320–325.

Remington, C.L. 1968. Suture-zones of hybrid interaction between recently joined biotas. *Evol. Biol.* 2:321–428.

Roose, M.L., and L.D. Gottlieb, 1976. Genetic and biochemical consequences of polyploidy in *Tragopogon*. *Evolution* 30:818–830.

Stebbins, G.L. 1947. Types of polyploids: their classification and significance. *Adv. in Genetics* 1:403–429.

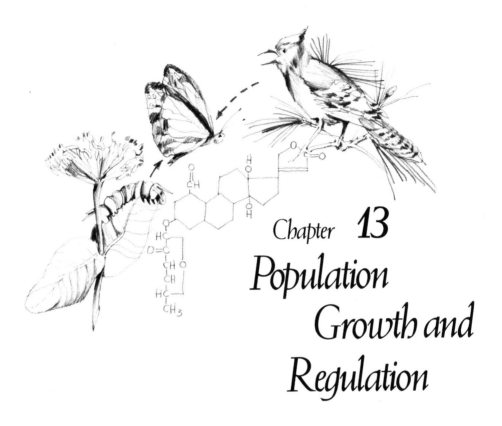

Chapter **13**

Population Growth and Regulation

The geneticist studying evolution looks for changes in the frequency or proportion of genes in the population. The geneticist investigates frequencies and not actual numbers, because the number of individuals in populations of plants and animals changes from year to year, or even within a year. These changes are sometimes very drastic. When the population is growing, most genes are also multiplying, and when the population is shrinking, most genes are becoming less abundant. But gene frequencies are statistical abstractions. To understand the relationships of the organism with its environment, we have to know how many individual organisms of each kind exist, what the yearly fluctuations in numbers are, how many die and are born annually, and when, since these parameters can affect the march of evolution.

The numbers of individuals of a given species in an area can change, sometimes very dramatically. For example, in 1536 the Spanish Adelantado, Pedro de Mendoza, brought 20 cows and 72 horses to America when he founded the city of Buenos Aires in what is today the Argentine pampas. Three years later the settlement was burned to the ground by the Indians and was abandoned by the Spanish. The horses and cows were left to their own devices. They multiplied in the pampas and by 1700 had grown into a population that numbered more than a million head of cattle and an equally large number of horses. The Spanish explorers of the sixteenth and seventeenth centuries also systematically introduced goats into oceanic islands to serve as food sources in case of possible shipwreck. One such explorer, Juan Fernandez, introduced a pair of goats into the islands in the South Pacific off the coast of Chile that today bear his name. In 1704, when Alexander Selkirk (Daniel Defoe's real-life Robinson Crusoe) was abandoned in the islands by the captain of his ship, the two goats had increased to a herd of more than 10,000, and that herd still lives there today. Every year new strains of influenza virus (Chapter 6) arise and in a matter of months spread throughout the world. Wheat rust fungi invade the northern wheat belt of the United States every year and there multiply severalfold in a period of weeks (Chapter 6). Whenever a piece of organic matter is left unrefrigerated, bacteria multiply at astonishing rates in hours; when bread is baked or beer is brewed, an inoculation of yeast cells quickly multiplies severalfold. The ability of organisms to increase in number under special conditions is tremendous.

But species also can become extinct, sometimes quite fast. In 1919 a fungus known as the "chestnut blight" was accidentally introduced into the United States. This fungus attacks and kills chestnut trees, including the native American chestnut, *Castanea dentata*. Within ten years of the introduction of the blight, the population of this majestic, dominant tree had been so reduced that the tree is practically extinct today. Before the nineteenth century, passenger pigeons were among the most abundant species of birds in the United States. However, hunting led to the extinction of the species. On the other hand, no amount of hunting has been able to halt the spread of starlings and blackbirds in the United States today.

Although examples of wide population fluctuations, both upward and downward, are easy to find, most species are in some sort of equilibrium, at least in the short term. We do not observe any obvious changes in the numbers of most species of plants and animals.

The study of changes in numbers of individuals in a population, called *demography*, is a highly specialized branch of population biology. Demography has derived its major impetus from the need of insurance companies to know the probability of death of an individual who wants to purchase a life insurance policy, in order to determine a reasonable premium. Consequently, a great deal is known about human demography and relatively little about the demography of plant and animal populations.

Observation tells us that there are great differences between species in regard to their demography. Some species of trees count their life span in tens or even hundreds of years, whereas most bacteria count their life span in hours or minutes. Most species of animals live a year or less, and most vascular plant species live more than a year. Nevertheless, there are ephemeral plants that live only a month or two, and a few species of animals live more than 50 years. The life span of most microorganisms is short (hours or days), but there is also a fair amount of variation among them.

The demographic characteristics of a species are related to the environment in which the species lives, and they constitute an adaptation of the species brought about by natural selection. In this chapter and the next two, we explore this aspect of evolution in greater detail.

POPULATION GROWTH

Every organism is born and dies.* If it lives long enough, it may also reproduce. Every individual's precise life span and number of descendants are unique and cannot be accurately predicted. However, a population of such organisms has statistical properties given by the proportion of individual organisms with similar characteristics. These patterns can be analyzed theoretically in mathematical terms. They have to do primarily with growth, reproduction, and death.

We will first describe some of the demographic properties of populations and follow with a discussion of some of the problems one encounters when studying real populations with some examples.

Exponential growth

Populations are said to have discrete or nonoverlapping generations when the offspring do not reproduce until after their parents have died. For example, in populations of annual plants, such as wild mustard, the plants die after producing seeds, and the new seeds do not germinate until the ensuing year. Populations are said to have continuous or overlapping generations when adults of different ages are reproducing simultaneously, as in human populations, many species of trees, and many animals. The mathematical tools necessary to analyze these two situations are slightly different.

Discrete Generations. Let us start with the simplest situation, namely, a population formed of individuals that are born, reproduce once, and then die. We will assume that no deaths whatsoever occur until the individuals have reproduced. The number of organisms in each generation will depend on the reproductive rate, b_0, which we will define as the average number of offspring that each individual leaves behind. If b_0 has a value greater than 1, it follows that the population is growing in numbers; if the value of b_0 is exactly 1, the population is stable; if b_0 is less than 1, the population is shrinking, and eventually it will become extinct. In species with two sexes in a one-to-one

---

* In organisms that divide by fission (such as bacteria) or bud off new ones (such as *Hydra*) or reproduce asexually (such as many plants) the question of what exactly constitutes "birth" or "death" is problematical.

ratio, the reproductive rate is usually calculated as the average number of female offspring produced by each female. Most species have reproductive capacities much in excess of the rate needed to replace the parent generation, so most populations, unless checked by some external factor, will tend to grow. For example, it has been determined that the female of the common house fly produces approximately 120 eggs per generation, of which 60 on the average will hatch into females. There are about seven generations in a year. If we assume that all the mortality of adult flies occurs after they have produced their eggs, a population of one male and one female fly at the beginning of the year will grow into 5,598,720,000,000 flies, or roughly 5.6×10^{12} flies by the end of the year! If we assume that each fly weighs 0.01g, the population will have increased in weight from 0.02g to 28,000 metric tons!

Putting this relation in the form of an equation, we have

$$N_t = N_0 b_0{}^t,$$

where N_t = number of individuals at generation t; b_0 = net reproductive rate of the females; N_0 = number of individuals in generation 0. That is, the number of individuals at a given time, t (measured in generations), is equal to the initial number multiplied by the replacement rate to the t power. In our example

$$N_7 = 2 \times 60^7 = 5{,}598{,}720{,}000{,}000.$$

This is called *exponential growth*.

It must be obvious that species cannot grow in this way, other than perhaps in very special circumstances and for very short periods of time.

Continuous Reproduction. Many species breed at all times and have no definable generations. For such species, in order to calculate the number of individuals at a given time, we must know the number that existed at some previous time and the rate at which the population has been growing in the interval. The rate at which the population grows is given by

$$\frac{dN}{dt} = bN - dN = rN,$$

where N = number of individuals in the population at some given moment; b = birth rate; d = death rate; and r = a species- and environment-specific constant called the *intrinsic rate of increase*—also called "little r"—which is equal to $b - d$, the birth rate minus the death rate. The rate of increase of the population is then given by a constant, r, multiplied by the number of individuals in the population, N. To find out the number of individuals in the population at a given time, we manipulate the formula:

$$\frac{dN}{dt} = rN$$

can be expressed as

$$\frac{1}{N}\frac{dN}{dt} = r$$

or

$$\frac{d \log N}{dt} = r,$$

which in the integrated form becomes

$$\log_e N_t - \log_e N_0 = rt,$$

and taking antilogs of both sides, we have

$$e^{(\log_e N_t - \log_e N_0)} = e^{rt}$$

and

$$N_t = N_0 \, e^{rt},$$

where N_0 = number of individuals at time zero; N_t = number of individuals at time t; and e = base of the natural logarithms.

Logistic growth

Exponential growth is not possible for more than a very short period of time because of the limitation of resources. Food and ultimately space run out. If every spring each pair of house flies multiplied into twenty-eight tons of flies, there would be no space for anything but flies on the earth. When the numbers are low, e.g., our initial two flies, growth can approximate exponential growth for a generation or two. However, as the population increases in size, it must reach some upper limit at which the number of individuals dying equals the number of individuals being born. This population size, at which the growth rate of the population equals 0, is called the *carrying capacity* of the environment and is represented by the symbol K. The carrying capacity is determined by the available resources.

Thus the growth of a population is controlled by two factors. The first, which we can think of as an inherent characteristic of the species, is its ability to reproduce at some maximum rate. The second, which we can think of as the resistance of the environment, is the limitation of resources in the environment that results in a greater death rate and lower birth rate as the number of individuals in the population increases. The formulas given before have to be modified to take into account the fact that the number of additional individuals that can subsist is a function of the number that are already there. The simplest way of doing this is to assume that the relationship is linear, and that it has an upper limit, when no new individuals can be added to the population unless some already there die. This upper limit is K, the carrying capacity of the environment.

For the more general continuous growth model, the rate of growth now becomes

$$\frac{dN}{dt} = rN\left(\frac{K-N}{K}\right).$$

This formula, often called the logistic equation or Volterra equation in honor of the Italian mathematician who proposed it, can be analyzed very easily. Note that when the number of individuals in the population, N, is small in relation to the carrying capacity, K, the term $(K-N)/K$ is close to 1 and the growth of the population is almost logarithmic. However, as N approaches K, $(K-N)/K$ approaches 0, reaching it when $N=K$. If we graph the growth of the population for successive values of N, it will yield a sigmoid curve (Fig. 13.1). This type of growth is called logistic growth.

Experiments have shown that certain microorganisms (e.g., *Paramecium*, Fig. 14.6, or yeast) in artificially controlled situations grow according to the logistic model. But does the logistic model accurately describe the growth of populations of all organisms or a majority of them?

In 1937 two male and six female ring-necked pheasants (*Phasianus colchicus*) were introduced into Protection Island in Puget Sound on the coast of the state of Washington. This population grew within five years to 1325 breeding birds, a 166-fold increase. There is some indication that this population grew in a logistic fashion (Fig. 13.2). The two cocks and six hens introduced in 1937 had grown to 30 birds by the spring of 1938, to 81 in 1939, to

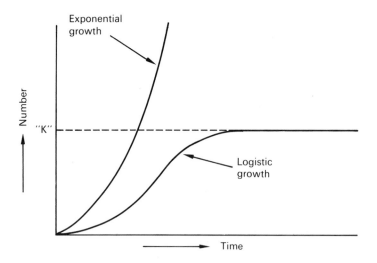

Figure 13.1
Exponential and logistic growth curves.

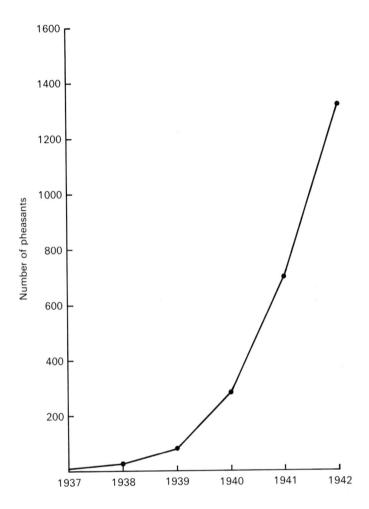

Figure 13.2
Population growth of pheasants on Protection Island. The initial population consisted of two cock and six hen pheasants. (From A.S. Einarsen, 1945. Some factors affecting ring-necked pheasant population density. *Murrelet* 26:39–44)

282 in 1940, to 641 in 1941, to 1194 in 1942, and to 1898 in 1943, when the Army arrived in the islands and soldiers began hunting pheasants. But it appears that the population was starting to level off by 1943. The pheasants during this six-year period were counted in the spring prior to reproduction, and again in the fall after the reproductive season had ended and the young were almost fully grown. It was observed that although the number of eggs laid per female pheasant did not change during the study, the percentage of yearly increase steadily decreased, because there was an increasing reduction

of nesting success and an increase in the mortality of the young. Consequently, there were fluctuations between spring and fall. In the winter months only deaths, no births, take place.

However, other studies indicate that the logistic model is a poor describer of population growth. A number of experiments were carried out by the Australian ecologist A.J. Nicholson with laboratory populations of the Australian sheep blowfly, *Lucilia cuprina*. In most of the experiments the limiting factor was the quantity of food supplied daily to each experimental culture. The blowfly larvae need to eat a certain level of food to grow large enough to pupate and develop into adults. The size at which the larvae pupate increases with the amount of food available to each larva. The larger the pupa, the larger the resulting adult, and the greater the number of eggs it can lay. A number of experiments were run in which the amount of food to adults and larvae was independently modified. It was found that even when conditions were quite adverse, each population succeeded in maintaining itself. In most situations there was a marked periodic fluctuation in the number of adults and/or of larvae (Fig. 13.3). The blowflies did not grow according to the logistic model. Another well-documented case of a population that did not grow according to the logistic model is that of the small crustacean *Daphnia pullex*, as depicted in Fig. 13.4.

Why is the logistic equation not sufficient in itself to describe the growth of a population in nature? The reasons are many. They undoubtedly vary from species to species, and not all are as yet understood. However, we can analyze some of them.

The growth rate given by the equation

$$\frac{dN}{dt} = rN\left(\frac{K-N}{K}\right)$$

changes in value whenever an individual is added or subtracted from the population. Let us assume the population is growing. In order for the growth curve to yield a smooth logistic curve, r has to be a true constant. But $r = b - d$, and we know from observation that in most species there are certain times of the year when the birth rate (b) reaches a peak and others when it levels off, and we also know that the death rate (d) is highest in some seasons of the year and lowest in others. Therefore r is not constant with time. This is the first reason for deviation from the logistic equation. The equation also implies that as an individual is added or subtracted from the population, there is an *immediate* effect on the growth rate. If we observe species, we will note that there usually is a lag between fertilization and birth. The number of births at time $t + 1$ is not determined by the number of individuals at time t, but by the number of females fertilized at time $t - g$ (where g is the gestation period) that survive to time $t + 1$.

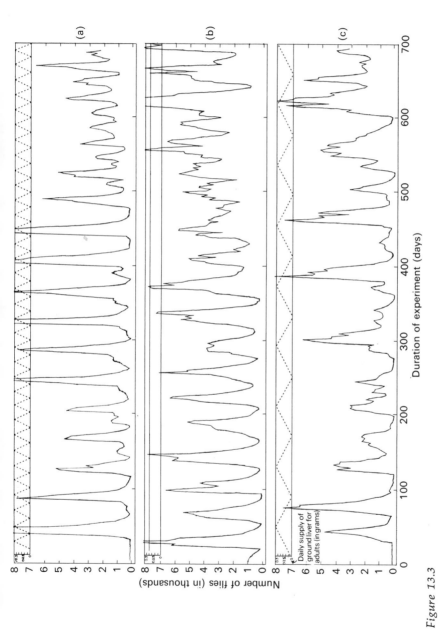

Figure 13.3

Population growth in the sheep blowfly, *Lucilia cuprina*. In (b) adults were fed the same amount each day (0.4g ground liver). In (a) and (c) the amount of food fluctuated, as indicated at the top of each graph. (From A.J. Nicholson, 1958. The self-adjustment of populations to change. *Cold Spring Harbor Symp. Quant. Biol.* 22:153–173)

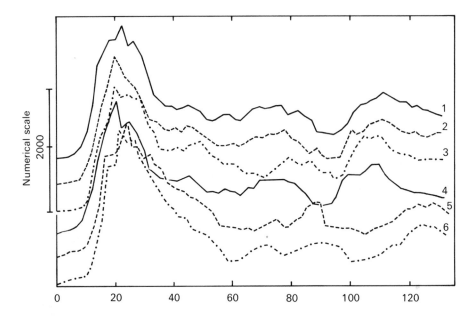

Figure 13.4

Numbers observed in six replicate populations with initial population of 25 *Daphnia* approximating the stable age distribution at that density. (The ordinate has been successively shifted by 400 for different replicates.) (From P.W. Frank, 1960. Prediction of population growth form in *Daphnia pulex* cultures. *Am. Nat.* 94:357–372)

To give an extreme example, imagine a population of mice that has $K - 1$ individuals with several females pregnant. When the next individual is added to the population, no wholesale abortion takes place among the rest of the pregnant females. Rather, the number of individual mice keeps growing even if the population size overshoots K. So the lag in the response of the population to increasing density is another reason why populations do not grow following exactly a logistic curve.

Finally, we have to explore the nature of K itself. The carrying capacity is given in our formula as a constant. We can think of it as the environmental resources that can sustain a certain number of individuals of a given species. But are the resources ever constant? Obviously not. Plants use light and water, whose availability varies throughout the year and from place to place. We all know how the forest changes from summer to winter, and even in the tropics it changes from the wet to the dry season. Since the plant biomass serves as the resource for herbivores, as it changes, the number of herbivores obtaining sustenance from it will change, and similarly with carnivores and decomposers. Furthermore, metabolic rate is a function of temperature, and temperature also oscillates.

In brief, for true logistic growth to occur, r and K must have constant values, and there must be immediate feedback. Such conditions are sometimes approached in the laboratory when a microorganism with a very short life span is used (preferably one that reproduces asexually) and when the resources can be kept perfectly constant. Under those conditions logistic growth is observed, but data from less ideal conditions indicate that as a describer of nature the logistic model is insufficient.

DEMOGRAPHY

How can we learn empirically how populations grow in nature? Very simply, by keeping track of the birth, growth, reproduction, and death of individuals in a population. This is what insurance companies and the Bureau of the Census do for the human species.

Survivorship schedule

The first statistic that has to be obtained is the survivorship schedule—that is, the number of individuals surviving to every particular age—which is usually

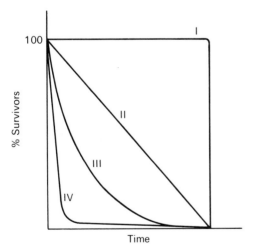

Figure 13.5

Kinds of survivorship curves. The type I curve describes those species with little mortality of the young, where all individuals reach a maximum age. The type II curve describes the case in which a constant number of deaths occur per unit time. Type III represents constant probability of death, regardless of age. In type IV most die very young, but once the organism is past a certain age, there is a good chance it will reach the maximum age for the species.

recorded as a proportion or frequency of those born at one given time. To obtain this figure, we determine how many individuals born at a certain time, *t*, survive the first interval of time, how many the second, how many the third, and so on until no more are alive. Graphing this information, we obtain a survivorship curve (Fig. 13.5). Each species has a characteristic survivorship curve, but we can recognize four general or basic types, which are usually called types I, II, III, and IV.

The type I curve corresponds to a population where most individuals survive to an old age for the species (which can in actuality be a few hours or days) and then die at similar ages. Very few species even approximate this survivorship curve. Human populations in developed countries are probably the closest example (Fig. 13.6). Type II curves correspond to populations where a fixed number of individuals die per unit of time. The type III curve

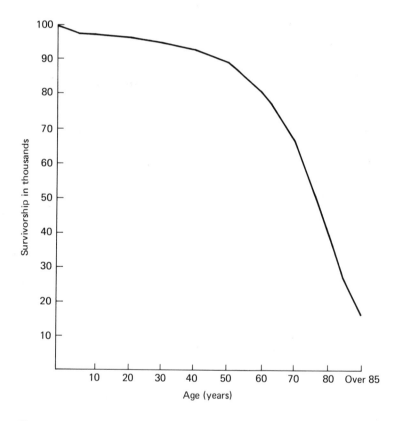

Figure 13.6

Survivorship curve for human males in the United States, 1964. (Data from R.D. Remington and M.A. Schork, 1970. *Statistics with Applications to the Biological and Health Sciences.* Englewood Cliffs, N.J.: Prentice-Hall)

corresponds to populations where a fixed proportion of the population die per unit of time, that is, when the probability of death is constant over time. Curves of types II and III are approximated by some populations of vertebrates and plants (Fig. 13.7). Type IV curves describe a population where there is a great deal of early mortality, and the probability of death decreases after some period of time. It is the most common situation and describes the survivorship curve of such varied organisms as maple trees, oysters, and some frogs (Figs. 13.8 and 13.9).

Fecundity schedule

The second statistic we need to know is the average number of offspring produced at each particular age. In most species, reproduction is confined to only some age groups. Furthermore, even during the time an individual is poten-

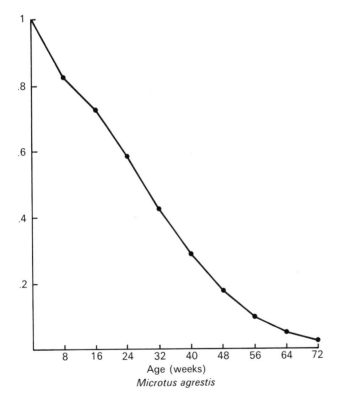

Figure 13.7

Mortality schedule for a laboratory population of *Microtus agrestis*. (Data from P.H. Leslie and R.M. Ranson, 1940. The mortality, fertility rate of natural increase of the vole (*Microtus agrestis*) as observed in the laboratory. *J. Anim. Ecol.* 9:27–52)

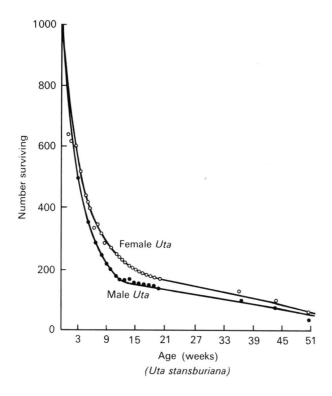

Figure 13.8

Survivorship curve for *Uta stansburiana*. (From D.W. Tinkle, 1967. The life and demography of the side-blotched lizard, *Uta stansburiana*. *Misc. Publ. Mus. Zool., Univ. Mich.* No. 132)

tially fertile, its probability of reproduction is not uniform. To obtain a fecundity schedule, we must record for each birth the number of female (by convention) offspring produced and the age of the mother. The fecundity schedule, which is the age-specific birth rate (Fig. 13.10), can then be constructed by calculating for each age group the average number of offspring born to individuals in that age group.

Net reproductive rate

Once we know the survivorship curve and the fecundity schedule of a population, the net reproductive rate (R_0), also called the replacement rate, of the population can be obtained. It tells us whether the population (assuming the survival and reproductive rates remain constant) is growing, remaining stationary, or declining.

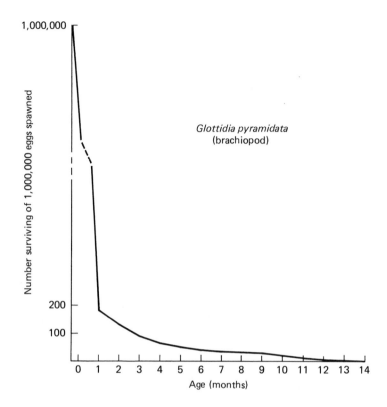

Figure 13.9

Survivorship curve of the brachiopod *Glottidia pyramidata*. (Data from R.T. Paine, 1963. Ecology of the brachiopod *Glottidia pyramidata*. *Ecol. Monogr.* 33:187–213)

To obtain the net reproductive rate we have to know: (1) the number of individuals in the population that were born at a given time (this group is called a *cohort*), what fraction will succeed in reproducing, and if they can reproduce more than once, what fraction will survive to each successive reproductive event; and (2) the average number of female offspring a female is expected to produce at each reproductive event.

Using the survivorship schedules, we calculate the probability of being alive after the first interval of time (usually a year, but the interval of time chosen will depend on the species). The probability is given by the number alive at time 1 divided by all of those born at time 0:

$$\lambda_1 = \frac{N_1}{N_0} = 1 - d_1,$$

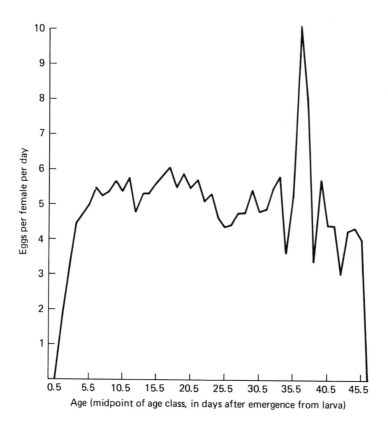

Figure 13.10

Fecundity curve for adult females of the human louse, *Pediculus humanus*. The data for this curve are from a laboratory population. The fluctuations at the end may be due to the small sample size (only nine of the original 400 females survived to the age class of 36.5 days). (Data from F.C. Evans and F.E. Smith, 1952. The intrinsic rate of natural increase for the human louse, *Pediculus humanus* L. *American Naturalist* 86:299–310)

where d is the death rate and is equal to

$$d_1 = \frac{N_0 - N_1}{N_0}.$$

The probability that an individual alive at the first interval will be alive at the second interval is given by

$$\lambda_2 = \frac{N_2}{N_1} = 1 - d_2,$$

and the probability that an individual born at t_0 will be alive at t_2 (which we will call l) is

$$l_2 = \lambda_1 \times \lambda_2 = \frac{N_1}{N_0} \times \frac{N_2}{N_1} = \frac{N_2}{N_0}.$$

Consequently, the proportion of individuals that survive from birth to age x is given by

$$l_x = \lambda_1 \times \lambda_2 \times \cdots \times \lambda_x = (1 - d_1) \times (1 - d_2) \times \cdots \times (1 - d_x).$$

Thus, to calculate the probability of surviving to a certain age, the only necessary statistic is the age-specific death rate.

Consider now the cohort of births numbering N_0 born in a given year. The number surviving to age x will be given by $N_0 \times l_x$, the number that were born multiplied by the probability of their surviving to age x. To know how many offspring will have been produced by this cohort at age x, we proceed as follows:

If b is the number of births, then at time x, the number of births in the cohort will be given by the number that are alive at time x multiplied by the age-specific birth rate, m_x, from the fertility schedule:

$$b_x = N_x \times m_x$$
$$= N_0 \times l_x \times m_x$$

At age $x - 1$: $\quad b_{x-1} = N_0 \times l_{x-1} \times m_{x-1}$

At age $x - 2$: $\quad b_{x-2} = N_0 \times l_{x-2} \times m_{x-2}$

$\cdots \qquad\qquad \cdots \qquad\qquad \cdots$

At age 1: $\qquad\qquad b_1 = N_0 \times l_1 \times m_1$

Total births $\qquad \Sigma b = \displaystyle\sum_{x=0}^{x=x_{max}} N_0 l_x m_x$

Thus, knowing the number of individuals born at a given time, we can calculate the probable number of offspring they will produce solely by knowing the survivorship and fecundity schedules of the population.

To obtain the net reproductive rate, R_0—that is, the average number of offspring per individual in the population—all that needs to be done is to divide the equation above by N_0, as follows:

$$\frac{\Sigma b}{N_0} = \frac{\displaystyle\sum_{x=0}^{x=x_{max}} N_0 l_x m_x}{N_0}$$

$$R_0 = \sum_{x=0}^{x=x_{max}} l_x m_x$$

The average number of offspring per individual in the population, R_0, is an important statistic. Its value will fluctuate around 1 in a population at equilibrium. If a population has an R_0 value larger than 1, the population is expanding (each individual is producing more offspring than necessary for replacement); if smaller than 1, the population is shrinking.

We can further manipulate the formula, remembering that

$$N_t = N_0 e^{rt}$$

and

$$\frac{N_t}{N_0} = e^{rt},$$

and making $t =$ one generation $= T$,

$$\frac{N_T}{N_0} = e^{rT} = R_0,$$

and we have connected the growth rate derived from the theoretical model of population growth with R_0. The importance of this equality lies in the fact that since a value for R_0 can be obtained empirically (through the $l_x m_x$ schedule obtained from censuses), r can be estimated:

$$R_0 = e^{rT},$$
$$\log_e R_0 = rT(\log_e e) = rT.$$

Consequently,

$$r = \frac{\log_e R_0}{T} \quad \text{and} \quad T = \frac{\log_e R_0}{r}.$$

Donald W. Tinkle of the University of Michigan has studied the demography of a small ground-dwelling lizard of the western United States and Mexico, *Uta stansburiana*. Populations of two subspecies, *Uta stansburiana stejnegeri* from Texas and *U. s. stansburiana* from Colorado, were investigated. Figure 13.11 shows the survivorship schedule (l_x) for the two populations. There is an interesting difference between the two populations. In the Texas population only 10 percent of the animals survive their first year of life, but in Colorado the figure is closer to 30 percent. In Texas during the breeding season each adult female lays at least three but probably four or more clutches averaging four eggs each. In Colorado, on the other hand, the young females usually lay only two clutches in their first year (occasionally three), but if they survive the winter, they lay three or four clutches during their second year of life. Many females survive for a third year, and their fecundity remains high.

The Mexican plant ecologist José Sarukhan has studied the demography of the tropical palm *Astrocaryum mexicanum*, a typical plant of the lower strata of the tropical evergreen forest of southern Veracruz, Mexico. Table

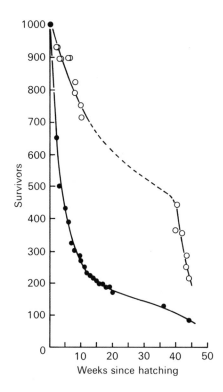

Figure 13.11
Survivorship curves for Texas (solid) and Colorado (open) *Uta stansburiana* based
on pooled data for several generations of lizards marked at hatching. Abscissa,
weeks since hatching; ordinate, number (of 1000) alive at each age interval. The
dotted portion of the curve for Colorado is an extrapolation in the absence of data
for the time period. (From D.W. Tinkle, 1969. Evolutionary implications of comparative population studies in the lizard *Uta stansburiana*. In *Systematic Biology:
Proceedings of an International Conference. Nat. Acad. Sci. Publ. 1692*, pp. 133–154)

13.1 and Fig. 13.12 show the survival probabilities from seedling to adult, the
average number of fruits per plant (a measure of fecundity), and the percentage of the total reproduction represented by each age category. Note that the
probability of surviving to the next life cycle stage is lowest during the seedling and juvenile stages of the life cycle (approximately the first 20 years of
the life of this palm). The plant does not start fruiting until it is approximately
40 years of age; after that fecundity increases with age, but the greatest contribution of seed is made by palms in the 65-year age category, because the
number of plants older than 65 declines steeply (Table 13.1).

These two examples illustrate some extreme differences in life histories.

Table 13.1

Survivorship probabilities and mean seeds per tree for individuals of the tropical palm *Astrocaryum mexicanum*.

Stage	Age (years)	Probability of Survival to Next Stage	Mean Seeds* per Tree
Seedling	8	0.37	0
Juvenile	15	0.32	0
Immature	27	0.81	0
	39	0.64	0
Mature	47.5	1.00	3
	56.0	1.00	13
	64.5	1.00	10
	73.0	0.75	28
	81.5	0.66	19
	90.0	0.75	25
	98.5	0.66	30
	107.0	—	34

From J. Sarukhan, 1979. Demographic problems in tropical systems. In O.T. Solbrig (ed.), *Demography and Evolution in Plant Populations*. Oxford: Blackwell Scientific Publications.

* Age approximate.

Stable age distribution

In most populations the reproductive capacity of its members changes with time. First there is a period when organisms are unable to reproduce (which may be only a few minutes, as in microorganisms, or several years, as in humans and many trees). That period is followed by an increase in reproductive capacity until a maximum is reached, then by a decline that eventually reaches zero. In most populations, individuals die before they reach the postreproductive period. However, humans and their pets and domesticated animals, as well as animals in zoos, which live in an environment more favorable than their natural surroundings, do reach that stage. If individuals of different ages contribute unequally to reproduction, it follows that the net reproductive rate, R_0, can vary in the population because of changes in the age structure without any changes in birth rate. Furthermore, populations can be *increasing* while the birth rate is decreasing, solely as a result of changes in the age structure, as occurs with the human population in many developing countries. Furthermore, changes in the parameters of the life history table, i.e., changes in age-specific birth or death rates, bring with them changes in the proportion of individuals of different ages in the population. Therefore any population in a steady environment eventually approaches some stable age distribution. This is true regardless of whether the population is increasing in size, decreasing,

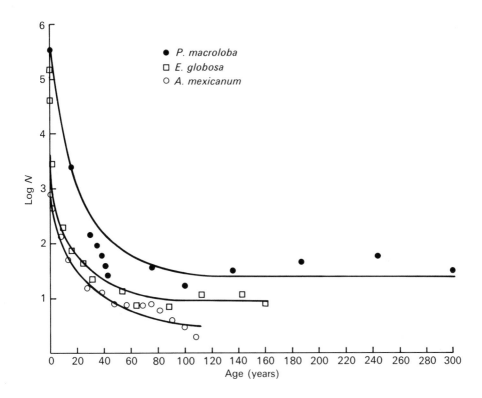

Figure 13.12

Population age structure for the tropical tree *Pentaclathra macroloba* (solid circles) and the palms *Astrocaryum mexicanum* (open circles) and *Euterpe globosa* (squares). (From J. Sarukhan, 1979. Demographic problems in tropical systems. In O.T. Solbrig (ed.), *Demography and Evolution in Plant Populations*. Oxford: Blackwell Scientific Publications)

or holding steady. One can readily obtain proof of this assertion by taking any real or imaginary life history table (i.e., a schedule of l_x and m_x for $x = 1$ to $x = n$, where n is the maximum possible age) and calculating the number of females in each age class for each time interval, starting with any arbitrary age distribution. In Table 13.2 the population reaches a stable age by the eighth year.

We have seen that populations in nature are constrained in their growth by a number of factors, so that on the average their net reproductive rate, R_0, must be 1 or close to it (it may oscillate around 1). To maintain an R_0 value of 1 requires some unique $l_x m_x$ schedule. It follows that in nature most populations are probably at or close to a stable age distribution. This means that the proportions of individuals belonging to different age groups are constant for

Table 13.2

Calculation of a stable age distribution.

An island has been devastated by a hurricane, and all but one two-year-old female of an endemic turtle has been killed. The growth of this imaginary population is determined by factors other than turtle density according to the following life table.

The Life History Table

Age	1 year	2 years	3 years	4 years	5 years
l_x	1	0.5	0.25	0.125	0
m_x	1	3	2	0.5	0

Age Distribution with Time

	1 year		2 years		3 years		4 years		5 years	
	No. of Turtles	% of Population	No. of Turtles	% of Population	No. of Turtles	% of Population	No. of Turtles	% of Population	No. of Turtles	% of Population
1st year	0	0	1	100	0	0	0	0	0	0
2nd year	3	100	0	0	0	0	0	0	0	0
3rd year	3	75	1	25	0	0	0	0	0	0
4th year	6	85	1	15	0	0	0	0	0	0
5th year	9	75	3	25	0	0	0	0	0	0
6th year	18	78	4	17	1	4	0	0	0	0
7th year	32	74	9	21	2	5	0	0	0	0
8th year	63	73	18	21	4	5	1	1	0	0
9th year	125	75	31	19	9	5	2	1	0	0
10th year	237	75	62	19	15	5	4	1	0	0
⋮										
15th year	·6215	74	1615	19	420	5	109	1	0	0

generation after generation in a steady environment. Consequently, if we take repeated censuses of the population at yearly intervals, we ought to obtain about the same proportion of individuals in each age category. If we do not, it means that the l_x's and m_x's are not constant.

Reproductive value

In a population with stable age distribution at equilibrium, where $R_0 = 1$, each female in the population produces one female offspring on the average. In reality, however, some females produce no offspring because they die before reproductive age, and consequently others produce more than one offspring. At birth each female has an equal probability of leaving one offspring. As time goes by and some females in the cohort die, the probability of leaving offspring rises, and it keeps rising with time. As the females start reproducing, the probability of producing additional offspring decreases, and it becomes zero once that individual reaches its postreproductive age. The measure of the probability of contributing to future generations of an individual of a given age is called its *reproductive value* (v), and it is defined as the age-specific expectation of future offspring. Mathematically,

$$v_x = \sum_{t=x}^{t=\infty} \frac{l_t}{l_x} m_t,$$

where the quotient l_t / l_x represents the probability of living from age x to age t, and m_t is the average birth rate at age t. Table 13.3 shows the reproductive value for a beetle, *Calandra oryzae*.

As there is a steady age distribution for each unique schedule of $l_x m_x$, there is also a unique schedule of reproductive values.

DENSITY

So far we have been discussing population growth, paying only indirect attention to the effect of crowding. The number of individuals per unit of space is referred to as the *density* of the population. It is usually measured and expressed as the number of individuals per unit area in a terrestrial species or per unit volume in an aquatic organism, especially when the organism is small, such as a unicellular alga. When it is difficult to estimate the number of individuals, the weight (or biomass) per unit area is measured instead.

Density is a measure of crowding. The crude density is the number (or biomass) of individuals per unit space (or volume) without any further considerations. The specific or ecological density is the number (or biomass) of individuals per unit area (or volume) that is actually suitable for the species—what is called the species habitat. Maximum density varies from organism to organism; many more dandelion plants or mice can live on an acre of land than redwoods or bison. But even among species that are similar in size, there

Table 13.3

Survivorship (l_x), fecundity (m_x), and reproductive values (v_x) for a laboratory population of the beetle *Calandra oryzae*.

Age in Weeks	l_x	m_x	$l_x m_x$	v_x
4.5	0.87	20.0	17.400	130.53
5.5	0.83	23.0	19.090	115.86
6.5	0.81	15.0	12.150	95.15
7.5	0.80	12.5	10.000	81.15
8.5	0.79	12.5	9.875	69.51
9.5	0.77	14.0	10.780	58.50
10.5	0.74	12.5	9.250	46.30
11.5	0.66	14.5	9.570	37.90
12.5	0.59	11.0	6.490	26.18
13.5	0.52	9.5	4.940	17.22
14.5	0.45	2.5	1.125	8.92
15.5	0.36	2.5	0.900	8.03
16.5	0.29	2.5	0.800	6.86
17.5	0.25	4.0	1.000	4.76
18.5	0.19	1.0	0.190	1.00

Data from L.C. Birch, 1948. The intrinsic rate of natural increase of an insect population. *J. Anim. Ecol.* 17:15–26.

are great differences in density. For example, certain species of birds, such as Antarctic penguins, herring gulls, and Pacific island boobies, have nesting colonies of thousands of birds per acre, whereas most species of song birds never attain even a tenth of such densities. More important, even within a species we note differences in density from place to place, and even within a locality, differences in density are sometimes observed from year to year. One of the best-documented examples of changes in population density is that of the snowshoe hare and the lynx in the Hudson Bay area of Canada. Because these species have been hunted for their valuable pelts, the records of the Hudson Bay Company can be used to calculate their density over the last hundred years (Fig. 13.13). Both species vary in a cyclical fashion, and over a ten-year period, tenfold changes in density can be observed. No other such extended record of density changes exists, but many species have been observed to fluctuate in density over periods of ten or more years; sometimes even within a year (Fig. 13.14).

Population number can change as a result of the direct interaction between individuals in the population. As more and more animals or plants pack into an area, they interfere with one another. Such interactions may have a destabilizing effect. For example, too many cows in a pasture may lead to

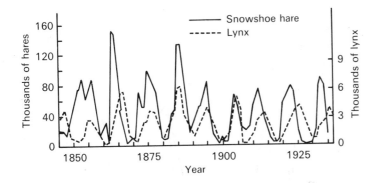

Figure 13.13

Population cycles in the snowshoe hare (*Lepus americanus*) and its predator, the lynx (*Lynx canadensis*). The data are based on furs acquired by the Hudson's Bay Company. (After D.A. MacLulich, 1937. Fluctuations in the numbers of the varying hare (*Lepus americanus*). *Univ. Toronto Studies, Biol. Ser. 43*)

fights and to the death of young calves. Population numbers may change also because of external factors of a physical nature: Storms, droughts, extreme cold, or extreme heat may lead to the death of individuals in the population. Finally, population density is affected by external biological factors, i.e., interactions with other species. Almost every species is a prey to some bacterium, fungus, or animal; all species but green plants and those animals that eat dead matter are predators on other species.

POPULATION REGULATION

We have seen that all species have the potential of multiplying in numbers and that under special circumstances they do (e.g., horses and cows in the Argentine pampas in the sixteenth and seventeenth centuries; pheasants in the Protection Islands). But it can easily be ascertained by observation that the numbers of individuals of most species are fairly constant. Consequently,

Figure 13.14

Density fluctuations in various species. (a) *Colinus virginianus*, the bobwhite quail, in Wisconsin. The upper curve (solid line) shows fluctuations in numbers based on the fall census. The lower curve includes the spring census and also shows seasonal changes. (From P.L. Errington, 1945. Some contributions of a fifteen-year local study of the northern bobwhite to a knowledge of population phenomena. *Ecol. Monogr.* 15:1–34) (b) Laboratory populations of *Paramecium aurelia* and *Saccharomyces exiguus*. *Paramecium* is a predator on *Saccharomyces*. (From G.F. Gause, 1935. Experimental demonstration of Volterra's periodic oscillations in the numbers of animals. *J. Exp. Biol.* 12:44–48)

▶

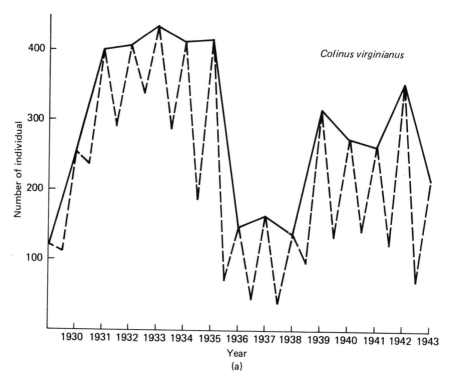

Number of individual

Colinus virginianus

1930 1931 1932 1933 1934 1935 1936 1937 1938 1939 1940 1941 1942 1943

Year

(a)

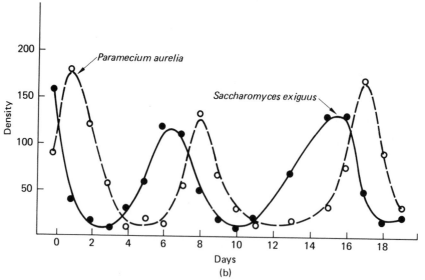

Density

Paramecium aurelia

Saccharomyces exiguus

0   2   4   6   8   10   12   14   16   18

Days

(b)

the rate of mortality must be in equilibrium with the birth rate (Fig. 13.15). Otherwise we would observe much greater fluctuations in the size of populations than we do. This apparent concordance of birth and death in most populations is referred to as *population regulation,* and it is an interesting and controversial subject. What we specifically wish to know is the mechanism or

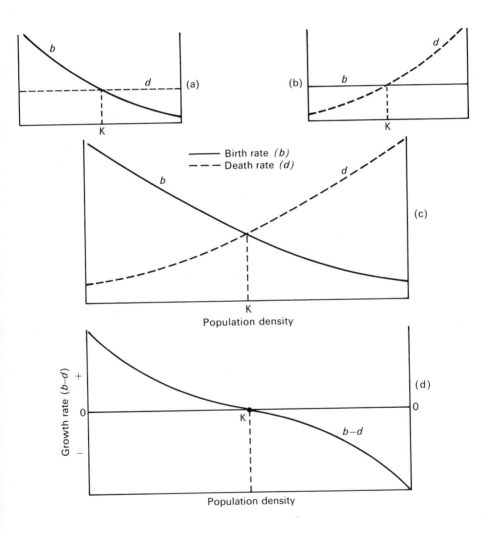

Figure 13.15
Models of population regulation. (a) Death rate constant, birth rate density-dependent. (b) Birth rate constant, death rate density-dependent. (c) Birth and death rates density-dependent. (d) Relation between growth rate and carrying capacity.

mechanisms that account for the apparent equilibrium of birth and death rates in most populations.

Two extreme explanations for population regulation are the following: (1) Death is due entirely to outside physical or biological factors—a fire, a flood, a predator—and affects the same proportion of individuals in the population, regardless of the population size. Mortality is therefore *density-independent* (Fig. 13.15a). Under this view populations regulate their numbers by adjusting their birth rate to equal the death rate. (2) Death is always a function of the number of individuals in the population: The larger the number of individuals in the population, the greater the risk of death. According to this view, individuals can find shelter or evolve defenses against a flood or a fire or a predator, but the number of shelters or the ways of escape are limited. Consequently, as the numbers of individuals increase, they start running out of resources—food, shelters, escape avenues—so that more and more are killed by predators or floods or starvation, and eventually the death rate equals the birth rate. Population regulation is always due to *density-dependent* mortality in this view (Fig. 13.15b).

Neither of these extreme positions is totally correct or incorrect. There are causes of mortality that are independent of population density (the eruption of the volcano on the island of Krakatau in 1883, to take an extreme example) and many others that are density-dependent (food shortage is the most obvious). The problem is to ascertain the relative importance of these two apparently different kinds of mortality, and to determine if species can adjust their birth rates and, if so, by which mechanism and under which conditions.

All species are subject to density-dependent and density-independent sources of mortality, but not all show the same demographic responses. Some species are fairly stable in numbers. Others are affected drastically by external factors and vary widely in density in what appears to be an irregular fashion. Finally, some species, such as the snowshoe hare, have wide fluctuations in density but in a cyclical fashion. What accounts for these great differences between species? A very likely possibility is that there are different ways in which a species can cope with the environment and with other species, and that these different ways are alternative evolutionary solutions called *adaptive strategies.*

DENSITY-DEPENDENT AND DENSITY-INDEPENDENT MORTALITY

When a devastating fire or flood affects an area, it kills every organism, regardless of how many of them are there. When there is a limited amount of food available, the number of organisms that will starve to death (if any) depends on the number present: If there are only a few, all will probably survive; if a large number are present, the number of deaths due to insufficient food can be considerable. Environmental factors that kill independently of the number of individuals present cause what is called density-independent mortality; those factors whose effects are influenced by density cause density-dependent mortality. Climatic and physical factors (earthquakes, floods, temperature, etc.) are usually classed as density-independent factors. Resources (food, shelter) are usually considered to be density-dependent. In reality, these are not two distinct classes but rather a continuum from primarily density-independent (a devastating fire, for example) to primarily density-dependent (such as food). Density-dependent and density-independent factors affect the parameters of the life history table differently.

Imagine a stable environment where no extreme catastrophes occur. Populations of organisms living in such an environment will be at or close to their carrying capacity most of the time. Most of the offspring that are produced will not survive to reproduction. In such an environment an individual that increases the number of offspring it produces affects its fitness only marginally, since each offspring has a small probability of surviving. However, mutations that increase the probability of survival from age t to $t+1$ will be favored since such individuals, by living longer, can reproduce over several seasons and leave considerably more offspring. On the other hand, in an unstable environment, where catastrophes occur frequently, the probability of surviving from age t to $t+1$ is reduced, regardless of phenotype, but by the same token the probability of the establishment of the offspring is enhanced. This can be shown mathematically in a very simple way.

Let us assume we have a mixture of two phenotypes, one short-lived that reproduces once and dies, the other living longer and reproducing twice. The difference between the phenotypes is genetically determined. We wish to know under what conditions one or the other phenotype is favored. By definition, selection will favor the phenotype that has the largest number of surviving individuals at time $t+1$:

$$N_{(t+1)} = \lambda \times N_{(t)}.$$

The number of individuals at time $t+1$ equals the number at generation t times the growth rate of the population (λ) (p. 331). For the short-lived phenotype, the number of individuals at $t+1$ equals the number of offspring produced per individual, multiplied by the probability that they will survive to reproductive age, and multiplied by the number of parents:

$$N_{(t+1)}{}^A = N_{off}{}^A \times S^A \times N_t{}^A,$$
$$\lambda_t{}^A = N_{off}{}^A \times S^A,$$

where $N_{off}{}^A$ is the average number of offspring, and S^A the survival probability. For the longer-lived phenotype, we must add to the calculation the number of individuals surviving the first year, as follows:

$$N_{(t+1)}{}^B = (N_{\text{off}}{}^B \times S^B \times N_t{}^B) + (N_t{}^B \times R),$$
$$N_{(t+1)}{}^B = N_t{}^B[(N_{\text{off}}{}^B \times S^B) + R],$$
$$\lambda^B = (N_{\text{off}}{}^B \times S^B) + R,$$

where R is the survival rate (l_x) of individuals of the B phenotype from t to $t + 1$. If we now equate the two growth rates,

$$\lambda^B = \lambda^A,$$
$$(N_{\text{off}}{}^B \times S^B) + R = N_{\text{off}}{}^A \times S^A,$$

and further assume that the survival rates of offspring (S^B and S^A) are equal for both phenotypes, we can predict when the long-lived phenotype will be favored over the short-lived individual, and vice versa, as follows:

$$(N_{\text{off}}{}^B \times S^B) + R = N_{\text{off}}{}^A \times S^A,$$
$$\frac{(N_{\text{off}}{}^B \times S^B) + R}{S^A} = N_{\text{off}}{}^A,$$

$$N_{\text{off}}{}^B + \frac{R}{S} = N_{\text{off}}{}^A.$$

Whether the long-lived individual is favored over the short-lived one is dependent on the ratio of its survival probability to that of its offspring. The quantity R/S will be a large number whenever the survival rate of adult individuals is high and that of newborn offspring is low; it will be a small number whenever their survival rate is about equal; and it will be an inconsequential number whenever S is larger than R.

We have assumed that λ^A and λ^B are the same. We want now to predict which environmental factors favor each phenotype, giving them the same degree of success. In stable environments, as we have seen, the establishment of young may be impeded by the lack of resources, since larger adults are better competitors. In such environments S will be small and R/S will be large. According to our formula, $N_{\text{off}}{}^A$ would have to be much larger than $N_{\text{off}}{}^B$ to maintain the same growth rate. The greater R/S is, the more likely it is that we will find the longer-lived genotype.

On the other hand, in an unstable environment, the probability of survival of the young may be the same as that of the adult. Then there is little advantage in having the potential to live another year, and R/S will be negligible. Looking back at our equation, we see that $N_{\text{off}}{}^B$ approaches the value of $N_{\text{off}}{}^A$. In other words, in an unstable environment the phenotype that is favored is A. Experimental verification of the prediction above is offered in a study by Solbrig and Simpson with different phenotypes of the common dandelion.

Dandelion plants were collected from three adjacent fields in Michigan. The plants were transplanted to a uniform garden, and their genotypic compositions were ascertained by means of the isozyme technique (Table 13.4).

Table 13.4

Percentage of each of four dandelion biotypes in three adjacent populations.

Habitat	Sample Number	Biotypes A	B	C	D
1. Dry; full sun; highly disturbed	94	73	13	14	0
2. Dry; shade; medium disturbed	96	53	32	14	1
3. Wet; semishade; undisturbed	94	17	8	11	64

From O.T. Solbrig and B.B. Simpson, 1974. Components of regulation of a population of dandelions in Michigan. *J. Ecol.* 62:473–486.

There were a minimum of four different genotypes out of 300 plants studied. Two of these, the so-called *A* and *D* biotypes, were most common. However, they were not evenly distributed across the three fields; the *A* biotype was most abundant in the first two fields, the *D* biotype in the third. The fields differed slightly in the moisture regimen and also in the amount of sun they received, but primarily they differed in the amount of disturbance, a cause of density-independent mortality, which was greatest in the first field (where *A* was dominant) and least in the third field (where *D* was prominent). Consequently, it was expected that the biotypes in the third field would have a greater survival rate but lower seed production, and that the inhabitants of the first field would be the opposite. Figures 13.16 and 13.17 and Table 13.5 show that these were indeed the findings when the performance of the *A* and *D* dandelions were studied experimentally.

The *intrinsic rate of increase, r,* is equal to the birth rate minus the death rate. The birth rate is species-specific within certain limits. Some organisms, such as the fruit fly *Drosophila*, lay approximately 300 eggs per female; others, such as an oak tree, produce thousands of seeds in a lifetime. Finally, some species, such as humans and elephants, have a very low birth rate of less than 20 offspring on the average. Since all these species are equally successful in that they are persisting in time, birth rate alone does not determine the evolutionary success of a species. Why are there such differences in birth rates?

DEMOGRAPHY AND NATURAL SELECTION

All individual organisms have only a limited amount of food, water, and other needed resources. With these they build their bodies and produce their young (eggs, spores, seeds, or live young). If they have strong bodies, well adapted

Table 13.5

Number of inflorescences per plant, number of seeds per inflorescence, mean seed weight, and average number of seeds for two dandelion biotypes in a greenhouse experiment.

	Biotype A (r-selected)*	Biotype D (K-selected)*
Average number of inflorescences/plant	0.24	0.04
Average number of seeds per inflorescence	105.5	204.3
Average seed weight	0.315 mg	0.437 mg
Total number of seeds/plant	25.3	8.2

* See p. 352.

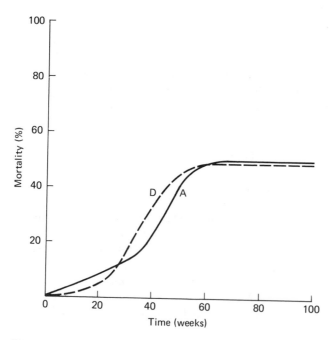

Figure 13.16

Mortality as a function of time for two dandelion biotypes grown separately in pure culture but under the same conditions. Note the great similarity of both curves. (Data from O.T. Solbrig and B.B. Simpson, 1974. Components of regulation of a population of dandelions in Michigan. *J. Ecol.* 62:473–486)

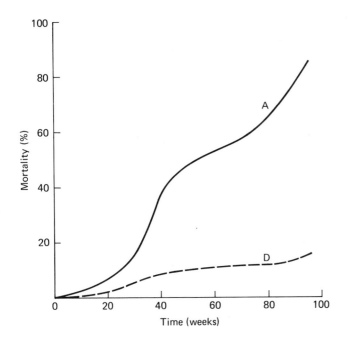

Figure 13.17

Mortality rate for two dandelion biotypes when grown together in a mixture initially made up of even numbers of both biotypes. (Data from O.T. Solbrig and B.B. Simpson, 1974. Components of regulation of a population of dandelions in Michigan. *J. Ecol.* 62:473–486)

to the environment, and the ability to defend themselves against enemies, their death rate will be lowered. Consequently, they will have a longer potential life span than other individuals in the population without those characteristics. But the resources needed to produce that large body are limited. Unless the large individual can gain access to new resources, the increase in body size may come at the expense of its fecundity, and that may result in lowered fitness. Two alternatives are possible. (1) The longer life span compensates for the lowered age-specific fecundity, so that total fertility over the life span of the individual is not reduced. Examples of such a situation are the Colorado populations of *Uta stansburiana* and the D biotypes of the dandelion. (2) The longer life span does not compensate for the lowered age-specific fecundity, but the offspring that are produced are larger (or heavier, or are fed by the parents more or for a longer period). If as a result of increased size an offspring has a higher probability of survival, this fact may compensate for the lower number of offspring produced, so that the contribution to the next generation may be greater than that of another individual in the population

Table 13.6

Post-fledging survival in relation to brood size for Swiss starlings (*Sturnus vulgaris*) banded as young.

Brood Size	Number of Broods	Number of Young	Recovered More Than Three Months after Fledging		Relative Productivity (Percent Recoveries Times Brood Size)
			Number	Percent	
1	65	65	0	—	—
2	164	328	6	1.8	3.7
3	426	1,278	26	2.0	6.1
4	989	3,956	82	2.1	8.3
5	1,235	6,175	128	2.1	10.4
6	526	3,156	53	1.7	10.1
7	93	651	10	1.5	10.2
8	15	120	1	0.8	6.4
9	2	18	0	—	—
10	1	10	0	—	—

From D. Lack, 1948. Natural selection and family size in the starling. *Evolution* 2:95–110.

that produces more offspring, but whose offspring have a lower probability of survival.

For example, consider a population of plants in which, as a result of mutation, individuals appear that produce a new chemical compound in their leaves that is poisonous to a predator. However, that compound is produced at the expense of the energy that would have been invested in offspring, so fecundity is reduced. Then the mutation will spread in the population only if the probability of survival of the young compensates for their lowered number.

These are just some of the trade-offs that can take place in the apportionment of energy between vegetative growth (which increases survival and the ability to garner resources), defenses (which reduce mortality from predation and disease), and reproduction. The organism that contributes the largest number of offspring to the next generation will be selected. To do so, it must maximize the product of survival and reproduction, that is, $l_x \times m_x$. A fine example of this solution is clutch size in birds studied by the English ornithologist David Lack.

Lack studied the clutch size in a number of species of birds. He was interested in finding out whether individuals that laid the largest number of eggs left the greatest number of descendants, or whether there were trade-offs between number of eggs and survival probability. One of the species he studied was the European starling. Each one of several thousand nestlings was marked with an aluminum band, and the size of the brood of each nest

Table 13.7

Nestling weights of starlings (obtained in England in May 1947 by R. Carrick).

Age in Days	Average Weight per Nestling in Grams in		
	Brood of 2	Brood of 5	Brood of 7
4	28.8	—	21.4
6	48.3	45.1	38.2
15	88.0	77.6	71.4

From D. Lack, 1948. Natural selection and family size in the starling. *Evolution* 2:95–110.
The ranges of weight on day 15 were for brood of 2:87.5–88.5; for brood of 5:72.5–83.0; and for brood of 7:66.0–77.0.

was recorded. The number of young older than three months (the period when most early mortality occurs) that were subsequently recovered was used as a basis to estimate reproductive success. Table 13.6 shows the results of Lack's studies. We observe several characteristics. (1) There is variation in brood size, going from one to ten eggs per brood. (2) The frequency of brood sizes is not evenly distributed. Approximately one-third of all nests have five eggs, deviations lower and higher being less abundant, the more they differ from five eggs per clutch. (3) Survival probability (as inferred from birds recovered three months after fledging) is essentially equal for broods with three, four, and five nestlings but decreases with broods larger than five. (4) When the brood size is multiplied by the probability of survival, broods of five contribute the largest number of fledglings. The weights of nestlings in broods of different size (Table 13.7) demonstrate that the failure of the larger broods to produce more offspring can be attributed to undernourishment of the young. Hence the brood size of five represents the perfect compromise. Nestlings are slightly lighter than nestlings from smaller broods but not enough so to decrease their probability of survival. Consequently fitness is maximized by a brood of five.

There are situations in nature, however, where death is inevitable, regardless of the characteristics of the phenotype. Landslides, floods, fires, earthquakes, human destruction of the environment, extremely and unusually cold, wet, hot, or dry years—all bring in their wake destruction of all or most of the organisms living in an area, regardless of phenotype. In situations where events leading to high density-independent mortality are frequent, investing energy in potentially superior phenotypes will result in little or no advantage. Consequently, organisms that channel a greater proportion of their energy into offspring, increasing their birth rate, should be favored in such a situation. Because populations in these environments are susceptible to a high degree of density-independent mortality, a high birth rate is necessary to avoid extinction.

The kind of environment in which an organism lives, then, determines whether selection will favor individuals that invest more energy in a superior phenotype at the expense of a lower birth rate or organisms that produce a large number of offspring. This trade-off is the inevitable consequence of the limited resources available to every individual. Organisms in stable environments with relatively fewer offspring are said to be *K-selected;* the ones in unstable environments with relatively many offspring are said to be *r-selected.*

Another trade-off is between few large offspring and many small eggs or seeds of low birth weight. The former will have a greater chance of success, provided density-independent mortality is not a factor. This trade-off is also reflected in the example of the dandelion (Table 13.5), where the *D* biotype produces fewer but heavier seed.

In general, such organisms as ephemeral plants in deserts, and plants and animals of temporary ponds or of natural or disturbed fields, which are specialized to exploit temporary resources, are said to be *r*-selected. Organisms of more stable situations, however, tend to be *K*-selected.

The demographic characteristics of a species—life span, age-specific fecundity, age of first birth, etc.—are shaped by the forces of the physical environment, by interactions with other species, and by intraspecific competition, and they represent an adjustment to maximize individual fitness. Natural selection favors the individual that leaves the largest number of offspring. However, the largest number of offspring is seldom achieved by maximizing fertility at the expense of survivorship. On the contrary, many successful species have very low fertility. Witness the success of the human species.

In this chapter we have investigated some factors that regulate the growth of populations. In a world of limited resources, populations increase their numbers basically only in two situations. The first is following a catastrophe, when population numbers have been reduced and there is a temporary excess of resources. Some of these "catastrophes" are periodic and predictable. For example, in middle and high latitudes, winter temperatures are too low for the growth of plants. As plants go dormant and drop their leaves in the fall, animal populations must go dormant, reduce their numbers, or both. With the advent of spring, all populations can increase in numbers again. The same is true in regions with a pronounced dry season. Other catastrophes are less predictable—hurricanes, extremely cold winters, floods, etc.—but the process of shrinkage and growth is similar, though more irregular.

The second situation in which a population increases its number is the acquisition by its members of phenotypic/genotypic characteristics that permit them to garner a greater share of the available resources. In previous chapters (4–8) we have seen how these changes come about. However, they will spread in the population only if the resultant phenotype can increase in numbers. Such an increase in most natural situations means that some other phenotype of the same species or some other species decreases in numbers. Traditionally ecologists have studied the interactions between species and the changes in numbers taking place in populations, whereas population geneticists have concentrated on the dynamics of the genetic changes. It must be understood, however, that these constitute *one process* and that the genetic and demographic changes take place simultaneously.

Species differ in a number of life cycle characteristics, the principal ones being (1) the number and size of propagules (seeds, eggs, hatchlings); (2) the average life span; (3) the form of the survivorship curve (i.e., the age-specific mortality); (4) the age of first reproduction; and (5) the age-specific fecundity (i.e., whether the individual reproduces more than once and how many offspring are produced at each age). These aspects of the life cycle constitute the life history of an organism. Natural selection shapes the life history of an organism in the same way that natural selection shapes phenotypic characteristics.

The study of life histories and the concept of life history strategy is intimately linked to the study of demography. Demographic studies provide the raw data that allow us to characterize a life history. The interpretation of the life history in terms of environmental selective forces is the "strategy." A life history strategy is therefore a hypothesis that explains the origin of a set of demographic characteristics. As such it must make testable predictions. In the present context, the word "strategy" carries no teleological implication. Even though an organism's "strategy" allows it to cope successfully with the diversity of environmental forces, with competitors, and with predators during its lifetime, the organism does not plan its "strategy" in the way a general plans a campaign. Instead, only those individuals that possess the appropriate set of morphological and behavioral characteristics succeed in the struggle for existence and are in turn able to transmit those characteristics to their descendants. We now investigate how interactions with other species help shape the life history characteristics of organisms.

SUGGESTED FURTHER READING

General references

Boughey, A.S. 1973. *Ecology of Populations*, 2nd ed. New York: Macmillan.

den Boer, P.J., and G.R. Gradwell (eds.). 1971. *Dynamics of Populations*. Waageningen: Centre for Agricultural Publishing and Documentation.

Harper, J. 1977. *Population Biology of Plants*. London: Academic Press.

MacArthur, R.H. 1972. *Geographical Ecology: Patterns in the Distribution of Species*. New York: Harper and Row.

May, R.M. (ed.). 1976. *Theoretical Ecology: Principles and Applications*. Oxford: Blackwell Scientific Publications.

Solbrig, O.T. (ed.). 1979. *Demography and Evolution of Plant Populations*. Oxford: Blackwell Scientific Publications.

Whittaker, R.H. 1975. *Communities and Ecosystems*, 2nd ed. New York: Macmillan.

Articles and reviews

Einarsen, A.S. 1945. Some factors affecting ring-necked pheasant population density. *Murrelet* 26:39–44.

Fox, S.F. 1975. Natural selection on morphological phenotypes of the lizard *Uta stansburiana. Evolution* 29:95–107.

Frank, P.W. 1960. Prediction of population growth form in *Daphnia pulex* cultures. *Am. Nat.* 94:357–372.

Harper, J.L., and J. White. 1971. The dynamics of plant populations. In P.J. den Boer and G.R. Gradwell (eds.), *Dynamics of Populations.* Waageningen: Centre for Agricultural Publishing and Documentation.

Harper, J.L., and J. White. 1974. The demography of plants. *Ann. Rev. Ecol. Syst.* 5:419–464.

Lack, D. 1948. Natural selection and family size in the starling. *Evolution* 2:95–110.

MacLulich. 1937. Fluctuations in the numbers of the varying hare (*Lepus americanus*). *Univ. Toronto Studies, Biol. Ser.*: 43.

McNaughton, S.J. 1975. r- and K-selection in *Typha. Am. Nat.* 109:251–261.

Mertz, D.B. 1972. The *Tribolium* model and the mathematics of population growth. *Ann. Rev. Ecol. Syst.* 3:51–78.

Nicholson, A.J. 1948. Competition for food among *Lucilia cuprina* larvae. *Proc. 8th Intern. Cong. Entom.*, pp. 277–281.

Nicholson, A.J. 1958. The self-adjustment of populations to change. *Cold Spring Harbor Symp. Quant. Biol.* 22:153–173.

Sarukhan, J. 1979. Demographic problems in tropical systems. In O.T. Solbrig (ed.), *Demography and Evolution of Plant Populations.* Oxford: Blackwell Scientific Publications.

Tinkle, D.W. 1969. Evolutionary implications of comparative population studies in the lizard *Uta stansburiana.* In *Systematic Biology: Proceedings of an International Conference. Nat. Acad. Sci. Publ. 1692*, pp. 133–154.

Chapter 14

Competition and Niche Theory

Populations and species are genetic units that owe their structure to a common descent and to their ability to cross and interchange genes. A forest, a grassland, a desert scrubland are not genetic units but ecological units called *communities*. A community is an assemblage of populations of plants, animals, bacteria, and fungi that live in a certain area with a characteristic physical environment, interacting with one another and forming a distinctive system. Each community has its own characteristic species, its own physical structure, given mostly by the plants (especially in terrestrial communities), and its own function.

The study of communities is the province of ecology. Nevertheless, in order to understand population biology and evolution it is necessary to understand some of the basic interactions that take place between organisms in communities since these interactions—competition, predation, mutualism—and the physical environment are the basic selective forces that act on populations and shape their evolution.

In the previous chapter we discussed the growth and regulation of single populations. We now extend that discussion to explore what happens when two or more populations of different species live in the same environment. When two different species in a community are in contact with each other, three basic interactions can take place. (1) The two species may interfere or inhibit each other; that is, species may *compete* with each other. Such competition typically takes place because they use a common resource that is present in limited amount. (2) One species may be eaten by or may eat the other; that is, a species may be a *prey* or a *predator*. (3) Both species may collaborate in using the environment in such a way that a net gain accrues to both.

Each of these interactions has implications, not only in terms of the morphological and physiological characteristics of the species but also in terms of its life history. In this chapter we take up competition, in the next one predator-prey and mutualistic interactions.

COMPETITION

Whenever two individuals interfere or inhibit each other and the presence of one reduces the fitness of the other, they are said to be *competing*. The competing individuals may be members of the same species, or they may belong to different species. Competition is usually for a resource that is in short supply in the environment. The resource can be a foodstuff, a nesting site, space to germinate seeds, etc. Despite the image conjured up by popular writers of fierce combat for scarce resources by competing organisms, competition is more likely a passive process, in which the individual that is more efficient at harvesting gets the bulk of contested resources. The result of competition is that the fitness of one individual participating in the interaction is increased and that of the other decreased. Otherwise the interaction cannot be viewed as competition in the evolutionary sense.

For example, Solbrig and Simpson grew plants of two biotypes of the common dandelion (*Taraxacum officinale*) side by side in experimental cultures (Chapter 13). After two years 65 percent of the plants of the D biotype had survived in lime (basic) soils, 90 percent in peat (acid soils), 69 percent in sandy soils, and 46 percent in normal loamy soils, against respective percentages of 35, 51, 87, and 40 for the A biotype (Table 14.1). In addition, plants of the D biotype were larger, and as Figs. 14.1 and 14.2 show, the gain in D came largely at the expense of the A plants. The D biotype showed an advantage in its ability to compete with A for the resources, but the advantage was a function of the soil type: the greatest advantage accrued in loam soils and the least in sandy soils, where fewer plants of D than of A survived, but where D plants were nevertheless slightly larger and captured more of the resources.

Competition can be quite direct, as when two birds compete for a territory. It is then called *interference competition*. More indirect competition,

Table 14.1

Survivorship, individual dry weight, and total dry weight of dandelion plants in a
1 : 1 competition experiment.

Soil Type	Biotype	Initial Number	Final Number	Survival (%)	Dry Weight Plant (g)	Dry Weight Total (g)
Sand	A	80	70	87.5	1.64	114.8
	D	80	55	68.8	2.80	154.0
Lime	A	80	28	35.0	1.38	38.6
	D	80	52	65.0	6.82	354.6
Peat	A	80	41	51.3	1.42	58.2
	D	80	72	90.0	6.19	445.7
Loam	A	80	32	40.0	4.83	154.6
	D	80	37	46.3	13.70	506.9

called *exploitation competition*, occurs when both populations use the same
resources. Because it is normally advantageous for individuals to avoid competition, natural selection presumably promotes the use, by competing individuals, of different resources or the use of the same resource at different times
or in different ways. This can result in an increase in the ways organisms use
the environment and a concomitant increase in species diversity. Exactly how
important competition is as a force in creating diversity is still a matter of
debate. Some ecologists feel that competition is the most important force in
promoting diversity; others feel that its effect depends on the environment,
the level of resources, and the density of the potential competitors. The feeling
that the outcome of competition is not necessarily an all-or-none phenomenon
but a matter of circumstances is becoming prevalent.

To assess the effect of competition, we can proceed in two ways. If we are
dealing with competition for a resource by members of the same species, we
can determine the loss in fitness of each genotype and proceed with the calculations outlined in Chapter 6. In this way we will learn how the gene and
genotype frequencies in the population are affected. An example of this approach is the calculation of the values of s and t for the homozygous ST/ST
and CH/CH chromosome arrangements of *Drosophila pseudoobscura* (p. 198).

Another way to approach the study of competition is to investigate the
effect that competition has on the growth and reproduction of the groups that
are competing. What is studied here is the effect of competition on the numbers of individuals of each phenotype in the population. This is the usual
approach, and it is of course the only one when the competitors belong to
different species. However, note that no direct information on gene and genotype frequency is obtained this way.

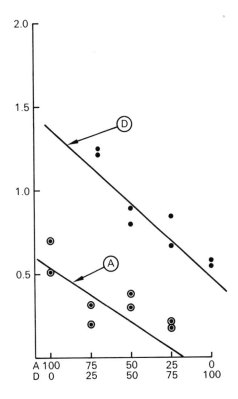

Figure 14.1

Average plant weight of dandelion plants of the *A* biotype and the *D* biotype grown in competition on sandy soils. Five different experiments were conducted, each with two replicates, with different initial mixtures: (1) all *A* plants; (2) 75% *A*, 25% *D*; (3) 50% *A*, 50% *D*; (4) 25% *A*, 75% *D*; (5) all *D* plants. Note that the weight of *A* and *D* plants when grown in pure culture is approximately the same, but that in mixtures *D* plants increase in size at the expense of *A* plants.

Models of competition and two-species growth

In the discussion of the logistic growth models (Chapter 13) we assumed that the growth of the population was regulated exclusively by resources. Usually many species in a community are competing for the available resources, so the growth rate of a species is normally also affected by the number of individuals of other species in the community that use that resource. Furthermore, although the interactions are mutual, they are seldom equal, some species affecting the growth of others strongly and others only minimally. The presence of one cow in a pasture affects the population of rabbits in the pasture (by consuming grass) more than one rabbit affects the population of cows.

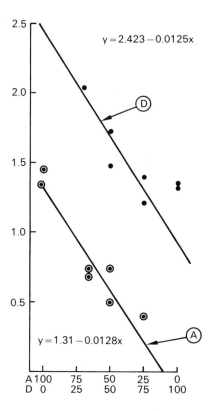

$$y = 2.423 - 0.0125x$$

$$y = 1.31 - 0.0128x$$

Figure 14.2

Average plant weight of dandelion plants of the *A* biotype and the *D* biotype grown in competition on loamy soil. Five experiments were conducted with different initial mixtures as in Fig. 14.1. The results are similar to those obtained when the plants are grown in sand. Compare with Fig. 14.1. The better soil (loam vs. sand) increases the advantage of *D*.

We now proceed to refine the logistic growth model to account explicitly for the presence of other species. The analysis was first presented by the American scientist A. J. Lotka in 1925 and independently and in greater detail by the Italian V. Volterra in 1926. We start with two competing species, which we will call A and B. From the previous discussion (Chapter 13) we know that when species are growing alone, their growth rates can be represented by

$$\text{species A:} \quad \frac{dN_A}{dt} = r_A N_A \left(\frac{K_A - N_A}{K_A} \right);$$

$$\text{species B:} \quad \frac{dN_B}{dt} = r_B N_B \left(\frac{K_B - N_B}{K_B} \right).$$

However, when the two species are together, the growth rate of A will be affected by how many individuals of B are present, and vice versa. An individual of B may be comparable in its demand on the resources to an individual of A, but it is more likely not to be so; it will then have either a greater or a smaller effect. We will represent by α_{12} the value of the effect of an individual of B on A and by α_{21} the value of the effect of an individual of A on B. It is now possible to write two simultaneous equations to represent the growth rates of species A and B when growing together, as follows:

$$\text{species A:} \quad \frac{dN_A}{dt} = r_A N_A \left(\frac{K_A - N_A - \alpha_{12} N_B}{K_A} \right);$$

$$\text{species B:} \quad \frac{dN_B}{dt} = r_B N_B \left(\frac{K_B - N_B - \alpha_{21} N_A}{K_B} \right).$$

We note that the growth rate of each species is modified by the average effect of each individual of the other species times the total number in the other species. Although one rabbit in a pasture affects minimally the growth rate of a population of cows, many rabbits have a very strong impact, as every rancher knows.

The model can be extended to n species, where it becomes

$$\frac{dN_i}{dt} = r_i N_i \left(\frac{K_i - N_i - \sum_{j \neq i}^{n} \alpha_{ij} N_j}{K_i} \right).$$

Here n is the number of species, r_i is the intrinsic rate of increase of species i, K_i is the carrying capacity of the ith species, N_i is its population density, and α_{ij} is the competition coefficient of species j on species i.

At equilibrium, all dN/dt must be equal to 0; that is, no species increases or decreases in number. The resulting population densities are

$$N_i^* = K_i - \sum_{j \neq i}^{n} \alpha_{ij} N_j^*,$$

where the asterisk represents the number at equilibrium.

This competition model greatly oversimplifies the process of interspecific competition. An imaginary example can give some insight into the process. We start in a field with a pair of buffalo and a pair of cows. Both eat grass, and we will assume that the amount of available grass is constant throughout the year. At the beginning of the experiment both species will be growing because there are plenty of resources for both. Cows, having shorter generations and bigger litters, have a greater intrinsic growth rate (r_B) than buffalo (r_A), and consequently their number increases faster than that of buffalo. However, each buffalo eats more grass and is known to be more efficient in digesting and converting it into meat, so α_{12}, the effect of one cow on one buffalo, is less than α_{21}, the effect of a buffalo on a cow. As the supply of grass becomes limiting, the interactions between the two species become more significant. Eventually competition results in one of three possible outcomes: (1) The buffalo win out and the cows become extinct. (2) The cows win out and the buffalo become extinct. (3) An equilibrium is reached where buffalo and cows coexist. The question is, can we predict with our model which of these outcomes will occur?*

If there are only individuals of species A in the community, that species will grow until it reaches its carrying capacity, K_A, when $dN_A/dt = 0$. However, if there are individuals of species B in the community, species A will reach its asymptote ($dN_A/dt = 0$) at a population size smaller than K_A. The precise number of A individuals when the growth rate of A is 0 is a function of the number of individuals of B and the competition coefficient α_{12}:

$$\frac{dN_A}{dt} = 0 = r_A N_A \left(\frac{K_A - N_A - \alpha_{12} N_B}{K_A} \right),$$

which implies

$$K_A - N_A - \alpha_{12} N_B = 0,$$

when r_A, N_A, K_A, and N_B have values greater than 0, so that at the asymptote, $N_A = K_A - \alpha_{12} N_B$. Likewise for B, $N_B = K_B - \alpha_{21} N_A$ at the asymptote.

We now make use of a graph (Fig. 14.3) whose coordinates are N_A and N_B, the number of individuals in species A and B. On this graph we will plot the zero growth rate line for each of the species, starting with species A. If we plot N_A on the abscissa and N_B on the ordinate, the N_A intercept for species A—that is, the value of N_A when $N_B = 0$—will be K_A. The N_B intercept—that is, the value of N_A when $N_B = K_B$ (and consequently theoretically N_A should be 0)—will be K_A/α_{12}. By the same reasoning, if the zero growth rates for species B are plotted on the same graph, the N_B intercept will be K_B, and the N_A intercept will be K_B/α_{21}. The zero growth line is obtained by drawing a line between the N_A and N_B intercepts of each species.

Both species will increase in numbers until they reach the zero growth isocline. The species represented on the abscissa meets the isocline horizontally, and the one represented on the ordinate will do it vertically (Fig. 14.3). The four possible types of relations between species A and B lead to four different outcomes (Fig. 14.4), depending on the values of K_A, K_B, α_{12}, and α_{21}.

Case 1. The carrying capacity of species A (K_A) is larger than the value K_B/α_{21}, and the carrying capacity of species B (K_B) is larger than K_A/α_{12}. Consequently, $\alpha_{12} > K_A/K_B$ and $\alpha_{21} > K_B/K_A$.

If the zero isoclines are graphed, they cross. At that point both species have zero growth and can coexist. If they are disturbed in any way, however,

---

* In real life the grass would become extinct first, resulting in the demise of both cows and buffalo. We assume that the grass is not trampled to death in this hypothetical experiment.

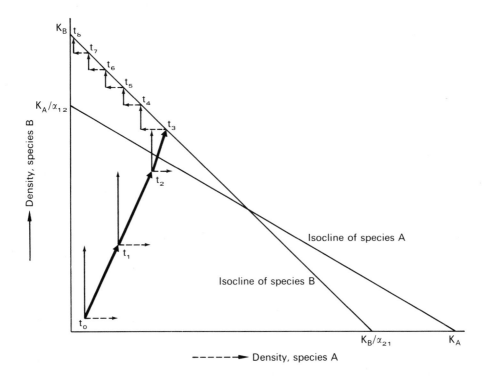

Figure 14.3

Species competition graph. The density of species A is graphed against that of species B. The growth of A during each time period is indicated by the horizontal arrows (dashed), that of B by the vertical arrows (solid). At t_1 there is a component of each, as shown. When one species, in this case A, reaches its isocline, it can no longer grow. The other species can continue to grow if it has not reached its isocline, as B has not. If the total number goes outside the isocline of A, as at t_3, the density of A decreases until it reaches its isocline. In this case B can still grow at the end of each time interval, so A becomes extinct.

one of them will increase in numbers while the other decreases. Case 1, then, is an unstable equilibrium that will lead to the extinction of one of the two species, depending on which of them first exceeds the zero isocline of the other. This depends in nature on the historical accident of which species arrived at the spot first.

Case 2. Here K_A is smaller than K_B/α_{21} and $K_B < K_A/\alpha_{12}$, and consequently, $\alpha_{12} < K_A/K_B$ and $\alpha_{21} < K_B/K_A$.

  In this case the isoclines also cross, but the equilibrium is a stable one. If B reaches its zero isocline before A, to the left of the equilibrium point, A will continue to grow. At higher numbers of A, B must decrease, until the equilib-

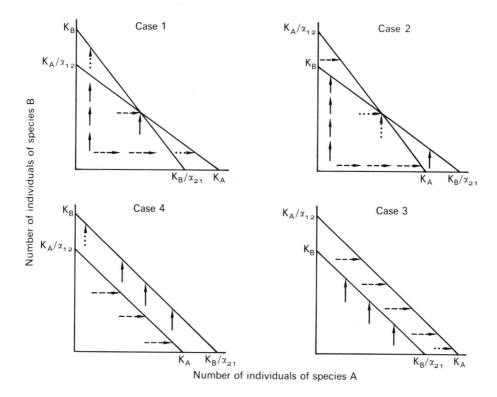

Figure 14.4

Two-species competition graphs. Case 1: $K_A > K_B/\alpha_{21}$ and $K_B > K_A/\alpha_{12}$. An unstable equilibrium may be reached, and either A or B eventually becomes extinct. Case 2: $K_A < K_B/\alpha_{21}$ and $K_B < K_A/\alpha_{12}$. A stable equilibrium is reached with both A and B present. Case 3: $K_A > K_B/\alpha_{21}$ and $K_B < K_A/\alpha_{12}$. The only survivor is A. Case 4: $K_A < K_B/\alpha_{21}$ and $K_B > K_A/\alpha_{12}$. The only survivor is B.

rium point is reached; A cannot increase beyond that point. Similarly, if A reaches its isocline first, B will continue to increase, forcing a decrease in A, until that equilibrium point is reached.

Case 3

$$\alpha_{12} < \frac{K_A}{K_B}\left(K_B < \frac{K_A}{\alpha_{12}}\right) \quad \text{and} \quad \alpha_{21} > \frac{K_B}{K_A}\left(K_A > \frac{K_B}{\alpha_{21}}\right).$$

Here A is the sole survivor under any initial conditions.

Case 4

$$\alpha_{12} > \frac{K_A}{K_B}\left(K_B > \frac{K_A}{\alpha_{12}}\right) \quad \text{and} \quad \alpha_{21} < \frac{K_B}{K_A}\left(K_A < \frac{K_B}{\alpha_{21}}\right).$$

Here B is the sole survivor under any initial conditions.

These models of two species growing in competition predict that two species using the same resource can coexist only under the condition that the minority species is always capable of growing at a faster rate than the majority species. This means that in each species in some way the effect on population growth by members of its own species is greater than the effect of individuals of the other species. Such a set of circumstances is unlikely to occur unless each species has resources not used by the other. In all other cases no coexistence is possible. The Russian population biologist G.F. Gause was the first to present this analysis. What is known as the "exclusion principle," sometimes called "Gause's exclusion principle," is attributed to him. This principle simply states that two species cannot coexist if they use *exactly* the same resource (Chapter 11).

The graphic model just presented is useful in gaining a feeling for the interactions of two species. However, as with the growth models of a single species, its validity depends on how realistic the underlying assumptions are. And not only are r_A, r_B, K_A, and K_B again treated as constants, but two new constants are added: α_{12} and α_{21}, the competition coefficients of A on B and B on A. The competition coefficients represent a composite of many other things, such as how efficiently one species uses resources, how well it physically competes, and how good it is at avoiding the effects of the other species. All these properties no doubt vary from one individual to the next in the population. They should therefore be influenced by selection. In that case the average value of the competition coefficient will change with time and with the size of the population. Furthermore, it will be affected by the spatial distribution of the two species and by the age of the individuals.

Nevertheless, these very simple models allow us to get some understanding of how the numbers of individuals in one population can affect the numbers of individuals and the probability of survival of members of a population of another species.

Whenever two species are made to grow together under constant conditions in the laboratory, an artificial situation is created, since in nature the environment is never constant and competition is rarely a one-on-one affair. However, observation of competition in nature is close to impossible, given the large number of variables. Nevertheless, so-called perturbation experiments, in which one species is artificially removed from or added to a community, can be very illuminating. In spite of all these reservations, a great deal of insight has been obtained from both laboratory and field experiments on competition.

G.F. Gause studied the growth rate and competitive interactions of two species of ciliates, *Paramecium aurelia* and *P. caudatum*, in a series of experiments published in 1934. Containers were inoculated with both species, and the amount of food added was kept constant. Both populations quickly expanded their numbers, but after a short time *P. aurelia* grew at the expense of *P. caudatum* until the latter disappeared from the culture (Fig. 14.5). In 1969

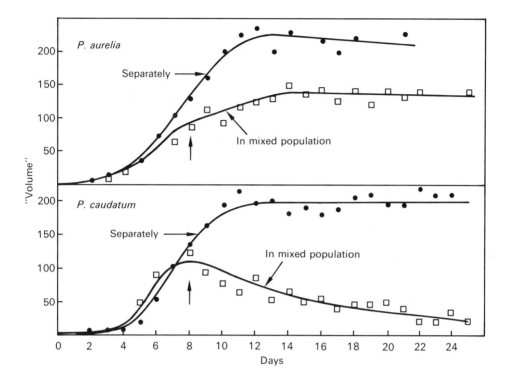

Figure 14.5

Gause's competition experiments with *Paramecium aurelia* and *P. caudatum*. The total volume (an estimate of biomass) for each species, grown in a medium with bacteria, changes with time. The growth curves for each species grown by itself and in combination are shown. (From G.F. Gause, 1934. *The Struggle for Existence*. Baltimore: Williams and Wilkins)

the American ecologist John Vandermeer repeated these experiments with some notable refinements, using three species of *Paramecium*, *P. aurelia*, *P. caudatum*, and *P. bursaria*, and a species of *Blepharisma*, another ciliate. As a first step, he grew the species singly in cultures, obtaining the results seen in Fig. 14.6. As one can observe, the species grew in a logistic fashion. From these experiments Vandermeer calculated the values of r, K, and N_0 for each species (Table 14.2). He then grew the species in pairs, obtaining the results depicted in Fig. 14.7. In these experiments one species always outcompeted the other. From these results Vandermeer calculated the values of α_{12} and α_{21} for each species in each experiment (Table 14.3). As a final step, he used the parameters of Tables 14.2 and 14.3 to predict the behavior of each species in a mixture of these four species under the assumptions that r, K, α_{12}, α_{21} are constants and that there are no higher-order interactions, i.e., effects due to the presence

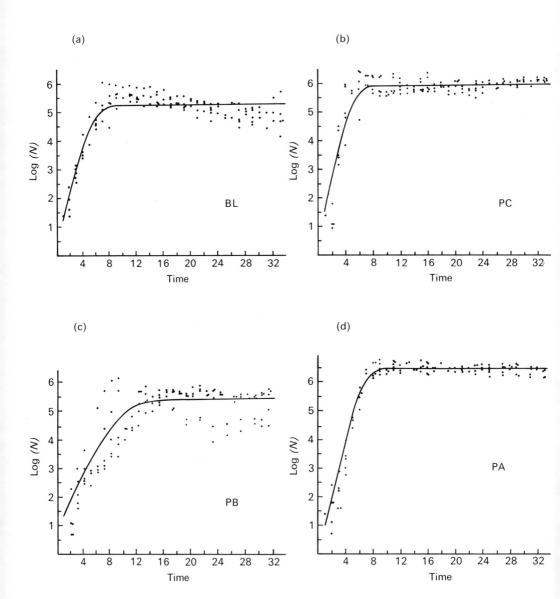

Figure 14.6

Growth curves for four ciliates grown in pure culture. (a) *Blepharisma* sp. (BL);
(b) *Paramecium caudatum* (PC); (c) *P. bursaria* (PB); (d) *P. aurelia* (PA). (From
J.H. Vandermeer, 1969. The competitive structure of communities: an experimental
approach with Protozoa. *Ecology* 50:362–371)

Table 14.2

Parameters of the logistic equation.

Species	r	K	N_0
Paramecium aurelia	1.05	671	2.5
P. caudatum	1.07	366	5.0
P. bursaria	0.47	230	5.0
Blepharisma	0.91	194	3.0

From J.H. Vandermeer, 1969. The competitive structure of communities: an experimental approach with Protozoa. Ecology 50:362–371.

of more than one competitor. The Gause equation extended to m species under these assumptions can be written as

$$\frac{dN_i}{dt} = r_i N_i \left(\frac{K_i - N_i - \sum_{j=1}^{m} \alpha_{ij} N_j}{K_i} \right), j \neq i,$$

where dN_i/dt is the growth rate of the ith species, m is the number of species in the community, α_{ij} is the effect of the jth species on the ith species, r is the intrinsic rate of natural increase, N is the number of individuals, and K is the carrying capacity. Vandermeer's results are indicated in Fig. 14.8. We see that the experimental data indicated by the dots in the figure are in very good accordance with the theoretical predictions indicated by the solid line. These results were somewhat surprising since some higher-order interactions were expected, but they are an elegant demonstration of the predictive power of Gause's competition equations, at least with unicellular organisms in laboratory cultures.

Competition in natural situations is much more difficult to study since, in addition to the competitive interactions between the species under study, there are competition effects from other species, as well as predation, that cannot always be controlled. The physical environment also has an effect. Nevertheless, there are a number of examples that very elegantly document competition in nature. One of the most elegant examples is that of the distribution of two barnacle species, Balanus balanoides and Chthamalus stellatus. These species are found growing in the intertidal zone on rocky shores. J.H. Connell, who studied these species in detail, found that although the larvae of both species showed a great deal of overlap in the area where they settled, the adults of Chthamalus were always found in the upper (and more exposed) part of the shore and those of Balanus in the lower (and less exposed) part. To test the relative effects of desiccation and of competition, Connell measured the growth rates of the two species in several places and the survival of young and older

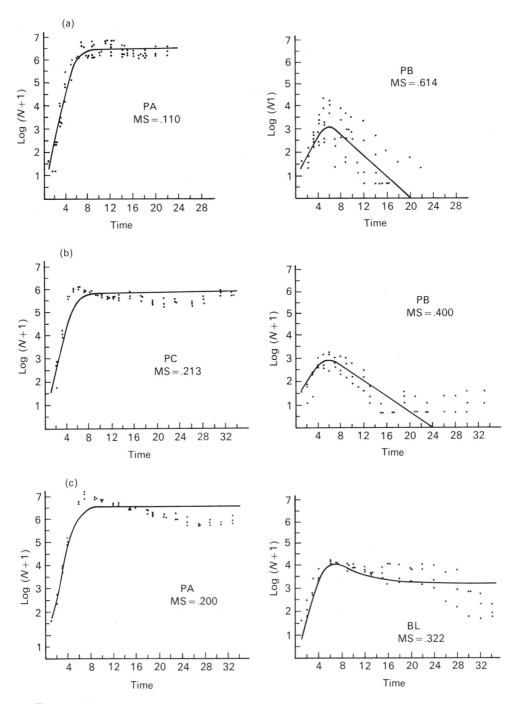

Figure 14.7

Growth curves for the species shown in Fig. 14.6, grown together in pairs. (a)
Paramecium aurelia (PA) and *P. bursaria* (PB); (b) *P. caudatum* (PC) and PB;

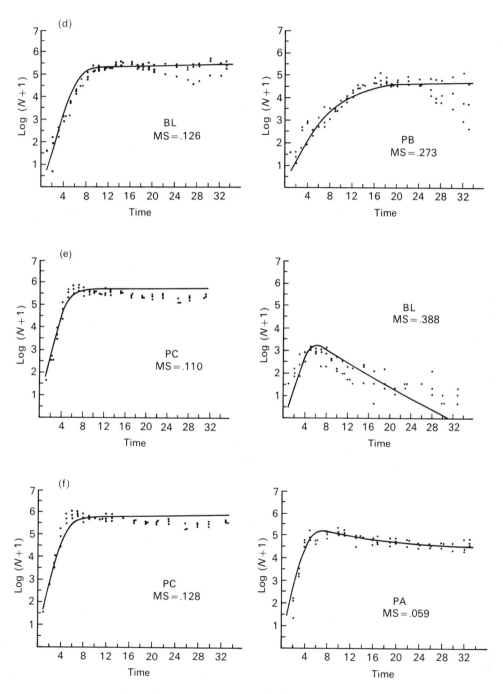

(c) PA and *Blepharisma* sp. (BL); (d) PB and BL; (e) PC and BL; (f) PC and PA. (From J.H. Vandermeer, 1969. The competitive structure of communities: an experimental approach with Protozoa. *Ecology* 50:362–371)

Table 14.3

Values of competition coefficients α_{ij}.

| | Second Species (j) | | | |
First Species (i)	Paramecium aurelia	P. caudatum	P. bursaria	Blepharisma
Paramecium aurelia	1.00	1.75	−2.00	−0.65
P. caudatum	0.30	1.00	0.50	0.60
P. bursaria	0.50	0.85	1.00	0.50
Blepharisma	0.25	0.60	−0.50	1.00

From J.H. Vandermeer, 1969. The competitive structure of communities: an experimental approach with Protozoa. *Ecology* 50:362–371.

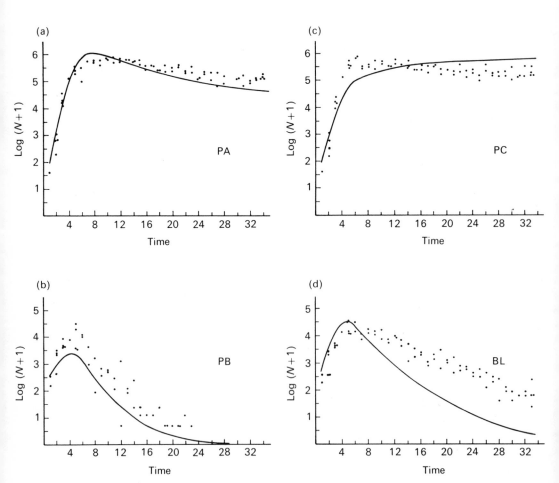

Table 14.4

Mortality rates of young and old *Chthamalus stellatus* on transplanted stones.

Level	Treatment	Number of Chthamalus Present in June 1954		Percent Mortality over One Year	
		1952 or Older	1953 Year Group	1952 or Older	1953
Mean tidal level in a tidal pool	Balanus removed	41	50	37	44
	Undisturbed	31	60	71	95
1 foot below mean tidal level	Balanus removed	3	51	0	35
	Undisturbed	16	69	31	90
2.5 feet below mean tidal level	Balanus removed	45	25	36	40
	Undisturbed	8	22	75	86

From J.H. Connell, 1961. The influence of interspecific competition and other factors on the distribution of the barnacle *Chthamalus stellatus*. *Ecology* 42:710–723.

barnacles on stones that had been placed at different levels. As Tables 14.4 and 14.5 show, *Balanus* cannot survive the desiccating environment of the high intertidal area. *Chthamalus*, however, can survive at the low levels of the intertidal shore in the absence of *Balanus*. *Balanus* grows faster than *Chthamalus* above the mean tide level and is able to overgrow or crowd out *Chthamalus* completely in the lower intertidal area. Consequently, the distribution of these two species is the result of the interplay of biological and physical factors. *Balanus* is able to outcompete *Chthamalus* in the lower and middle intertidal, but since it cannot resist the desiccation of the upper intertidal, it leaves an area where the more desiccation-resistant *Chthamalus* can grow (Fig. 14.9).

A similar situation is found among certain species of plants studied by the German ecologist Ellenberg. In Europe certain species of plants are encountered only in acid soils. They are known as acidophylls, "acid-loving"

Figure 14.8

◄ Growth curves of the four species in Fig. 14.6, when grown all in one culture. (a) *Paramecium aurelia*; (b) *P. bursaria*; (c) *P. caudatum*; (d) *Blepharisma* sp. (From J.H. Vandermeer, 1969. The competitive structure of communities: an experimental approach with Protozoa. *Ecology* 50:362–371)

Table 14.5

Growth rates of 25 *Chthamalus stellatus* and 27 *Balanus balanoides*.

	Size		Absolute Growth
	June 11, 1954	Nov. 3, 1954	Rate per Day × 100
A. *Chthamalus*	2.49	4.24	1.21
B. *Balanus*	1.87	4.83	2.04

From J.H. Connell, 1961. The influence of interspecific competition and other factors on the distribution of the barnacle *Chthamalus stellatus*. *Ecology* 42:710–723.

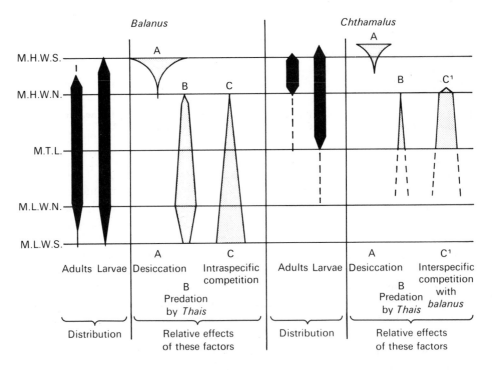

Figure 14.9

The intertidal distribution of adults and newly settled larvae of *Balanus balanoides* and *Chthamalus stellatus* at Millport, with a diagrammatic representation of the relative effects of the principal limiting factors. (From J.H. Connell, 1961. The influence of interspecific competition and other factors on the distribution of the barnacle *Chthamalus stellatus*. *Ecology* 42:710–723)

plants. Ellenberg grew a number of acidophyll species along a gradient of soils from acid to basic and measured their growth. He found that plants growing on neutral soils (pH = 7) grew faster and were larger than individuals of the same species on acid soils. He then repeated the same experiment with plants normally found in neutral soils and obtained similar results. However, these species were not able to grow on the acid soils or grew very poorly. He repeated the experiment, this time with mixtures of both types of species. He discovered that, when the two types were grown together, the plants from neutral soil outcompeted the acidophylls in the neutral soils, but the opposite happened in the acid soils (Fig. 14.10). In other words, acidophylls are not really "acid-loving"; rather, they tolerate the acid soils and are excluded from the more favorable neutral soils by competition.

Even though it is very difficult to study competition in nature, observation tells us that it is a potent, ever-present force. Whenever a species is removed from an area (without concomitant alterations of the environment), some other species immediately occupies the area. The conclusion is inevitable that the invading species had been kept out by competition. But competition is also a potent evolutionary force. Consider two populations of different species that use a common resource, such as insect-eating lizards. Assume that individuals of one species are slightly larger than individuals of the other species and can consume larger insects. However, there is still a large potential overlap in the insects they eat, and consequently, there is competition for them. If we analyze Fig. 14.11, we see that the small animals of species A and the large ones of species B do not have to contend as much with competitors as do those of both species that are of more intermediate sizes. Consequently, the fitness of the former animals will be slightly higher, they will tend to increase in number, and the populations will diverge until competition is minimized or eliminated. This process is similar to character displacement between related species, discussed in Chapter 12. We will call this *ecological divergence*.

THE CONCEPT OF THE NICHE

The examples of the barnacles and of the "acid-loving" plants point out that specialized phenotypes that harvest resources efficiently may gain competitive advantages and eventually exclude less efficient phenotypes. This idea is embodied in the concept of the niche. The definition given here is attributed to the great American ecologist G. Evelyn Hutchinson from Yale University.

Imagine a hypervolume (a volume with many dimensions) with an axis for every relevant variable in the life history of organisms, such as food, time, space, temperature, and light. Any species can be mapped into such a hypervolume by the values of all the variables that permit the organism to exist.

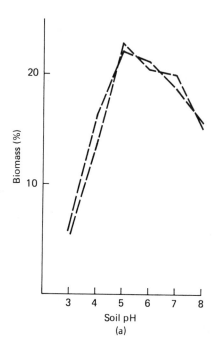

(a)

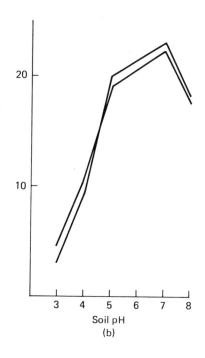

(b)

Figure 14.10

Percent relative biomass of plants of wild radish (*Raphanus raphanistrum*, dotted line) and wild charlok (*Sinapsis arvensis*, solid line), under different soil pH's in pure culture (a and b) and mixture (c). Equal numbers of plants were planted in all soils, but as the graphs show, plants grew more in soils of certain acidities. Note that without competition both species did well in the pH 5–7 range, but with competition they did worst in that range. (From H. Ellenberg, 1952. Physiologisches und ökologisches Verhalten derselben Pflanzenzarten. *Ber. Deut. Bot. Ges.* 65:350–361)

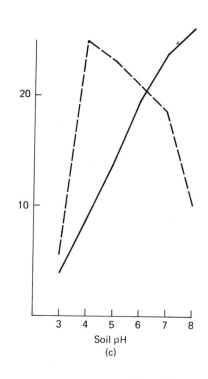

(c)

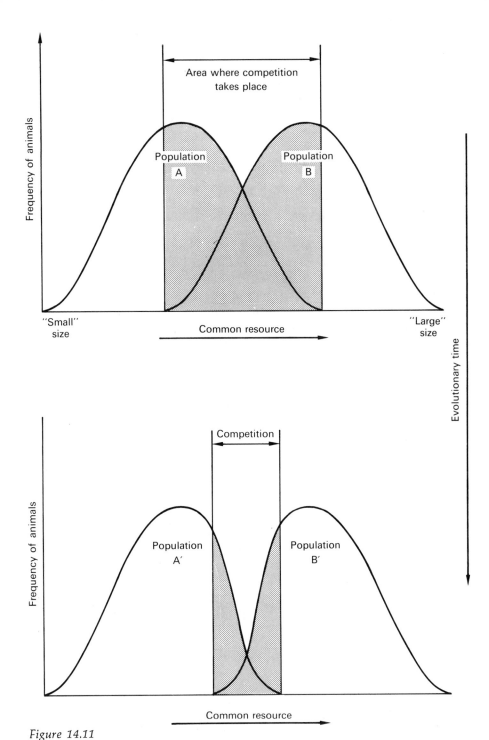

Figure 14.11

Schematic representation of ecological divergence. Since individuals of both species in the nonoverlapping areas do not have to compete with members of the other species, their fitness is greater than that of the individuals in the area of overlap. As time goes by, both species will tend to occupy nonoverlapping ecological niches.

For example, a species may be able to remain alive only at temperatures between 0° and 20°C, at humidities between 80 and 90 percent, in air with more than 5 percent oxygen, with leaves of certain species of plants available to eat, and so on. The hyperspace bounded by the values of the variables that the organism requires to live is the *fundamental niche* of the species. However, more than one species may potentially occupy parts of that hypervolume (Gause's principle of competitive exclusion says that no two species have *exactly* the same fundamental niche). When more than one species is present, each species will occupy a smaller volume than their fundamental niches because of competition. The space a species occupies when in competition with others is called its *realized niche* (Fig. 14.12). The realized niche of a species is equal to or smaller than the fundamental niche but never larger. The fundamental niche has absolute boundaries, but the realized niche can change as a function of the number and abundance of competitors and predators.

The concept of the niche is an abstraction that allows us to visualize better the evolutionary and ecological factors affecting the biology of a species, and to define better the resource axes. If the fundamental niches of species in a

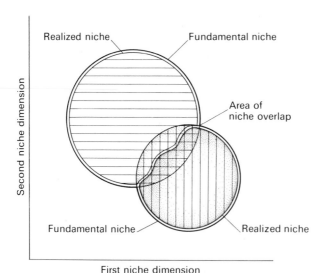

Figure 14.12

Realized vs. fundamental niche. The outer circle indicates (in two dimensions) the areas of niche space where the two species can *potentially* exist. In the area of overlap, when *both* are present, they displace each other into a smaller area of their respective niche space. The smaller area is the realized niche.

community are known, predictions regarding the intensity of competition and possible outcomes can be made. However, that state has not yet been reached. But ecologists are in the process of describing accurately the resource axes of many organisms. A good example of resource partitioning among species, which illustrates the usefulness of the niche concept, is a study with lizards undertaken by Thomas W. Schoener of the University of Washington.

Four species of lizards of the genus *Anolis* live in the small (5 square miles) island of South Bimini in the Bahama Archipelago. These four species live together on the island, which has very little relief and a rather uniform vegetation. What allows these related species with similar diets to coexist? Schoener studied quantitatively several dimensions of the niches of the four species: (1) preferred vegetation type; (2) diameter of the perch on which the lizard rests and height above the ground of the perch; (3) color of the bark of the perch; (4) frequency of time spent in leaves; (5) size of prey taken; and (6) species of prey taken. He found that all species overlap somewhat in each of the niche dimensions: There was no one axis used exclusively by one species. But if all axes were taken together, each species occupied a defined and unique part of the niche hypervolume. For example, *Anolis sagrei* is partly terrestrial and occurs more often on small rather than large perches near the ground; *A. distichus* prefers to perch on trunks and large branches of medium-to-large trees; *A. angusticeps* perches preferentially on small twigs high above the ground; and *A. caroliniensis* is found mostly on leaves or on the adjacent twigs or branches. Although all four species are of about the same size, the species differ somewhat in the maximum head length that they can attain: 19.7 mm for *A. caroliniensis* (average 18 mm); 16.4 mm for *A. angusticeps* and 15.5 mm for *A. sagrei* (averages 15.7 and 14.4 mm, respectively); and 13.5 mm for *A. distichus* (average 12.6 mm). The head length is important since the largest prey a lizard can take depends on its head size. These lizards eat a variety of insects and spiders, as well as some plant material, such as small fruits. However, although there are some differences in the maximum size the males attain, there is considerable overlap among smaller animals (as well as females and juveniles). By analyzing the stomach contents of a large number of lizards, Schoener determined that within the same species, the larger animals usually eat fewer but larger items of food, and that they have a greater range of food sizes. He also found that *A. distichus* prefers ants over other insects, ants constituting 75–90 percent of its diet, whereas *A. sagrei* eats considerably more fruit than the other three species. The diets of *A. caroliniensis* and *A. angusticeps* are fairly similar. However, when Schoener determined the relative abundance of these lizards in relation to different types of vegetation (Table 14.6), he found that species that were found most commonly in the same type of vegetation (*angusticeps-distichus; caroliniensis-sagrei*) differ most in their diets, and that the pair with the most similar diets (*angusticeps-caroliniensis*) prefer different kinds of vegetation. The last point is particularly

Table 14.6

Abundance of *Anolis* species in vegetation types of South Bimini. A = abundant relative to other areas where the species was seen; R = present but rarely seen; 0 = not observed.

Habitat	Species			
	A. angusticeps	*A. caroliniensis*	*A. distichus*	*A. sagrei*
Blackland (interior)	A	R	A	A
Coccothrinax shrub	0	A	R–A	A
Mangrove (*Rhizophora*)	0	0	0	R
Mangrove (*Avicennia*)	0	R	0	R
Salicornia tide flats	0	0	0	R
Uniola strand	0	R	0	R
Coastal rock	0	0	0	0
Incipient blackland	A	A	A	A
Number of habitats where seen	2	5	3	7

From T.W. Schoener, 1968. The *Anolis* lizards of Bimini: resource partitioning in a complex fauna. *Ecology* 49:704–726.

interesting since it points out that two species can use the same kind of resource if they are separated physically or temporally, as would be predicted from the concept of the multidimensional niche, where time and space are different axes.

The concept of the multidimensional niche allows the quantification of the activities of a species (feeding, caring for young, reproducing, sleeping) in time and space and allows the precise determination of how species interact in a community. Each species can be represented as a cloud in the hyperspace of the community (Fig. 14.13). With this information in hand, it is possible to ask how many species can theoretically coexist in a given area. The number and relative abundance of species in an area is the species diversity. Referring back to the multidimensional niche, we can see that species diversity is a function of the number of resource axes and the variety and abundance of each resource, called the *length* of the resource axis. For example, if nutrients are uniformly low in an area, the resource cannot be divided in as many ways as they can when there is a great richness of nutrients. The diversity of species is also a function of niche width, that is, the variety of resources used by a species, which in turn is a function of the relative abundance of the resource in time and space. Niche theory is an area of ecology that is being developed intensively both mathematically and experimentally, and that promises to combine the theoretical and empirical results of studies in population biology with the more classical aspects of community ecology.

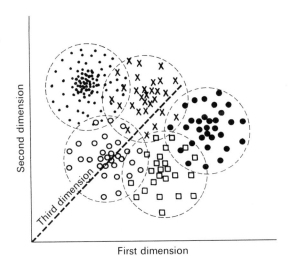

Figure 14.13

Species niches as "clouds" in niche space. Each point represents a population of a species in a three-dimensional niche space. Note that the borders are not sharp; rather, they interdigitate because of varying outcomes of local competition due to such facts as initial occupation, varying numbers, environmental fluctuations, and chance.

SUGGESTED FURTHER READING

General references

Emlen, J.M. 1973. *Ecology: An Evolutionary Approach*. Reading, Mass.: Addison-Wesley.

Gause, G.F. 1934. *The Struggle for Existence*. Baltimore: Williams and Wilkins.

Hutchinson, G.E. 1965. *The Ecological Theater and the Evolutionary Play*. New Haven: Yale University Press.

McNaughton, S.J., and L.L. Wolf. 1973. *General Ecology*. New York: Holt, Rinehart and Winston.

May, R.M. (ed.). 1976. *Theoretical Ecology: Principles and Applications*. Oxford: Blackwell Scientific Publications.

Ricklefs, R.E. 1973. *Ecology*. Portland, Ore.: Chiron Press.

Slobodkin, L.B. 1966. *Growth and Regulation of Animal Populations*. New York: Holt, Rinehart and Winston.

Articles and reviews

Colwell, R.K., and E. Fuentes. 1975. Experimental studies of the niche. *Ann. Rev. Ecol. Syst.* 6:281–310.

Connell, J.H. 1961. The influence of interspecific competition and other factors on the distribution of the barnacle *Chthamalus stellatus*. *Ecology* 42:710–723.

Ellenberg, H. 1952. Physiologisches und ökologisches Verhalten derselben Pflanzenarten. *Ber. Deut. Bot. Ges.* 65:350–361.

MacArthur, R.H., and R. Levins. 1967. The limiting similarity, convergence, and divergence of co-existing species. *Am. Nat.* 101:377–385.

Roughgarden, J. 1974. Niche width: biogeographic patterns among *Anolis* lizard populations. *Am. Nat.* 108:533–540.

Schoener, T.W. 1968. The *Anolis* lizards of Bimini: resource partitioning in a complex fauna. *Ecology* 49:704–726.

Solbrig, O.T., and B.B. Simpson. 1974. Components of regulation of a population of dandelions in Michigan. *J. Ecol.* 62:473–486.

Solbrig, O.T., and B.B. Simpson. 1977. A garden experiment on competition between biotypes of the common dandelion (*Taraxacum officinale*). *J. Ecol.* 65:427–430.

Stearns, S.C. 1977. The evolution of life history traits. *Ann. Rev. Ecol. Syst.* 8:145–172.

Steiner, W.M. 1977. Niche width and genetic variation in Hawaiian *Drosophila*. *Am. Nat.* 111:1037–1045.

Vandermeer, J.H. 1969. The competitive structure of communities: an experimental approach with Protozoa. *Ecology* 50:362–371.

Vandermeer, J.H. 1972. Niche theory. *Ann. Rev. Ecol. Syst.* 3:107–132.

Wilbur, H.M. 1977. Propagule size, number, and dispersion pattern in *Ambystoma* and *Asclepias*. *Am. Nat.* 111:43–68.

Yoshiyame, R.M., and J. Roughgarden. 1977. Species packing in two dimensions. *Am. Nat.* 111:107–121.

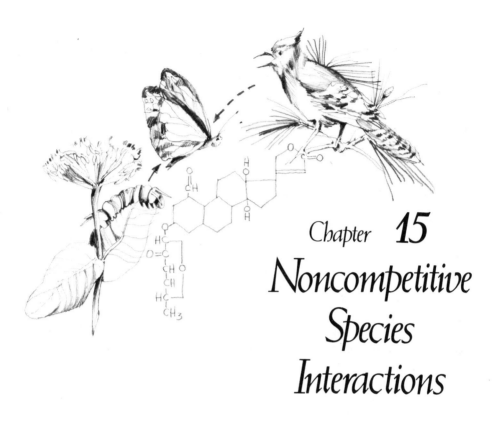

Chapter **15**

*Noncompetitive
Species
Interactions*

Most species of plants and animals are potential food to some animal (and a few plants). Consequently, plants and animals have evolved a large variety of defenses. Poisonous or distasteful flesh, hard and horny skins, spines, stinging hairs, the ability to inflict poisonous or painful stings or bites, the ability to run or fly swiftly or hide (aided by cryptic colors or shapes)—these are just some of the many chemical, morphological, and behavioral adaptations of plants and animals to avoid predation. Predators in turn have evolved their own adaptations to break down the prey's defenses: detoxifying mechanisms; powerful mandibles and digestive systems; strong, powerful claws, arms, or jaws to overpower the prey; extremely good eyesight or hearing; the ability to run or fly faster than the prey; etc. The defenses of the prey and the weapons of the predators point out a basic and important biological fact, namely, that capturing food and escaping from predators are two basic occupations of organisms, and they are two of the most powerful selective pressures molding the characteristics of organisms.

But not only do species compete with or eat each other; they also collaborate with each other. Many plants require insects to effect pollination. Rumi-
nants (cows, deer, camels) require bacteria in their guts to be able to digest
cellulose and would starve without them, and the same applies to termites.
Even the human species requires the colon bacterium *Escherichia coli* to obtain vitamin K. Consequently, collaborative interactions also act as strong
selective forces in shaping the features of organisms. In this chapter we explore these noncompetitive interactions.

PREDATOR-PREY INTERACTIONS

A predator is an organism that consumes all or part of another live organism
for his food needs. There are a whole series of such interactions, from the
large carnivore, such as a lion or a wolf, that hunts and kills his prey to the
small insect, such as a mosquito, that sucks blood and has a very small effect
on the prey. The prey need not be another animal; it can be a plant. In that
case the predator is called a herbivore. Normally herbivores do not kill their
prey but limit themselves to consuming some tissues (leaves, fruits, roots).
Nevertheless, herbivores can have devastating effects. Consider, for example,
the Japanese beetle's effect on the vegetation of New England. If the predator
develops a very close association with the prey, we refer to it as a *parasite* or
parasitoid, depending on the level of damage inflicted on the prey.

In order to simplify the theoretical presentation of predator-prey interactions, we restrict the discussion to the animal that kills its prey (and therefore removes it from the population of prey).

The interaction between predator and prey is much more complex than
that between competitors. In effect, individual selection will favor the efficient
predator. However, an efficient predator can totally destroy the prey and bring
about his own demise when the prey becomes extinct. This fact is depicted in
Fig. 15.1, where the growth of populations of *Paramecium caudatum* and of
Didinium nasutum, a predator on *Paramecium* growing in the same culture, are
graphed. When the predator is introduced, the population of *Paramecium*
decreases and that of the predator increases. As the prey population is reduced
and the number of predators increases, a point is reached when there are not
enough prey. After that both populations decrease in number and eventually
become extinct.

Predator-prey interactions can be modeled mathematically by a pair of
equations as follows:

$$\frac{dN_1}{dt} = r_1N_1 - p_1N_1N_2,$$

$$\frac{dN_2}{dt} = p_2N_1N_2 - d_2N_2,$$

where N_1 is the prey population density, N_2 is the population density of the
predator, r_1 is the instantaneous rate of increase of the prey population (per

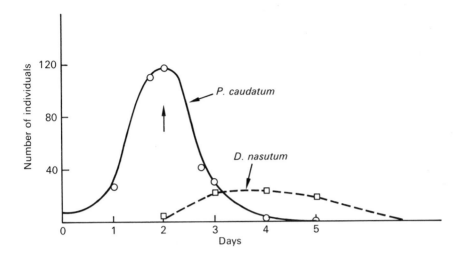

Figure 15.1

Example of the outcome of species interactions, where *Paramecium caudatum* is the prey and *Didinium nasutum* is the predator (From G.F. Gause, 1934. *The Struggle for Existence.* Baltimore: Williams and Wilkins)

individual), d_2 is the death rate per individual of the predator population, and p_1 and p_2 are predation constants. Each population is limited by the other, and there are no regulating density effects (note the absence of K). These equations are known as the Lotka-Volterra equations.

The growth rate of the prey population (N_1) is equal to its intrinsic growth rate (individual birth rate minus individual death rate) times the number of individuals at time 1 in the population (exponential growth; see p. 319) minus the product of the number of prey and predators, $N_1 \times N_2$ (which represents the expected encounters between them) multiplied by a constant, p_1, depicting the probability that the prey will be eaten by the predator in such an encounter. The growth of the predator is reflected by the number of prey and the encounters between prey and predator (N_1N_2) multiplied by a factor that indicates how much the predator gains from eating a prey (p_2), and reduced by the death rate of the predators due to all causes.

If we graph the growth rate of predator and prey on a pair of cartesian coordinates, we obtain Fig. 15.2. This system is cyclic, with both the prey and the predator populations oscillating. However, we have seen that in simple laboratory experiments prey and predator populations do not oscillate. Instead, the predator searches out every prey until it drives the prey population to extinction, and then it starves itself to death. The problem is that the equations are too simple. They make unrealistic assumptions of continual growth of the prey, and furthermore, they assume that encounters between prey and

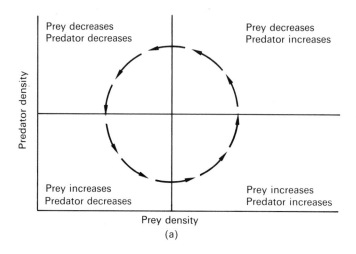

Prey decreases
Predator decreases

Prey decreases
Predator increases

Predator density

Prey increases
Predator decreases

Prey increases
Predator increases

Prey density

(a)

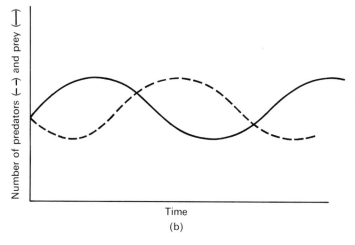

Number of predators (←–) and prey (—|)

Time

(b)

Figure 15.2

Theoretical numbers of predators and prey in a perfectly oscillating system. (a) Relation between the prey and the predator densities. (b) Time course of the prey and predator densities.

predator are random. As is well known, predators actively search out prey, and prey actively hide from predators.

In nature we observe an equilibrium of sorts between prey and their predators. Cycles are known, such as that of the lynx and its principal prey, the snowshoe hare in Canada (Fig. 13.13), or the lemming cycle in Alaska studied by the California ecologist Frank Pitelka near Point Barrow, but they are rare.

Lemmings are herbivorous animals that consume the low shrubs and herbs in the Arctic tundra. They are year-round dwellers of the tundra, feeding and breeding even in the winter months under the snow. During the summer months the lemmings are hunted by a variety of animals, especially such predatory birds as jaeggers and owls, which breed in the Arctic tundra during the summer months. We start the description of the cycle with a low lemming population. The low lemming population means that the number of young the predatory jaeggers and owls can raise is limited, given that availability of food determines the number of chicks raised. Consequently, the number of predatory birds will decrease. It also means that there will be plenty of vegetable material to feed the lemmings without impeding healthy plant growth. This situation—lots of food and relatively few enemies—creates the best combination of circumstances for the reproduction of lemmings during the following winter and summer. However, as lemmings increase in numbers, plant food supplies decrease and the predator population increases. This leads to some starvation by lemmings during the second winter of the cycle, and as a result of the growth of the population of jaeggers and owls during the second year, it leads to a high rate of predation during the third summer of the cycle. Consequently, by the fourth year the lemming population is down to its original size and the cycle can be repeated once more (Fig. 15.3). We note that the state of the vegetation is an important component of the lemming cycle, as are other aspects of the physiology of the animal that we have not considered.

The lemming cycle, however, is not typical of predator-prey interactions. Although the numbers of most predators and their prey fluctuate some from year to year, they rarely oscillate drastically, as in this case, and to our knowledge seldom in regular cycles.

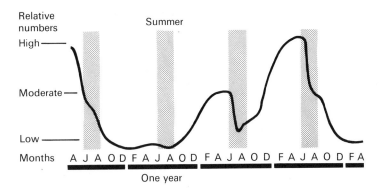

Figure 15.3

A generalized curve of a single oscillation in the short-term cycle of *Lemmus trimueronatus* (lemming) near Barrow, Alaska, drawn to show the important subordinate fluctuations or changes in trend and also to show their approximate timing relations to the summer seasons of successive years. (From F.A. Pitelka, 1958. Some aspects of population structure in the short-term cycle of the brown lemming in northern Alaska. *Cold Spring Harbor Symp. Quant. Biol.* 22:237–251)

Predator-prey interactions can take several forms. A simple graphic demonstration, provided by Michael Rosenzweig and Robert MacArthur, is given in Fig. 15.4. It consists of graphing on a cartesian set of coordinates the number of predators and prey, as was done in Fig. 15.2. However, we now make the following assumptions. In the absence of any predators, the prey population will reach a maximum density K_1. The prey's carrying capacity is the

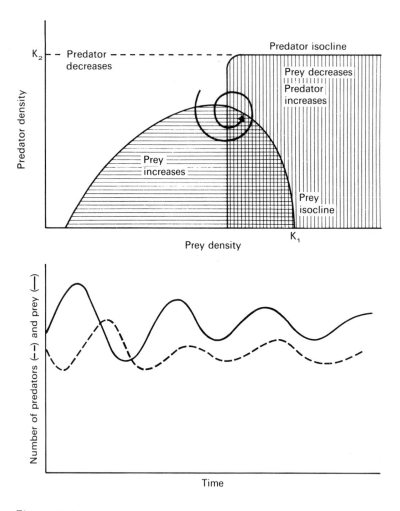

Figure 15.4

A predator-prey model with damped oscillations. (a) Model. (b) Time course. This model depicts the case of the inefficient predator. (After M.L. Rosenzweig and R.H. MacArthur, 1963. Graphical representation and stability conditions of predator-prey interactions. *Am. Nat.* 97:209–223)

result of factors (food, shelter, etc.) other than predation that limit the prey's growth. There must also exist a lower limit below which the prey population cannot maintain itself, because there are not enough possibilities for mating to occur. The prey population will therefore increase only if it is below the prey growth curve shown in the figure. Similarly, for the predator there must be some threshold below which predators cannot find enough food to maintain themselves, as well as an upper limit, or predator carrying capacity, beyond which the predator population cannot grow because of other factors (such as lack of nesting sites, for example). If predators do not compete among themselves, the predator growth curve will be close to a straight line until the predator carrying capacity, K_2, is reached. If there is some interference between predators, the zero growth curve will deflect to the right with a decrease in the predator population (Fig. 15.4).

Let us now imagine three examples, which we will call the case of the inefficient predator, the case of the highly efficient predator, and the case of the just predator, and we will analyze them with the graphic method of Rosenzweig and MacArthur. The inefficient predator needs a high prey density to survive. Consequently, the lower threshold corresponds to a high prey density (Fig. 15.4). The prey population will grow until it reaches its carrying capacity. When it reaches the predator threshold, the predator is able to invade the population and starts growing itself. As predators increase, the prey population starts decreasing. Eventually, the number of prey is not sufficient to maintain the high number of the predators, and they start decreasing too. If we follow the combined trajectory of both populations (see arrow) we note that it forms an inwardly coiling spiral, which eventually comes to rest at the intersection of both curves. An inefficient predator can live in happy equilibrium with its prey. When that equilibrium is disturbed, a damped oscillation ensues until the equilibrium is restored. Just the opposite is true in the case of the highly efficient predator. The interaction leads to a very unstable cycle, with the predator eventually destroying the prey and itself (Fig. 15.5). This is what happened with the chestnut blight in the United States during the decade after its introduction in 1915. The chestnut blight population grew until all trees were infected, but as the chestnuts died, so did the blight. Finally, in the case of the just predator, a stable cycle ensues. Such a situation requires that the two zero isocline curves intersect exactly at the point where the prey isocline is at its maximum (Fig. 15.6).

The Rosenzweig-MacArthur model is much more realistic than the Lotka-Volterra equations. It acknowledges that the growth of both predator and prey are affected by factors other than the interaction between them. It obtains also more realistic results and shows that, for predator and prey to coexist, natural selection has to operate in such a way that the predator efficiency is not exceedingly high. Paradoxically, an overefficient predator brings its own demise unless the prey acquires mechanisms to avoid predation. It is possible that some form of group selection is operating here: Only popula-

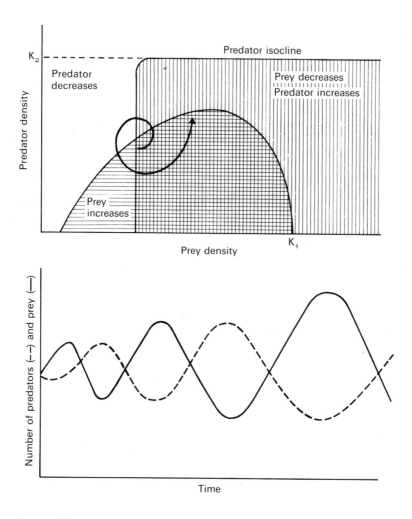

Figure 15.5

A predator-prey model with increased oscillations, according to Rosenzweig and MacArthur. Parameters are as in Fig. 15.4. This model shows the case of the highly efficient predator.

tions of "just" or "inefficient" predators survive. Other factors are the existence of "safe" or "hiding" places for the prey; immigration and emigration of both predators and prey; the ability of some species of predators to survive periods of absence of prey in some resting stage, such as cysts or eggs.

Competition and predation are important components of selection, and they determine to a large extent the evolution of the phenotypic characteristics and life history of organisms. Some examples will illustrate.

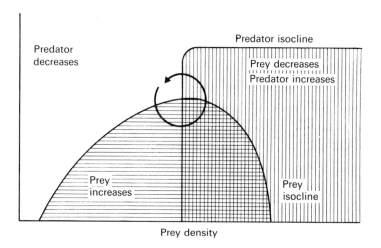

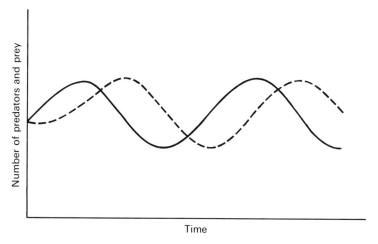

Figure 15.6

A predator-prey model with regular oscillations, according to Rosenzweig and Mac-Arthur. This model shows the case of the just predator.

The tissues of plants are eaten by a very large and diverse group of animals and fungi and even a few plants (such as dotter, *Cuscuta* spp.). Plants cannot escape or hide from their predators (known by the generic name of herbivores). However, individual plants with mutations that make their tissues unpalatable or poisonous will experience less predation and therefore should be favored by natural selection. It therefore should not come as a surprise to learn that plants produce a fantastic diversity of chemical substances. A whole branch of chemistry, called natural-products chemistry, is devoted to the study of the compounds that plants produce. Every month new compounds are discovered. Many of these, such as the antibiotics (poisonous to bacteria), are poisonous to predators; other substances, such as tannins, are undigestible. The existence of these compounds in plants acts as a strong selective agent on herbivores. In effect, among the herbivore population, individuals with mutations that allow them to detoxify the poisonous or undigestible compound that the plants produce will be quickly selected. In turn, the existence of immune herbivore species will act as a selective agent on the plants, favoring new mutations. This process takes place constantly within each plant population and within each herbivore population in every community. The appearance of a mutation being a question of chance, it is unlikely that all the plants in an evolving community will develop the same antiherbivore compound, or that the detoxifying mechanism that the herbivore acquires will work on all plants in the community. Consequently, specialization ensues, since most herbivores can digest only one or a few species of plants, and within a plant often only one type of tissue, such as leaves or wood or roots.

Herbivores are in turn prey to carnivores. Unlike plants, herbivores can escape from their predators and hide. Within the herbivore community, selection will favor those individuals that are least conspicuous. One way of becoming inconspicuous is to acquire the same color and even the shape as the plant the herbivore eats. Figure 15.7 depicts a variety of insects that have acquired the shape of their background. Another way for the herbivores to escape being eaten is to develop certain types of behavior, such as feeding at night when predators cannot rely on visual cues. However, if too many herbivores look like a leaf or a twig and are active at night, the predator learns to look more carefully at leaflike or sticklike objects, or it changes its own behavior to hunt at night. Since each different herbivore population in a community acquires its own peculiar escape mechanism, predators cannot hunt all herbivores with the same degree of efficiency, and they will tend to specialize on certain kinds. For example, species of birds and lizards that feed on insects tend to hunt preferentially insects of certain sizes living in certain places. They also show preference for hunting at certain times of day. An added effect of predation is to keep the herbivore population to a reasonably low level, and that helps to preserve the vegetation. In turn, among the herbivore population, those that are most cryptic or are best at hiding will be most likely to survive, thereby decreasing the efficiency of the predator and avoiding destabilizing cycles. When the number of individuals of one herbivore species decreases and they are hard to find, the predator switches to another herbivore species, allowing the first to recover, and so on.

An interesting experiment that shows the complexity of the interactions in a community was performed by Robert Paine from the University of Washington. On the rocky shores of the Pacific Coast of North America, there is a community consisting of mollusks, barnacles, and one starfish (Fig. 15.8). The starfish, *Pisaster ochraceus*, is a carnivore that feeds on all the other species; *Thais emarginata*, a snail, feeds only on bivalves and barnacles. All

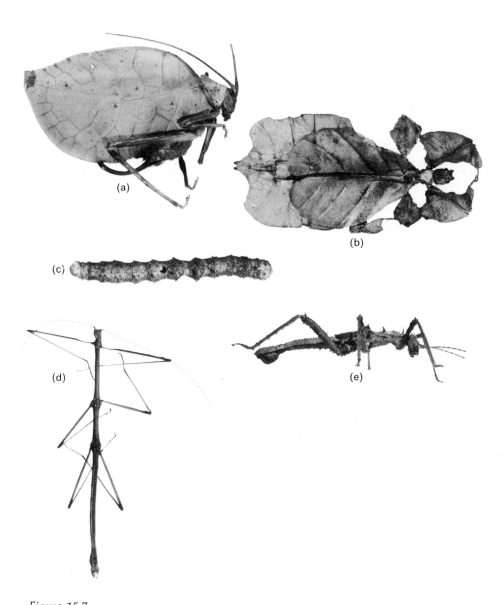

Figure 15.7

Insects that resemble leaves or twigs. (a) A katydid, family Tettigoniidae (Orthoptera). (b) *Phyllium siccifolium* (Phasmida). (c) *Boarmia hispidarius* cater-pillar (Lepidoptera). (d) Walking stick, tribe Lonchodini (Phasmida). (e) *Haaniella echinata* (Phasmida). Photo by Alfred Coleman.

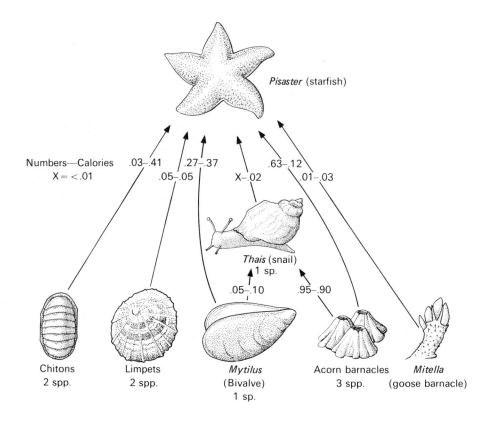

Numbers—Calories
X = <.01   .03–.41   .27–.37   .63–.12
               .05–.05   X–.02   .01–.03

Thais (snail)
1 sp.
.05–.10   .95–.90

Chitons   Limpets   *Mytilus*   Acorn barnacles   *Mitella*
2 spp.   2 spp.   (Bivalve)   3 spp.   (goose barnacle)
                          1 sp.

Figure 15.8

Feeding relationships in a marine tidal community. There are two predators, *Pisaster*, a starfish, and *Thais*, a snail. The proportion of each predator's diet made up of each prey type is shown. The decimal fraction on the left is in terms of the total number of prey taken, and that on the right is in terms of the total caloric intake. (After R.T. Paine, 1966. Food web complexity and species diversity. *Am. Nat.* 100:65–75)

the other species, which are primarily herbivorous, feed on a variety of algae and planktonic organisms. Paine artificially removed the starfish from an experimental area of the intertidal rocky shore. Soon afterward he noted that one species of acorn barnacle (*Balanus cariosus*) started outcompeting the other species so that in a period of three months it occupied from 60 to 80 percent of the space. With time, however, a bivalve (*Mytilus californianus*) and a goose-necked barnacle (*Mitella polymerus*) in turn displaced the *Balanus*. Eventually the *Mytilus* took over most of the space. The original community of more than a dozen species was reduced to essentially two. What happened here? The best competitor, the bivalve *Mytilus*, also happens to be a good morsel (in terms of calories) for the starfish. With the starfish present, it is subject to strong predation and consequently cannot build up a large population to outcompete the other species. As a result, more species can coexist. Predation by starfish in this case keeps the best competitor from winning.

Warning coloration and the evolution of mimicry

Another example illustrating the complexity of food webs and the complicated interactions that take place in communities is provided by what we may call the "monarch butterfly system," studied by a variety of scientists but primarily by Lincoln Brower. The monarch butterfly, *Danaus plexippus*, occurs widely in North America. It is a large, conspicuously colored butterfly with bright orange wings bordered with dark brown. It is usually avoided by birds, who find the butterflies very repugnant to eat. When naive birds (birds that have been hand-reared and have never seen a monarch butterfly) are given a butterfly, they will attack it but will quickly spit it out. The reason for this very strong reaction became apparent when the feeding habits of the monarch butterfly were studied. The larvae of the monarch butterfly feed on milk-weeds, which are very poisonous plants because they have highly toxic alkaloids in their tissues. However, the monarch butterfly larvae are capable of isolating the alkaloids and storing them in their tissues from where they are passed to the adult butterflies. Consequently, this species has been able to break the chemical defense of the plant and turn it into its own protective armor! (Fig. 15.9) But that is not the end of the story. For the chemical protection to be effective, would-be predators must recognize it *before* they strike. This explains the very conspicuous coloration of this butterfly, which warns potential predators. That birds learn to associate the conspicuous coloration of the monarch butterfly with its distastefulness was demonstrated experimentally by Lincoln Brower. In one series of tests Brower fed monarch butterflies and the palatable swallowtail butterflies to hand-reared birds (Florida scrub jays) that had not tasted either kind before. Whereas the swallowtail butterflies were always eaten, the monarchs were usually not eaten, but if swallowed, they were soon regurgitated by the birds, who from there on avoided monarch butterflies. Evidently the jays learn by experience to avoid monarch butterflies as food, recognizing the butterfly by its conspicuous colored wings.

Many animals have poisonous or distasteful flesh that protects them from their would-be enemies. Bad-smelling or unpalatable bodies are found in many groups of bugs, beetles and butterflies, shrews, salamanders, and skunks. Bees, wasps, sting rays, and some snakes can inflict poisonous stings on their attackers.

Possession of poisonous or distasteful flesh is not of much use if the attacker learns of it only after an attack that results in the death or maiming

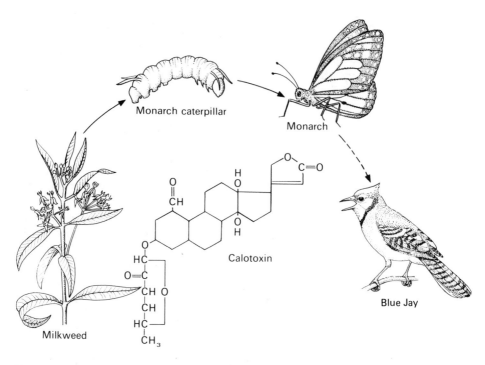

Milkweed

Calotoxin

Monarch caterpillar

Monarch

Blue Jay

Figure 15.9

Unpalatability in the monarch butterfly (*Danaus plexippus*). Milkweeds (*Asclepias* spp.) produce several compounds toxic to vertebrates. The formula for one of these, calotoxin, is shown. The caterpillar of the monarch eats the milkweed leaves and assimilates the compound, which is retained when the caterpillar becomes an adult butterfly. The compound makes the butterfly distasteful to bird predators.

of the victim. However, if the poisonous species is conspicuously colored, the potential enemy soon learns to associate the conspicuous color with the noxious property of the species. Experiments conducted by Carpenter substantiate this. He classified 244 species of insects subjectively as conspicuous or cryptic. He fed them to a *Cercopithecus* monkey in Africa and then classified the insect species as edible or distasteful on the basis of the monkey's reactions. The results were as follows:

	Edible	*Distasteful*	*Total*
Cryptic	83	18	101
Conspicuous	23	120	143
	106	138	244

Note that most of the cryptic species were edible, and most of the conspicuous ones were distasteful.

Protection is not operative until the potential enemy has at least one encounter with a distasteful species and *learns* to associate the conspicuous color with the noxious property of the animal. But since many poisonous or distasteful species of different families and orders have the same basic colors—black, red, orange, and yellow—to advertise their noxious properties, most would-be enemies learn to associate that general color or pattern with obnoxiousness. That there is convergence in warning coloration among distasteful species is in itself an adaptation to reduce the "mistakes" of the would-be predator, since such mistakes are usually fatal to the attacked individuals. For example, many unpalatable tropical butterflies of the family Nymphalidae that occur together in parts of Brazil have similar shape and color pattern in their wings. They are species of *Lycorea, Melinaea, Mechanitis, Heliconius,* and *Eveides,* which belong to three different subfamilies of the Nymphalidae.

After a bird or a toad has learned to associate a certain bright color or pattern of colors with an obnoxious property and avoids animals with those characteristics, not only poisonous individuals but all animals with those colors will be avoided. This allows the evolution of *mimicry*—that is, of a palatable animal's resembling another that is poisonous or distasteful. The evolution of mimicry is basically not different in principle from the evolution of crypsis: In one the model is nature, in the other some conspicuous obnoxious animal. In both cases, the closer the match with the model, the greater the protection. In both cases, too, if the mimics become too numerous, the predator learns to associate their color or background with good food rather than with unpalatable food (whether it is an obnoxious animal or a lichen or a rock).

Mimics are known in many groups of animals. For example, ants are unpalatable to most predators because of their sting (but not to all animals—witness at least three unrelated mammals specializing in a diet of ants). Yet ants are mimicked in various parts of the world by spiders, grasshoppers, beetles, flies and bugs. The same is true for bumblebees, mimicked by various flies and moths. But the best known and studied cases of mimicry are among butterflies, and among these, probably the best example is furnished by the mimics of the monarch butterfly.

Another species of butterfly, the viceroy (*Limenitis archippus*), occurs in eastern North America (Fig. 15.10). In its superficial visible characters, especially the color and patterning of the wings, the viceroy resembles the monarch. Since viceroys are not very distasteful, it has long been assumed that the viceroy is a mimic of the monarch butterfly. To test this idea, Brower fed only viceroys to a set of jays who had no previous experience with these butterflies. Most birds ate the viceroys that were offered, and none showed any adverse effects. Another set of jays were offered monarch butterflies instead. They soon learned to reject them, as we have seen. They were then offered viceroys, and for the most part they rejected them, too. Viceroy butterflies are indeed protected from predation by their resemblance to the distasteful monarch.

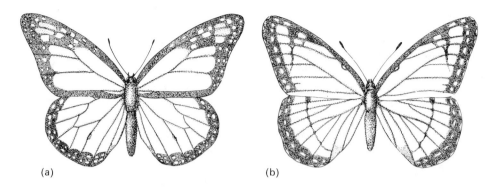

(a)   (b)

Figure 15.10

Batesian mimicry. (a) The distasteful model, the monarch butterfly (*Danaus plexippus*). (b) The mimic, the viceroy butterfly (*Limenitis archippus*).

Since the protection that a mimic butterfly obtains depends on the potential predator's learning to associate its color or shape with a distasteful characteristic of the model, the fate of the mimic is closely associated with the model species. If mimics are much more numerous than the model, the predator may conceivably learn to associate their color with the good properties of the mimic, and given their conspicuous color, they could become very vulnerable. Such a situation would be harmful not only to the mimic species but to the model as well. Furthermore, if during the learning process the predator encounters many mimics, the learning process will be slowed down. Finally, because of its conspicuousness, the mimic is restricted to live only at the same time and in the same place as the model.

In an attempt to quantify the relationship between mimics and models, Brower fed starlings (*Sturnus vulgaris*) mealworms (*Tenebrio molitor*). Some mealworms were dipped in quinine, making them highly unpalatable. Together with these unpalatable worms (the "models") he fed the starlings edible mealworms that had been dipped in water (the "mimics"). Starlings soon learned to associate mealworms with unpalatability and avoided them. When the proportion of models to mimics was in the order of 9:1, the birds rejected 80 percent of all edible mealworms; when the ratio was 2:3—that is, slightly more mimics than models—the birds still rejected about the same 80 percent of the edible mealworms. When the ratio was 1 model to 9 mimics, the birds consumed a large number of the mimics, but some 17 percent of the edible mealworms that were offered were taken for models and rejected. Mimics can apparently still derive some protection, even when they outnumber the models. However, 83 percent predation can probably not be sustained easily by a species.

Common warning coloration is denoted as *Müllerian mimicry* after Fritz Müller, a German naturalist who correctly explained the phenomenon in 1879.

The resemblance of an unprotected animal to an unrelated species is called *Batesian mimicry*, after the British naturalist Henry Walter Bates, who explained this type of mimicry in 1862.

The number of palatable mimics of a distasteful model cannot be too large; otherwise the predator will learn to identify the color with edible food, and both species will lose protection. The evolution of mimicry is a direct result of predator-prey interactions. The detailed investigation of many of these relations is being actively pursued at this time and constitutes a fascinating aspect of the interface between evolutionary and ecological studies.

SYMBIOSIS

Not all interactions between species are to the detriment of one of the species participating in the interaction. Often both species benefit; a good example is flowering plants and their pollinators. The pollinator obtains needed food in the form of nectar and/or pollen; the plant is fertilized as the result of the insect's visit. Unless there was an alternative source of food, the insect would starve without the plant, and likewise without a pollinator the plant would become extinct. Such interactions are known broadly as *symbiosis* or *mutualism*. Some authors divide mutualistic relations even more finely (Table 15.1).

Table 15.1

Types of interactions where both participants benefit and/or where neither is hurt.

A. Within Species

1. *Mating*

2. Some *social interactions* (e.g., schooling in fish; bird flocks).

3. *Division of labor* among members of a colony (e.g., social insects; *Volvox* algae).

B. Between Species

1. *Commensalism*. One species benefits; the second neither benefits nor suffers (e.g., commensal spiders that live in the web of another spider and eat its leavings; mosquitoes in pitcher plants).

2. *Mutualism*. Both species benefit.
   a) *Nonobligatory* (e.g., nonspecialized flowers and their pollinators; cattle and cattle egret).
   b) *Obligatory* (e.g., *Yucca* species and their necessary pollinator, *Tegeticula* sp.).

3. *Symbiosis*. Extreme cases of obligatory mutualism, where both species occur in close associations. The classical example is the lichens.

The degree to which two species benefit in a symbiotic interaction varies. At one extreme is the situation in which two species cannot exist without each other, e.g., the *Yucca* plant and its obligatory pollinator, the moth *Tegeticula* (Fig. 15.11). At the other extreme are pollinators that visit many types of flowers and plants that are pollinated by many such insects. In the latter case, the relationship between any two species is a loose one.

An interesting example of a symbiotic relation is that of ants and some plant species, of which the so-called bullhorn acacias of Central America, studied by the American ecologist Daniel Janzen, are the best known. Bullhorn

(a)　　　　　　　　　　　　(b)

(c)　　　　　　　　　　　　(d)

Figure 15.11

Pollination of *Yucca* spp. by the moth *Tegeticula*. (a) View of *Yucca* flowers with moths collecting pollen from anthers. (b) Detail of moth collecting pollen from anther. (c) Cross section of ovary and adjacent anther. After gathering pollen, the moth oviposits on the ovary. Offspring will develop in the ovules of *Yucca*, destroying a few seeds. (d) After ovipositing, the moth crawls up to the stigma and deposits there the pollen it has collected, presumably to feed the larvae when they emerge, but in fact pollinating the flower. (Photographs of the glass flower models at Harvard University, courtesy C.E. Wood)

Figure 15.12
Thorns of *Acacia cornigera,* the bullhorn acacia.

acacias get their name from their large swollen spines (Fig. 15.12). The spines are hollow and have a patch of thin tissue to one side. Certain small stinging and very aggressive ants of the genus *Pseudomyrmex* perforate that thin tissue and use the hollow interior of the spine to nest. In addition, bullhorn acacias produce nectar glands at the base of their leaves, and at the tips of the leaves they produce small packets of tissue, called Beltian bodies, that are rich in protein and other nutrients. The ants eat both the nectar and the Beltian bodies. Consequently, the ants can complete their entire life cycle on the *Acacia.* The ant in turn attacks any insect that tries to eat the leaves of the *Acacia.* They also cut shoots of neighboring vegetation that come in contact with the *Acacia.* These activities by the ant ensure ample light for the *Acacia* and freedom from predators (Table 15.2). The acacia-ants therefore act as antiherbivores, and the *Acacia* does not need to invest in chemical defenses (but has to invest in shelter and food for the ant). Not all species of *Acacia* have ants (actually only a minority do), nor do all species of *Pseudomyrmex* live on acacias. But the "ant-acacias" and the "acacia-ants" have developed a close interdependence not unlike that of the *Yucca* plant and the *Tegeticula* moth so that their future evolutionary histories are closely interlocked.

SPECIES INTERACTIONS AND LIFE HISTORY PARAMETERS

The study of the adaptations of plants and animals to their physical and biological environment is tightly linked to the science of physiology and functional morphology, often referred to as ecophysiology or physiological ecology.

Table 15.2

Effect of the ant *Pseudomyrmex ferruginea* on *Acacia cornigera*.

	Ants Present	Ants Removed
Weight of suckers	41,750 gm	2,900 gm
Number of leaves	7,785	3,460
Average growth in 45 days	72.86 cm	10.23 gm
Percent mortality	28%	56%
Percent shoots with insects: Day	2.7%	38.5%
Night	12.9%	58.8%
Mean number of insects per shoot: Day	0.039	0.881
Night	0.226	2.707

After D.H. Janzen, 1966. Coevolution of mutualism between ants and acacias in Central America. *Evolution* 20:249–275.

The particular ways that organisms interact with their environment have demographic implications. This fact is demonstrated by a number of independent studies on different aspects of the life cycle of goldenrods (*Solidago* spp.) in the northeastern part of the United States, studies performed by the American ecologist Warren Abrahamson and some of his colleagues.

In a first study six populations belonging to four different species were researched. Two populations grew in a forest, two in a wet meadow, and two in a highly disturbed field. The three sites were very close to one another (within a one-mile radius) so that no major macroclimatic differences were present. By harvesting plants at regular intervals, dividing them into stem, leaves, and flowers, and then drying and weighing the various parts independently, the researchers ascertained the differences in the relative proportion of these parts in the different populations through the yearly cycle of the plant. Plants in the disturbed field had a relatively high proportion of biomass in flowers, followed by the populations from the wet meadow, while those from the forest had the smallest proportion of energy in flowers. In turn, plants from the wet meadow had a great proportion of their biomass in stem tissue, whereas those in the forest had the highest proportion of leaf biomass (Fig. 15.13). The differences (which are genetically inherited) can easily be interpreted as responses to the environment in which these populations grow. The disturbed field is the most uncertain environment, and consequently (see Chapter 13) one can predict that the plants there will invest heavily in seed production. The wet meadow is intermediate in terms of disturbance. It is a sunny environment, but the vegetation is dense and plants can easily be overtopped and shaded. Goldenrod plants compete for the light by growing tall and trying to avoid being outshaded. In doing so, they divert energy to stem production. The forest floor, on the other hand, is darker than the field because of the shading effect of the trees. But the vegetation is less dense and there is

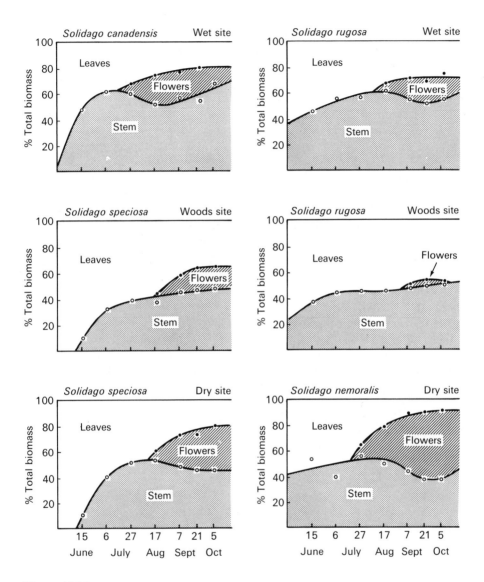

Figure 15.13

Apportionment of biomass in *Solidago* species in different environments. The percentages of the total biomass in leaves, stem, and flowers are shown as a function of time during the growing season. (From W.G. Abrahamson and M. Gadgil, 1973. Growth form and reproductive effort in goldenrods (*Solidago*, Compositae). *Am. Nat.* 107:651–661. Used by permission of The University of Chicago Press)

much less interference between the herbaceous components. Since goldenrods can never grow taller than the trees and capture their light, and there is little interference with other goldenrods, they do not grow tall in forests and consequently have more energy to invest in leaf production. Goldenrods in these three fields have morphologies and life histories that are directly related to their environment. Furthermore, since these environments are so close that seeds from one environment can regularly be blown into the other, the absence of forest types in meadows and fields, and vice versa, implies that these morphologies and life histories are of direct adaptive value.

In another study Abrahamson and his collaborators repeated the study with one of the same species (*Solidago canadensis*), using populations in disturbed fields. This time they included underground structures (roots and rhizomes) and ascertained not only the weight but also the caloric content of the tissues. In these fields they took three kinds of samples: (1) plants with no parasites; (2) plants that had been attacked by gall-forming herbivorous insects; and (3) plants that had been attacked by the gall-forming insects, which in turn had been attacked by a wasp that is a natural enemy of the gall insect. Plants without parasites were larger than those with herbivorous parasites, since the gall-forming insect diverts a significant proportion of goldenrod calories into galls, as well as into the body of the insect. However, when the gall-forming insects are parasitized, they are no longer capable of diverting as many goldenrod calories to their use as before (Fig. 15.14), and some of those calories go into the wasp (Fig. 15.15). The wasp therefore has

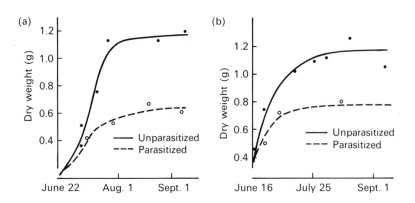

Figure 15.14

Growth curves of stem galls on the goldenrod *Solidago canadensis*. The galls are caused by two insects: (a) *Gnorimoshema gallaesolidaginis*, a moth, and (b) *Eurosta solidaginis*, a fly. The gall insects are in turn affected by insect parasites. (From B.R. Stinner and W.G. Abrahamson, 1979. Energetics of the *Solidago canadensis*–stem gall insect–parasitoid guild interaction. *Ecology*, in press)

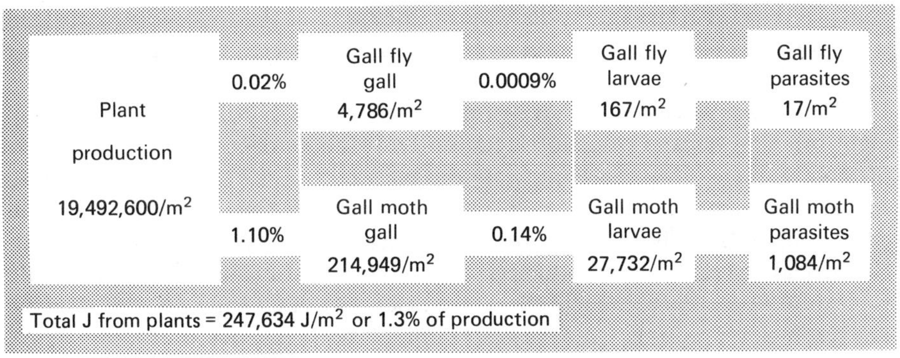

		Gall fly gall 4,786/m²		Gall fly larvae 167/m²		Gall fly parasites 17/m²
Plant production 19,492,600/m²	0.02% / 1.10%		0.0009% / 0.14%			
		Gall moth gall 214,949/m²		Gall moth larvae 27,732/m²		Gall moth parasites 1,084/m²

Total J from plants = 247,634 J/m² or 1.3% of production

Figure 15.15

Energy flow model (in joules) for *Solidago*–gall insect–parasite chains expressed on a per area (m²) basis. The percentages in the "channels" of the model indicate the ratio of each component to plant production per m². Note that the parasites are not a cost to the plant. (From B R. Stinner and W.G. Abrahamson, 1979. Energetics of the *Solidago canadensis*–stem gall insect–parasitoid guild interaction. *Ecology*, in press)

a beneficial effect on the goldenrod, and it constitutes a natural biological pest control for the goldenrod.

Abrahamson was able to ascertain not only that the insects altered the total energy available to the plant, but that the plant responded by changing its relative allocation of energy to the various plant tissues. When attacked by these herbivores, goldenrod plants reduce the amount of reproductive tissues (flowers and rhizomes) proportionally more than they reduce vegetative tissue. This is interpreted as an adaptive response of the plant to ensure its survival since plants that are smaller (with less stem and root, and fewer leaves) have a lower probability of survival.

From this study it was also determined that there is a successional pattern involving plants and insects in an old field. After an agricultural field is left fallow or a plot of forest has been clear-cut and the trees have been removed, goldenrod seeds are blown in, and eventually they germinate and establish a population. In the early stages the new population is usually free of gall-forming insects. Sooner or later an adult fly or moth (the most common gall-forming insects that attack goldenrods) finds the field and lays its eggs in the growing tips of the goldenrods. Larvae hatch and bore into the plant and, by chemically affecting the tissues, induce the plant to form galls. Inside each gall a larva develops into an adult insect. With time, the herbivore spreads in the goldenrod population, but eventually the population of gall-forming insects is found by the parasitic wasps, and they in turn lay their eggs on the galls, so that their larvae can develop on the body of the herbivore. This leads to a reduction of the population of gall-forming insects and a recovery of the goldenrods.

This example clearly illustrates the complexities of the organism-environment interactions and their interrelation with the demographic characteristics of the organism. One can appreciate the plasticity of the goldenrod responses to changes in the environment. In each physically distinct environment, goldenrod populations have evolved that apportion their energy differently between leaf, stem, and reproductive tissue, presumably in order to maximize their total fitness over their lifetime. As discussed in Chapter 13, this does not always mean producing the largest number of seeds per unit time. It can involve changes in longevity, or diversion of resources so as to reduce mortality. In addition, within each population, plants have been selected with the ability to alter their basic pattern of allocation under special circumstances—for example, when grazed by herbivores. The response is of course not a conscious one, but it is probably mediated by hormones in a manner not yet understood.

So far we have been referring to organisms with an open type of growth (such as plants, many invertebrates, fishes, reptiles, and amphibians). But most animals do not have the ability to change their growth rate or body proportions. However, trade-offs of the same kind are also operative in animals, but they involve other aspects of the life cycle. An example is the evolution of parental care of the young.

Adaptive advantages of parental care

Parental care means that adults provide their young with food and invest time and take risks to protect them from predators. The adults may directly present food they have harvested for the purpose, as many birds do, or they may supply food indirectly by provisioning the egg with an ample supply of yolk. Parental care involves a diversion of potential parental resources to the offspring, in the same way that blooming involves a diversion of potential resources from the mother plant to the offspring. Since these resources (energy, time) are always in short supply, the greater the care parents lavish on their offspring, the fewer offspring they can care for. This inverse relationship has been documented for marine invertebrates (sea urchins, starfish, worms, mollusks, and crabs), for fishes, and for birds. On the other hand, the probability of survival of the offspring increases with parental care. Consequently, although the number of eggs that are produced is smaller, the number of surviving offspring may be larger in a species that provides parental care as compared with one that does not. Furthermore, if young are fed and protected by their parents, they do not need to be born with characteristics necessary for independent life. This allows for a more precocious birth and a faster growth rate after birth. This condition is illustrated by a comparative study between two species of birds performed by the American ecologist Robert Ricklefs.

One of the species is the common starling (*Sturnus vulgaris*). In this species the young hatch after only 12 days of incubation. When they are born, they have no feathers, their eyes are closed and they are totally helpless. They must be fed and brooded by their parents in the nest for the next three weeks if they are to survive. On the other hand, since much of their body is digestive apparatus, they grow very fast. The other species studied was the European quail (*Coturnix coturnix*). In this species eggs take 18 days to hatch, and the young are born open-eyed, covered with down, and fully able to run about gathering their own food. The parents still care for the offspring by showing them food items and warming them during cold weather, but not to the extent that starlings take care of their young. The quail lays more eggs per clutch, but mortality among the young is greater since, as a result of their slower developmental time, they spend more time in the vulnerable early stages of development. And the quail must provision the egg with more nutrients than the starling.

Many more examples can be presented of how different characteristics of plants and animals increase their survival in a given environment by increasing their share of resources, by allowing them to outcompete other species, by decreasing the danger from predators or increasing their efficiency in catching prey, or by making them better partners in symbiotic interactions. All adaptations have in common that they somehow increase the survival probability of the individual or its offspring, that they entail the efficient use of resources within certain constraints, and that they involve some trade-off. Something is done better, but something is given up. But the rigorous description of individual adaptations alone does not by itself explain how they came about, nor do they help us predict future behavior of species or populations. What are needed are predictive and testable theories.

In the opening pages of this book, we stated that more than half a million species of extant plants and more than a million species of animals had been described to date. Despite the magnitude of these numbers, they perhaps represent only half the real number of living species. We also pointed out that the number of species that once existed and are now extinct is possibly an order of magnitude higher. In the last five chapters, we have discussed some of the processes and mechanisms involved in creating and maintaining this great diversity, but of necessity we have left out many other topics. We have not asked how energy flows through an ecosystem, or whether the diversity is a stable condition or just a transient phenomenon. Likewise, we have not addressed ourselves to the question of ecological succession, to theories regarding optimal harvesting of resources, or to the number of species that can coexist in a given area. These and other subjects constitute some of the many interesting and fascinating subjects that concern population biologists and ecologists today.

In a world threatened by human destructiveness, it is important to learn as quickly as possible exactly how species interact and how they respond to the disturbance created by us if we are to keep from committing irreversible ecological damage.

SUGGESTED FURTHER READING

General references

Cott, H.B. 1940. *Adaptive Coloration in Animals.* New York: Oxford University Press.

Gilbert, L.E., and P.H. Raven (eds.). 1975. *Coevolution of Animals and Plants.* Austin: University of Texas Press.

May, R.M. 1973. *Stability and Complexity in Model Ecosystems.* Princeton, N.J.: Princeton University Press.

Maynard Smith, J. 1974. *Models in Ecology.* Cambridge: Cambridge University Press.

Articles and reviews

Abrahamson, W.G., and M. Gadgil. 1973. Growth form and reproductive effort in goldenrods (*Solidago*, Compositae). *Am. Nat.* 107:651–661.

Brower, J.V.Z. 1958. Experimental studies of mimicry in some North American butterflies. Part 1. The monarch, *Danaus plexippus,* and viceroy, *Limenitis archippus archippus. Evolution* 12:32–47.

Brower, L.P. 1969. Ecological chemistry. *Scientific American* 220 (2):22–29.

Brower, L.P., and J.V.Z. Brower. 1964. Birds, butterflies and plant poisons: A study in ecological chemistry. *Zoologica* 49:137–159.

Carpenter, G.D.H. 1921. Experiments on the relative edibility of insects, with special reference to their coloration. *Trans. Ent. Soc. London.*

Connell, J.H. 1972. Community interactions on marine rocky intertidal shores. *Ann. Rev. Ecol. Syst.* 3:169–192.

Ehrlich, P.R., and P.H. Raven. 1967. Butterflies and plants. *Scientific American* 216 (6):104–113.

Janzen, D.H. 1966. Coevolution of mutualism between ants and acacias in Central America. *Evolution* 20:249–275.

Oaten, A., and W.W. Murdoch. 1975. Functional response and stability in predator-prey systems. *Am. Nat.* 109:289–298.

Paine, R.T. 1966. Food web complexity and species diversity. *Am. Nat.* 100:65–75.

Paine, R.T. 1974. Intertidal community structure: Experimental studies on the relationship between a dominant competitor and its principal predator. *Oecologia* 15:93–120.

Pitelka, F.A. 1958. Some aspects of population structure in the short-term cycle of the brown lemming in northern Alaska. *Cold Spring Harbor Symp. Quant. Biol.* 22:237–251.

Ricklefs, R.E. 1968. Patterns of growth in birds. *Ibis* 110:419–451.

Rosenzweig, M.L., and R.H. MacArthur. 1963. Graphical representation and stability conditions of predator-prey interactions. *Am. Nat.* 97:209–223.

Stinner, B.R., and W.G. Abrahamson. 1979. Energetics of the *Solidago canadensis*–stem gall insect–parasitoid guild interaction. *Ecology* (in press).

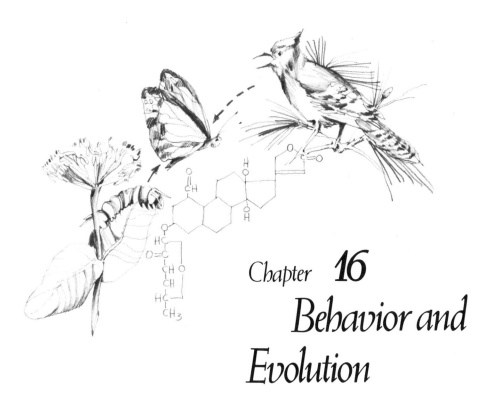

Chapter *16*

Behavior and Evolution

In previous chapters, we have discussed genes and their resulting morphological traits in a populational context. We will now extend the discussion to behavioral traits.

In a given situation, an animal can react in a number of ways. Naturally, not all reactions are favorable to the animal. Some are immediately fatal, and some decrease the animal's fitness in the longer term. Other behaviors can increase the animal's chances of survival and reproduction. We will address two questions in this chapter: Is behavior subject to natural selection? If so, how does natural selection work to produce various patterns of behavior?

The behaviors an animal is capable of are limited by its morphology and physiology in three respects: the movements it is capable of making, the information it can obtain from the environment, and the control it has over its movements in response to information about the environment. There is a wide diversity throughout the animal kingdom in all three respects.

That behavior is limited by physical form is simple to understand. An amoeba is capable of only a narrow range of behavior. Animals need wings of some kind to be able to fly, lungs or other special structures to live on land rather than in water, special glands in order to nurse young.

For behavior to be effective, the animal must be able to recognize things and events in the environment. This information is obtained through its senses. The kind and complexity of information varies greatly. Many kinds of external stimuli can be recognized by animals (Table 16.1). Organs to detect temperature, light, heat, pressure, electricity, gravity, and various chemicals are known. The information an animal receives depends on its sensory equipment. Dogs have senses similar to ours, but they depend more on scent. Bees use light to see, but they see further into the ultraviolet and not so far into the red as we do. They also can use polarized light. These capabilities affect their behavior.

If an animal is to survive, it must be able to use the information collected by its senses to stimulate appropriate behavior. There are two main systems in the animal that regulate behavior. The first is the nervous system, which carries information from the senses to the brain to the motor muscles. In order for an animal to be capable of complex behavior patterns, it must have a highly developed nervous system.

The second control system is chemical, and it is composed of hormones.

Table 16.1

Kinds of external stimuli recognized by animals.

Type of Stimulus	Stimulus	Example of Animal	Animal's Receptor Organ
Temperature		Pit viper	Pit organs
Light	Polarized light	Bee	Compound eye
	Color	Birds	Eye (with cones in retina)
Sound	Airborne sound	Butterflies	Tympanal organ
	Airborne sound	Insects	Johnston's organ
	Airborne sound	Man	Ear
	Substrate vibration	Bee	Subgenual organ
Chemical	Airborne (smell)	Man	Nose
	Airborne pheromones	Moth	Antennal receptors
	Taste (contact between stimulus source and receptor)	Blowflies	Chemosensory hair
	Taste	Dog	Taste bud
Mechanical	Touch	Mammals	Meissner's corpuscle
	Pressure	Cat	Pacinian corpuscle
Gravity		Crustaceans	Statocysts

Some hormones can cause prompt changes in behavior, such as the eclosion hormone of the *Cecropia* moth, which stimulates preparation for emergence of the adult from the pupal cocoon. Other hormones have longer-lasting effects. For example, in migrating birds, increased day length in early spring stimulates the release of hormones that cause physiological changes leading to the accumulation of fat, which then sustains the bird during its migratory flight. Similarly, hormones coordinate a long series of physiological and behavioral changes in preparation for reproduction. Depending on its hormonal state, an animal may respond in different ways to the same stimulus. The interaction between hormones and behavior is complex. In general, hormones influence an animal's behavior, and many types of behavior can cause hormonal change.

ADAPTIVENESS OF BEHAVIOR

We have discussed the constraints of an animal's body on its behavior. It is clear that the morphology and physiology of an animal are subject to natural selection, but what about the behavior patterns themselves? If behavior is subject to natural selection, it must meet two criteria: It must be adaptive, and it must be heritable. We shall first discuss the adaptiveness of behavior.

There are many ways in which behavior patterns can increase the fitness of an organism. One overall category, which concerns the maintenance of the individual, includes those behaviors that tend to increase the efficiency and prolong the life of an individual. By their behavior, animals can take advantage of favorable aspects of the environment, and to some extent they can protect themselves from unfavorable aspects. A lizard does so when it regulates its body temperature by moving in or out of the sun. Behavior patterns can also form a defense against predators. Making a safe home (such as a burrow), running at the sight of a predator, and feigning death are some of the forms predator defense can take. Another important category of maintenance behaviors is concerned with feeding—both the search for food and the handling of it once it is found. Some of these behaviors are simple, such as the chemotaxic response of many protozoans to certain chemicals, and some are quite complex, such as the coordinated hunting of a pride of lions.

The other large category of behavior patterns that increase fitness is concerned with reproduction. This category can be subdivided into behavior patterns of attraction between male and female, courtship, and care of young.

The basic function of all reproductive behaviors is to ensure fertilization of the eggs by the sperm. The first step toward the fulfillment of this function is the meeting of potential mates. There are various mechanisms to attract one sex of a species to another, such as the release of chemical attractants called pheromones by some insects, light flashes by fireflies, or the croaking of many species of frogs. These signals are often species-specific (i.e., performed only by members of a given species), and an individual of one species does not respond to a signal from an individual of another species. For example, two closely related species of tree frogs of the *Hyla versicolor* species complex are very similar morphologically. The two groups differ in the mating calls of the males, which are a fast trilling call in one group and a slower trill in the other. The calls can be recorded and played back from two speakers. If a female of one of the species is placed equidistant between the two speakers, she will move toward either the speaker with the slow trill or the one with the fast trill. If the experiment is repeated, she will move consistently toward the same call. Thus the females can discriminate between different calls.

Behavior from the time of meeting until the time of gamete release is called *courtship*. The simplest kind of courtship involves recognition of a member of the opposite sex (of the same species) and the simultaneous discharge of gametes. This kind of behavior occurs in many invertebrates. In many species courtship displays are more developed, and some are very complex, each individual of a species (or at least of one sex) following an elaborate pattern, as exemplified by many birds. Courtship may involve responsive behavior by both partners or display by one. An example of courtship primarily involving display by the male is the fruit fly *Drosophila melanogaster* (Fig. 16.1). The male first stands near the female, facing her, and follows her as she walks away. One of his wings, held perpendicular to the body, is vibrated. The male comes up behind the female and attempts to mount. The first few times he is usually rejected, and the pattern is repeated until he succeeds or gives up.

A common phenomenon in courtship (as well as in other kinds of behavior, such as territorial defense) is *ritualization*. Ritualization is the modification, through evolution, of a behavior pattern whose original function has become subordinated to communication. Ritualized patterns in aggressive encounters usually give information concerning the individual's likelihood to fight or to flee. In courtship the use of these behaviors signals a readiness to mate; frequently it specifies the sex of the animal; and since the patterns are usually species-specific, they communicate whether the two involved are members of the same species. In some cases it is possible to trace the evolution of a pattern by comparing closely related species. Signals can originate from many possible behavior patterns, such as blushing, feeding, defecation, etc. In the course of ritualization, the original behavior tends to become simplified and exaggerated, and thus it becomes a more effective signal. The animal's anatomy may also change so that it emphasizes the signal. An example of ritualized courtship display can be seen in the dabbling ducks, such as the mallard. Figure 16.2 shows 10 courtship poses found in various species of these ducks. In the mallard the head-up–tail-up posture (e) in particular shows off the color patterns of the male's plumage. The same movements can be observed in related species with different markings. The birds of different species that adopt these poses do not necessarily display them in the same order.

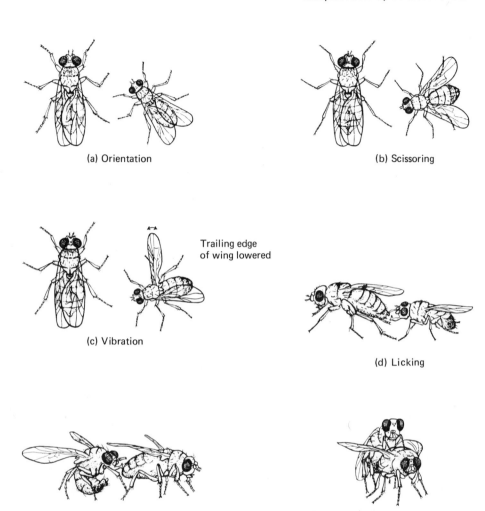

(a) Orientation

(b) Scissoring

(c) Vibration

Trailing edge
of wing lowered

(d) Licking

(e) Attempted copulation

(f) Copulating pair

Figure 16.1

Courtship in *Drosophila melanogaster*. (a) Orientation. The male approaches the female and follows her. (b) Scissoring. The male aligns his wings parallel down his back, then spreads them. (This step is uncommon in this species, but it occurs more frequently in other species of *Drosophila*.) (c) Vibration. One wing, held perpendicular to the body, is vibrated. (d) Licking. The male licks the abdomen of the female. (e) Attempted copulation. (f) Copulating pair. (From A. Manning, 1965. *Drosophila* and the evolution of behavior. In J.D. McCarthy and C.L. Duddington (eds.), *Viewpoints in Biology*, vol. 4. London: Butterworths)

Figure 16.2

Courtship poses of the mallard: (a) bill-shake, (b) head-flick, (c) tail-shake, (d) grunt-whistle, (e) head-up–tail-up, (f) turn toward female, (g) nod-swimming, (h) turning the back of the head, (i) bridling, (j) down-up. These poses can be combined in several sequences, both in the mallard and in other surface-feeding ducks. (From Konrad Z. Lorenz, 1958. The evolution of behavior. *Scientific American* 199(Dec.):67–78. Copyright © 1958 by Scientific American, Inc. All rights reserved)

Courtship behavior serves several functions. It stimulates simultaneous arousal in the male and the female, so that both are ready to breed at the same time. It inhibits aggressive behavior between mates that might otherwise be expressed. And it allows an individual to recognize another as a member of the same species. This is important, particularly in areas where two closely related species exist. If individuals from different species attempted to breed, they might not be able to produce offspring, or if they did, the offspring would most likely be sterile or inviable. Hence there would be a selective advantage to those individuals who did not mate with individuals of the other species.

The third category of reproductive behavior is the care of the young. According to present theory, certain environmental factors can lead to the evolution of parental care. E.O. Wilson gives four. (1) High predator activity can mean that a parent increases its fitness by protecting the young. (2) Specialization on difficult food sources can mean that the young cannot cope for themselves until they have reached a certain physical size or until they have learned enough skills. Wolves and the great cats attack large prey, and a certain minimum of size and stalking skill is needed for success. In these animals the young are cared for and taught until they can fend for themselves. (3) If a species lives in an unusually physically stressful environment, young may not be able to survive without help. (4) Species adapted to predictable environments leading to *K*-selection (see Chapter 13), will also tend to have more developed parental care. *K*-selection tends to favor large size and longer life, with reproduction occurring over several seasons rather than only once. These conditions favor broods of relatively few offspring whose chances of survival are improved by parental care.

HERITABILITY OF BEHAVIOR

We have discussed the adaptive value of various behavior patterns, including behavior supporting individual maintenance and reproductive behavior. But in order for natural selection to work on behavior, behavioral characteristics not only must have adaptive value but must be passed from parents to offspring.

To what extent can behavior be transmitted genetically? Acquired characteristics cannot be inherited and hence cannot be selected, but it is possible that there is selection for genetic characters that facilitate or predispose the animal toward certain behaviors. There is some evidence for genetically controlled behavior. Some behavior patterns appear suddenly, before the individual exhibiting them can have observed conspecific individuals performing them, such as hatching behavior in birds. The hatching behavior observed in most birds involves first a change to the prehatching position, in which the head is tucked under the right wing and the bill points to the air space at the blunt pole of the egg. The bill then penetrates the membrane separating the air space, and breathing with the lung begins. This is followed by pipping, the formation of the first cracks in the eggs. Then there is a series of movements in which the beak thrusts strongly against the shell followed by leg movements that turn the embryo. This results in a gradually extended hole in the shell. You can see, then, that hatching involves a fairly complex and stereotyped series of movements that cannot have been learned.

Other evidence that behavior can be inherited comes from studies of interspecific hybrids, especially in birds. If the parents have identical behavior patterns, the hybrid has the same patterns. For those behaviors in which the parent species differ quantitatively, the hybrids are usually intermediate. Where the parents differ qualitatively, behavior of the offspring varies considerably. In some cases behavior is intermediate. (It may tend to resemble one parent's behavior as the animal gains experience.) In other cases one parent's traits may be suppressed. Working with dabbling ducks, Lorenz obtained an interesting hybrid between the Chiloe teal and the Bahama pintail, *Anas bahamensis*. Neither parent displays the head-up–tail-up courtship pose characteristic of some related species (Fig. 16.2), but the hybrid does! These experiments indicate that there is a genetic component to behavior.

Many species have stereotyped behaviors that are alike in all members of the species (or at least in all members of one sex) and alike in each repetition by an individual. Certain acts are performed only in specific situations. Many of these stereotyped behaviors are species-specific. This does not necessarily mean that the behavior is inherited, however. It could also be due to similar experiences in similar environments by all individuals of the species.

In addition, there is evidence for behavioral differences controlled by single genes. One example is the mutant allele "yellow" in fruit flies, which in a homozygous male produces a yellow body and a slower rate of wing vibration in courtship. These males are significantly less successful than normal males in mating, because they are rejected more often by the females.

All this evidence supports the contention that genes have a strong influence on behavior. They are not the sole influence, however. Individuals also modify their behavior on the basis of experience, and some learned behavior is passed to offspring culturally (see Chapter 10). Individuals can learn appropriate behaviors from their parents or from other members of the species. Prairie dogs learn the boundaries of their colonies from older colony members.

Bird song seems to be heritable in varying degrees, depending on species. Some species, such as the canary, sing normally as adults, even when they have been reared in isolation from others of their species. Others, such as the chaffinch, sing a simplified form of the normal song when reared in isolation. Variations and extensions of this "blueprint," including a characteristic terminal flourish that most hand-raised birds do not sing, are evidently learned from other chaffinches under natural conditions.

A great deal has been said about which behavior patterns are transmitted genetically and which are learned. The consensus today is that the "nature-nurture" dichotomy has been overemphasized. Genes and the environment interact extensively, and very few behaviors can be attributed exclusively to one or the other. It is more appropriate to ask to what extent and in what manner the genetic and the environmental components contribute to the observed behavior pattern.

SOCIAL BEHAVIOR

We have discussed the behavior of individual animals in terms of their adaptation to their physical environment and to their need to reproduce. Let us now look at how they interact with other individuals. Most animals spend part or all of their life cycles in contact with others of the same species. In some species this contact may be restricted to the time of mating; in others the whole life may be spent as a member of a group of some sort.

If we look over the animal kingdom, social groupings appear in many taxa, and they are tremendously varied in form. They can differ in such qualities as size, cohesiveness, the relative proportions of different demographic classes, and differentiation of roles within the group. What are the forces that give rise to the various kinds of groupings?

A population is constrained in the social forms it can take by its history, which determined its present genetic make-up. It can evolve only at rates and into forms within the limits of its genetic flexibility. These constraints, however, can be quite broad, particularly when considered over a long time span. Given a range of genetic possibilities, there are a number of environmental factors that constitute the principal selective pressures that determine the extent of sociality and the form of society of a species. The following list is modified from one given by sociobiologist E.O. Wilson.

1. *Predator defense.* Aggregations of individuals make it more difficult for a predator to approach undetected. Thus an individual in a group has a higher chance than an individual alone has of surviving a predator approach. As a corollary, an individual in a group can spend less time on the alert for predators and more time on other activities, such as feeding. Predators may also find a group more dangerous to attack than an individual. In many birds, a flock will "mob" an intruding predator. Mobbing behavior also occurs in some social mammals, such as baboons.

2. *Increased competitive ability.* A group of social animals, by its size alone, may intimidate potential competitors of other species. Competitive ability can also be increased by cooperation. For instance, army ants, which forage in swarms, flush out and capture prey more efficiently than related species of ants that forage individually. Army ants also tend to capture larger prey. (This example can also be considered under the next category.)

3. *Increased feeding efficiency.* Social animals can increase their efficiency by observing other individuals and profiting from their experiences. In addition, some kinds of animals feed more efficiently in a group than alone. For instance, a flock of insectivorous birds stirs up more insects than the same number of birds flying separately, so each individual has more opportunities to capture food.

The pattern of food distribution is theoretically an important factor in determining the kind of social organization. If food is dispersed fairly evenly, animals will set up territories and defend them against other individuals of the species (provided the food is abundant enough to make defense worthwhile). If food occurs in patches and is available all the time or at predictable intervals, animals will live as close as possible to it and collect it individually. If food occurs in unpredictable patches, organized groups are favored for feeding efficiency.

4. *Increased reproductive efficiency.* Mating swarms, composed of individuals who have traveled over relatively long distances for courtship displays and mating, are known in many groups of insects. Presumably swarming is of particular value to rare species, individuals of which might otherwise have difficulty finding a mate. Other possible functions are to reduce inbreeding and to prevent hybridization with other species (premating isolating mechanism).

5. *Modification of the environment.* Animals in groups can modify the environment in their favor. An example at the most primitive level is the huddling together of rodents in winter for warmth. The other extreme is demonstrated by the hives of some higher social insects, such as the honeybee and the termite, which have a high degree of temperature and humidity control (Fig. 16.3).

SOCIAL ORGANIZATION

Now let us examine some specific patterns of social organization and, where possible, analyze the ecological and evolutionary factors that produce them and their adaptive value. The phenomena we will examine are territory, dominance, and mating systems.

Territory

The *territory* of an animal or group is defined by E.O. Wilson as "an area occupied more or less exclusively by the animal or group of animals by means of repulsion through overt defense or advertisement."

The concept of territory comes from studies of birds, and it is in this group that the clearest classic examples occur. In many species, such as the European robin, a male establishes a territory in the spring. He sings to advertise his presence, both to warn off other males and to attract a female. A pair forms, and throughout the rest of the breeding season, the male continues to defend the territory. In some species the female will also defend it.

Defense of the territory involves confrontation with intruders. Sometimes there is actual fighting; more often there is an aggressive display between the territory holder and the intruder. Usually the resident drives out the intruder, and the closer he is to the center of his territory, the more likely he is to succeed.

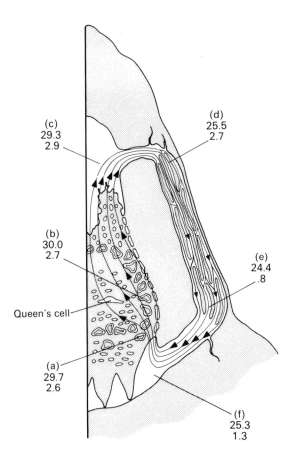

(c)
29.3
2.9

(d)
25.5
2.7

(b)
30.0
2.7

(e)
24.4
.8

Queen's cell

(a)
29.7
2.6

(f)
25.3
1.3

Figure 16.3

Temperature control and ventilation in a termite nest (*Macrotermes*). Termites require warm temperature and high humidity. *Macrotermes* nests can measure up to 16 feet high and up to 16 feet in diameter. The central nest is supported by pillars rising from the cellar (f), and contains the brood chambers (b), surrounded by fungus combs (a). Air heated in the central nest rises by convection to the attic (c). It then passes to a series of channels located in ridges on the nest surface. Here (d,e) CO_2 is passed to the atmosphere, and the oxygen content rises. The upper number at each location is the temperature (°C); the lower number is percentage of CO_2. Humidity is 98–99%. (From M. Lüscher, 1961. Air-conditioned termite nests. *Scientific American* 205(July):138–145. Copyright © 1961 by Scientific American, Inc. All rights reserved)

The size of a territory depends on the available resources and the population density. Territories are larger when the limiting resource is scarce than when it is plentiful. When there are relatively few individuals of the species, their territories will probably be larger than when there are many. However, there is a minimum size that depends on resources, and if there are too many individuals to hold territories in favorable areas, some set up territories in marginal areas, where success is less likely, and others form a nonbreeding floating population without territories. The latter birds may replace territory holders that die.

The term "territory" covers a variety of functions (Table 16.2). The area defended is usually a nest site, as in many birds that nest in groups, each defending its own nest, but the area can also or instead be a feeding ground. Or it may be a *lek*, a site for courtship display, where males gather, each defending a particular plot, and where females pass through and choose a mate. The animal may defend a fixed territory, making regular rounds to ensure that there are no intruders, or it may defend only the area where it happens to be (when the whole area is too large to monitor). The territorial defenders can be individuals or groups, male (more frequently) or female. Animals may be territorial at one stage of their life cycle or at one season, and solitary or colonial at others.

The advantage of territoriality for the individual seems to be that it gives ensured access to a desired resource, such as a nesting site, shelter, food, access to females, or some combination. One possible disadvantage of territoriality is that it can increase exposure to predators. But the chief disadvantage is the energy cost of defending the territory. If an animal spends time and energy in monitoring its territory and driving off trespassers, it cannot use that time and energy for activities contributing more directly to fitness, such as searching for food or care of young. Given this energy drain, it is advantageous for a territorial animal to recognize its neighbors, not to waste time in confrontations with them. Many birds can recognize and ignore a neighbor's song but respond aggressively to a stranger's.

Thus we expect to find territoriality where the resources are present over a reasonably long period, to make it worth staying in one place. Another necessary condition is that resources be fairly evenly distributed. If they are patchy, other individuals will also find it necessary to live near a patch, and territorial encounters will become very frequent, so frequent that they occupy too large a proportion of the animal's time. Under these conditions, the "scramble" strategy—each individual's gathering resources for its own needs—is more advantageous.

Table 16.2

Functions of territory.

Resource Defended	Example of Species
Food supply	Rhesus monkeys, song sparrows
Space for sexual display	Bullfrogs, prairie chickens
Nest site	Iguanas
Shelter	Spiny lobsters

Dominance

Territoriality is an interaction between two individuals (or two groups), whereas dominance concerns interactions within a group. The *dominance order*, or social hierarchy, is the set of sustained aggressive-submissive relations among a group of animals coexisting in one territory. There are many forms that the dominance order can take (Fig. 16.4). There may be one animal dominant over all the others, with no distinctions among the subordinates. It may be a linear order, with one at the top, a second submissive to the first but dominant to all others, and so on down the line. Or there may be a more complex network, with some circular elements.

The dominance order is formed through aggression, sometimes with actual fighting. The exact qualities that determine the dominance of an individual are not known. Some of the factors are sex (males are usually dominant over females); age (adults are dominant over juveniles, and age seems to be one of the determining factors in the ordering of adult males in various species of some ungulates, birds, and fish); size; aggressiveness; the status of the mother (in some monkeys); and previous history of defeats and victories.

When a group first forms, there is a good deal of aggression as the dominance order is being established. Once it is established, there is relative quiet. The animals go about their business peaceably except for occasional squabbles over food or the like. An individual can change rank, but once the order has been set up, such changes are much more difficult than they were before it was settled. In many cases, if the top individual is removed, aggressive encounters increase, as previously equal subordinates compete for the top position.

The selective advantage of being dominant is intuitively clear. The dominant animal has priority in food, nest sites, and mates. In addition, the top ranker is less subject to stress and hence to endocrine upset. *No!*

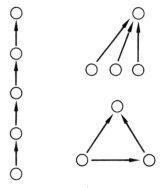

Figure 16.4

Unit relationships in a dominance order. Arrows indicate greater dominance. Relationships within a society consist of one or more of these unit relationships, with many combinations possible.

The advantage of the dominance order to the subordinate animals is less obvious. In some cases, particularly in the social insects, kin selection (see Chapter 10) may play a role. If an animal has only a small chance of surviving and rearing offspring on its own, it may increase its fitness by helping a relative instead. For instance, in *Polistes*, the paper wasp, groups of fertile females, many of them sisters, cooperate in founding a nest. One of them becomes dominant and lays the eggs, while the others function as workers. It has been suggested that, since many new nests fail, and since the subordinates generally have less developed ovaries than the dominant, the subordinates are unlikely to be able to rear their own offspring. Under such circumstances it may be more profitable for them to aid a sister.

In most societies, animals at the bottom of the hierarchy have the option of emigrating. Particularly when there is crowding, this is a viable option. A low-ranking individual in a situation where there is a shortage of some resource may be as likely to survive outside the group as in it. Species that live in transient habitats produce many emigrants. The meadow vole *Microtus pennsylvanicus* is an example. It lives in wet meadows, such as those appearing when a beaver dam is abandoned, which are rapidly invaded by forest trees. (The vole is also very common now in agricultural lands.) The population fluctuates from season to season. The higher the population, the greater the pressure is on subordinate individuals to emigrate. They are driven out by aggressive encounters with dominant individuals. The emigrants have a poor chance of survival, but some of them discover new places with a favorable environment and become dominant there. In addition, J.J. Christian thinks that the subordinate animals forced to disperse are an important factor in the evolution of mammals. Many of these animals emigrate to areas less well suited to their species. Of these, many die, but there may be some with mutations that are favorable in these new environments. Under appropriate conditions, they may give rise to new species.

There are some species that produce few emigrants. The hazards outside the group are so great that there is a big incentive to remain, even in a subordinate position. Baboons are an example. Subordinates eat well and occasionally mate, whereas loners do not live long. There is also the possibility of a change in the hierarchy when the dominant animals die. In many systems, rank is determined at least in part by age, so the longer an animal lives, the more likely he is to arrive at a high rank.

Dominance systems are similar to territorial systems in that both are organized by aggressive interactions. There is a certain amount of overlap between the two. A society with a hierarchy can be territorial. Some groups show a mixture. Black lizards are territorial, but some are dominant and are allowed to trespass. Some cats have a dominance order, but subordinates will defend their resting site against their superiors in rank. Many species are territorial under normal conditions, but they switch to a dominance hierarchy when crowded.

SEXUAL SELECTION AND MATING SYSTEMS

Natural selection can be separated theoretically into two aspects: survival selection, which maximizes the organism's competitive ability and hence its longevity (the l_x of a population); and reproductive selection, which acts on the other component of fitness, the reproductive success (the m_x of a population). Reproductive selection acts on all characters pertaining to reproduction, from mate selection to the rearing of the young, but the aspect that has attracted most attention is the differential ability of individuals of different genetic types to acquire mates, which Darwin called sexual selection (Chapter 10).

Two types of sexual selection have been distinguished. *Epigamic selection* involves the choice between males and females. *Intrasexual selection* results from competition among members of the same sex, usually males. Darwin believed that both types could account for the evolution of secondary sexual characteristics (differences between the two sexes apart from organs directly involved in reproduction). Epigamic selection would lead to ornamentation, odors, display behavior, etc., which would attract the opposite sex. Intrasexual selection would tend to produce armament, aggressiveness, etc., which would be advantageous to an individual competing with others of the same sex. We find in nature both kinds of secondary sexual characters. Extreme examples, such as the elaborate plumage of the bird of paradise or the impressive antlers of some deer, may seem so overdone as to be detrimental to the animal possessing them. Thus a bird's plumage, if more showy than that of other males, may be more attractive to females, but it can also make the bird more conspicuous to predators. Modern theory says that selection operates to maximize fitness, taking both sexual selection and survival selection into account. Sexual selection will lead to extreme characters only insofar as they increase overall fitness.

The fact that sexual selection leads to sexual dimorphism is an indication that selection can operate differently on males and females. A given character or strategy need not be equally advantageous to both sexes. This potential difference accounts in a large part for the variety of mating systems we observe in nature.

Let us define *parental investment* as any behavior toward an offspring that increases the chances of the offspring's survival at the cost of the parent's ability to aid survival of other offspring. Females in general produce gametes with considerable stored energy, whereas males produce gametes with genetic instructions but little stored energy. Therefore egg production is limited compared with sperm production, and a female's parental investment in a single gamete is greater than a male's. A male can increase the number of his offspring by mating many times; a female usually has all her season's eggs fertilized in a single mating. Thus errors in mate selection are much more serious to a female than to a male. A mating with a member of another species or an undesirable individual of her own species may mean that a female produces no viable young in a season, but such an error in a male may mean the loss of only a little time and a few sperm. Thus discrimination in choosing a mate is strongly to the female's advantage, whereas it is not so important to the male.

Gordon Orians has developed a theory of mating systems based on mate selection by the female that maximizes individual fitness. He gives two main bases for female choice. (1) The genetic quality of the male. The more fit the father, the more fit the female's offspring. Therefore the female should be responsive to characters of the male reflecting fitness. (2) The likelihood of receiving support from the male in reproduction and caring for the young. In some species the male provides a defended territory with its resources, food for the young, incubation for the eggs, etc. In territorial species, the female's choice is likely to be governed by the quality of the territory.

When female choice is based on genetic quality, the mating system is likely to be polygyny (one male with several females) since it makes no difference to the female if the male has mated before. There is likely to be strong sexual dimorphism, with emphasis in the male on characters attractive to the female or useful in intramale competition.

Where the male offers more than his genes, the mating system is frequently monogamy (one male and one female). A female will be likely to prefer a somewhat inferior unmated male to a male already mated, whose help she will have to share. Polygyny can occur where the female's fitness will be greater by sharing a male on a high-quality territory than by accepting an unmated male on a low-quality territory.

Under what conditions, then, is the male likely to aid the female to rear offspring? In some cases, conditions are so severe that the female usually cannot raise offspring by herself. Then the male's fitness will be greater if he helps provide food, shelter, protection from predators, etc., instead of trying to father more offspring. For example, defense of scarce nesting sites or food sources by both parents may be necessary. Another factor favoring monogamy applies to species where mating is lifelong, such as the kittiwake, a cliff-dwelling gull. In these species breeding can start earlier in the season, since the two sexes are already paired and do not have to go through courtship. In the kittiwake, the already paired gulls, on the average, raise more young than those that must first find a mate.

CONCLUSION

As we have seen, behavior is subject to natural selection. Behavioral characteristics are derived from and cannot be isolated from an animal's anatomy and physiology. Within the context of the organism as a whole, behavior patterns are usually adaptive, and frequently they can, at least to some extent, be transmitted either genetically or culturally.

Some specific patterns of behavior can be found in many species. One of the main thrusts of current work on the evolutionary aspects of behavior is to pinpoint the ecological pressures that promote these patterns. We have discussed some of this work in this chapter.

SUGGESTED FURTHER READING

General references

Brown, J.L. 1975. *The Evolution of Behavior.* New York: Norton.

Hinde, R.A. 1970. *Animal Behavior: A Synthesis of Ethology and Comparative Psychology,* 2nd ed. New York: McGraw-Hill.

Thorpe, W.H. 1961. *Bird Song: The Biology of Vocal Communication and Expression in Birds.* Cambridge: Cambridge University Press.

Wilson, E.O. 1975. *Sociobiology: The New Synthesis.* Cambridge, Mass.: Harvard University Press.

Wilson, E.O. 1978. *On Human Nature.* Cambridge, Mass.: Harvard University Press.

Articles and reviews

Alexander, R.D. 1974. The evolution of social behavior. *Ann. Rev. Ecol. Syst.* 5: 325–384.

Christian, J.J. 1970. Social subordination, population density, and mammalian evolution. *Science* 168:84–90.

Hamilton, W.D. 1972. Altruism and related phenomena. *Ann. Rev. Ecol. Syst.* 3: 192–232.

Littlejohn, M.J., M.J. Fouquette, and C. Johnson. 1960. Call discrimination by female frogs of the *Hyla versicolor* complex. *Copeia* 1960:47–49.

Lorenz, Konrad Z. 1958. The evolution of behavior. *Scientific American* 199 (Dec.): 67–78.

Luscher, M. 1961. Air-conditioned termite nests. *Scientific American* 205 (July): 138–145.

Manning, Aubrey. 1965. *Drosophila* and the evolution of behavior. In J.D. McCarthy and C.L. Duddington (eds.), *Viewpoints in Biology,* 4. London: Butterworths.

Orians, G.H. 1969. On the evolution of mating systems in birds and mammals. *Am. Nat.* 103:589–603.

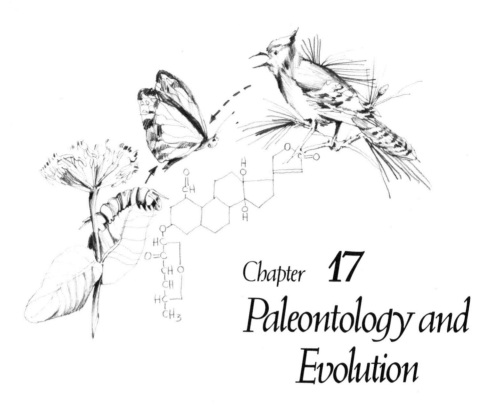

Chapter **17**

Paleontology and Evolution

In our discussion so far we have analyzed three major components of the theory of evolution. First we discussed genetic variability, which provides the raw materials for evolution. The ultimate causes for the existence of genetic variability are the physical and chemical properties of the substances of which organisms are formed, in particular the nucleic acids, DNA and RNA, and the proteins. These substances assembled into genes and chromosomes provide the continuity of genetic information from generation to generation. At the same time, the way chromosomes divide makes possible the occurrence of errors of duplication, called mutations, as well as the rearrangement of genes, so that new syntheses of genetic information appear. Selection is the second dimension of evolution. Due in part to the overabundance of offspring in terms of the available supply of food, selection will favor the type that is capable of using resources in the most efficient way and as a result leaves the most offspring. The third dimension is provided by a diversified environment, which allows for the selection of more than one type of organism over a given area. The final dimension, time, allows an interplay between genetics, selection, and a variable environment. At any one moment a series of diverse environments are found over the surface of the earth. Given enough time, a series of diverse environments will succeed each other at any one place. Looking at environmental fluctuations in terms of space and time, one can observe a veritable crazy quilt of ever-changing conditions. At each point existing organisms reflect in their genetic constitution and morphological makeup their adaptation to the then-present environmental conditions, as well as the conditions of their immediate ancestors and, to a lesser degree, of their more remote ancestors. Consequently, in order to get a true picture of evolution, it is essential not to overlook this fourth dimension.

IMPORTANCE OF FOSSILS IN EVOLUTIONARY STUDIES

The great diversity of past and present plants and animals is a manifestation of the response of organisms to the fluctuations of the environment over the millennia of time. The origin of living forms from simpler, nonliving substances is in itself a response to changing conditions in the environment. In effect, if the earth had not cooled and the elements had not become distributed over its volume as they were, or if the conditions of temperature and energy had been such that water could not have been formed, life would not have originated—at least not in the form familiar to us. The ultimate causes responsible for changes in the environment are extraterrestrial and not too well understood. Nevertheless, we have a certain comprehension of the types of changes that have occurred on the earth since its inception, particularly those that have taken place in more recent periods. This evidence is provided by the remains of plants and animals that once occupied the surface of the earth and the waters of the sea. These remains are known as fossils (Fig. 17.1).

The branch of biology involving the study of fossils and the past conditions of the earth is paleontology. If the shapes and forms of the organisms that inhabited a certain part of the globe in a certain period are known, inferences can be made about the climate of that period by analogy with what we know about the requirements of present-day forms. For example, we know that monkeys are restricted to tropical regions and that they will die if exposed to cold winters. Fossil remains of monkeys have been found in beds corresponding to the geological era called the Tertiary of Patagonia in southern Argentina, a region that today has very cold winters and is unsuited for monkeys. Two explanations are possible: The monkeys of the Tertiary were cold-adapted, or Patagonia was much warmer in the Tertiary than today. The presence of other fossil remains of tropical animals, such as anteaters, and of tropical trees indicates that in all likelihood the second interpretation is the correct one. In the same fossil beds, remains of other animals are found that are now extinct, such as marsupial saber-toothed tigers and giant sloths. By analogy and extrapolation, we can conclude that these animals, too, preferred

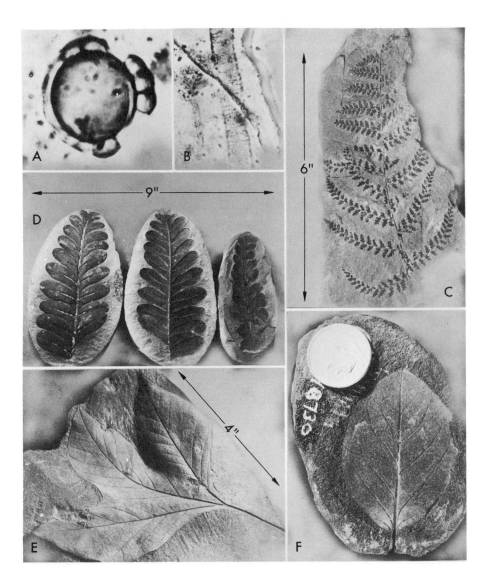

Figure 17.1

Some plant fossils of different ages. (A) *Eosphaera tilerii* (Precambrian). (B) *Ammikaea septata* (Precambrian). Both are among the oldest remains of living organisms known. (C) *Sphenopteris* sp. (Mazon Creek, Ill., upper Carboniferous). (D) *Alethopteris* sp. (Mazon Creek, Ill., upper Carboniferous). (E) *Sassafras* sp. (Dakota sandstone, middle Cretaceous). (F) *Diospyros rotundifolia* (Dakota sandstone, middle Cretaceous). Note how the plants become more familiar as we move forward through time. (Photos courtesy Professor E.S. Barghoorn)

warmer climates. Likewise, the type of plants with which they were associated can tell us something about their feeding habits and the type of environment they lived in.

The paleontologist, then, tries to reconstruct the life and the environments of the past, on the basis of the assumption that the response of living things to the environment was the same then as it is today. This inference is in part conjectural, but it has its basis of fact in the study of chemical substances extracted from fossil beds. Analysis of fossils, as well as of meteorites and the light that comes to us from stars millions of light-years away—light that emanated in many cases before the earth was formed—shows that the chemical composition of these bodies is the same as that found today and that the chemical substances display a similar behavior. These results back the conclusion that the building blocks of the universe are the same and that there is only one kind of matter. It follows that natural laws have always operated in the same way they do today, and if so, the behavior of living things in the past must have been similar to that of present-day organisms. We know that some species have evolved with little or no superficial morphological modification but with considerable internal and presumably physiological modification. However, it is highly unlikely that most species have evolved in such a manner.

HOW FOSSILS ARE FORMED

The word "fossil" is derived from the Latin word for digging up. We may define a fossil as any recognizable organic structure, or impression of such a structure, preserved from prehistoric times. This includes bones, wood, skin, and pollen, as well as tracks, dung, burrows, or any other remains formed by a prehistoric organism, or any that gives an indication of the presence of such an organism.

The bodies of most dead plants and animals decompose sooner or later. If this were not so, the surface of the earth would be covered several miles thick by the remains of dead organisms, and actually life would probably have ceased long ago. All available mineral matter would have become encased in the bodies of organisms. Decomposition is caused by several agents. Scavengers of all sorts, from vultures to bacteria and fungi, take care of the soft parts very rapidly. Harder tissues, such as bones and wood, are more resistant. Nevertheless, enzymes secreted by certain fungi and bacteria, acids occurring naturally in the soil, and the combined action of such environmental agents as water, wind, and temperature destroy even the hardest organic remains with time. Under very special circumstances, however, some parts may be preserved, with varying amounts of modification, to form fossils (Fig. 17.2).

For preservation to take place, rapid burial after death in a substance that protects the organism from destruction by scavengers and weathering is needed. The most common substance is some heavy silt or mud, and burial

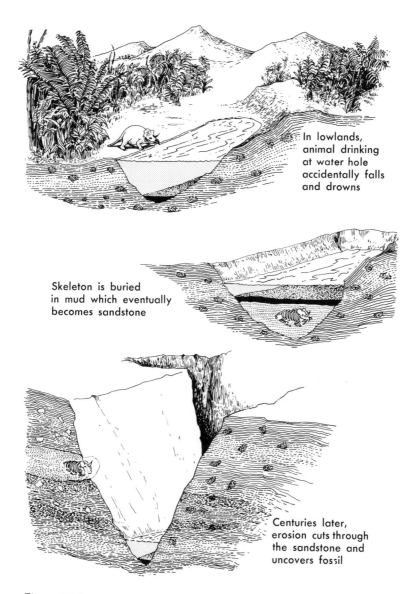

In lowlands, animal drinking at water hole accidentally falls and drowns

Skeleton is buried in mud which eventually becomes sandstone

Centuries later, erosion cuts through the sandstone and uncovers fossil

Figure 17.2
Schematic representation of the process of fossilization. (From O.T. Solbrig, 1966. *Evolution and Systematics.* New York: Macmillan)

will occur more frequently in water than on land, and in colder areas than in warm ones. But other substances can serve as well. Some of the best fossils are found in amber, which is a fossil resin. Particularly small insects, which probably got stuck in the resin and could not free themselves, are found in amber. Some very well preserved remains of woolly mammoths and rhinoceroses from the Pleistocene age have been found frozen in Siberia. Volcanic ash also can create good conditions for preservation. But even when an organism is rapidly buried, its soft parts decompose, and in most cases only the hard parts remain. Shells, which are inorganic in nature, are usually preserved completely unaltered; bones and wood, on the other hand, are often mineralized (commonly referred to as petrified)—that is, the free spaces in the tissue are impregnated with mineral substances carried by ground water. Sometimes the plant or animal structure is totally destroyed, leaving an empty space that nevertheless faithfully reflects the exterior aspects of the dead organism; such a fossil is a mold. If the space is subsequently filled by a foreign substance, the result is a cast (Fig. 17.3).

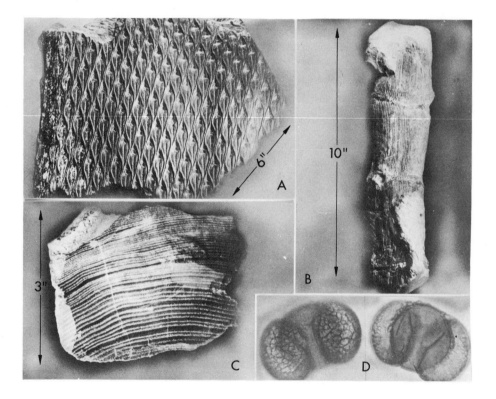

Figure 17.3

Examples of different types of fossil remains. (A) Part of trunk and bark of *Lepidodendron* sp. (Carboniferous). This is an impression made by the original plant on the rock. (B) Stem fragment of *Calamites* sp. (Carboniferous). This is an original pith cast. (C) Section of silicified wood of *Gleditchia montanense* (Miocene). (D) Fossil pollen grains of pine (*Pinus* sp.) from the Miocene of Oregon. (A–C: Photos courtesy Professor E.S. Barghoorn; D: courtesy Dr. A. Graham)

After burial the fossil may be very much altered. The commonest type of alteration is flattening due to pressure as more and more layers of sediment are piled on top of the bed containing the fossil. The bed may be further altered by folding and lateral movements. Eventually it has to be uncovered by erosion if we are to get possession of the fossil, and it may be further damaged in this process. Some fossils have been exposed in the past and reburied. Consequently, the chances of recovering an intact specimen are very slender, particularly a specimen of great age.

THE PROBLEM OF THE MISSING PARTS

The paleontologist is presented in most, if not all, cases with a fossil that represents a highly modified remnant, usually incomplete, of life from the past. The goal is to learn as much as possible about it and the conditions in which the organism lived. The first thing to do is to classify the fossil and relate it if possible to some known group. This is no minor problem with an incomplete and modified fragmentary fossil remain of an extinct organism. The fossil may have key structures preserved, such as teeth, jaws, or skull, but in many instances only a femur or a rib or a tarsus has been found. Since these structures can give at best only an approximate idea of the animal, paleontologists use the device of *form species*. Each remain is given a specific name and recorded as a different species, and an educated guess is made as to its possible affinities. If a more complete skeleton is found later on, and two or more form species are shown to be parts of the same organism, the form species names are abandoned in favor of the correct name of the animal, which can now be classified more precisely. Form species permit the paleontologist to deal in a scientific way with incomplete remains that otherwise would be useless. The paleobotanist in particular is aided by this device. In effect, vertebrate animals by and large can be identified by the remains of their bones and teeth, which are the structures most likely to be preserved. Not so with plants. Wood and leaf impressions, which are the most common plant fossils, are often insufficient material for identifying with any certainty the group to which they belonged.

DATING FOSSILS: RELATIVE AND ABSOLUTE AGES

If fossil specimens are to be of any value, their age must be known. To determine the age of fossiliferous beds, paleontologists use fossils! This may seem like a paradox, but actually it is not. William Smith, the father of modern geology, as early as 1790 made the observation that different geological strata were characterized by unique assemblages of fossils, and that consequently particular strata could be identified by their fossil contents, even when their physical characteristics were changed. In areas where little or no disturbance had occurred, the order of deposition of the strata could be ascertained, since obviously the overlying beds must have been deposited after the ones below them. By then comparing the fossils of beds from different areas, paleontologists were able to establish a relative chronology, known as the *geological column* (Table 17.1). A harder task was to establish time intervals in years, and until recently there was no accurate way of doing it. Since 1950 methods of obtaining more exact ages based on radioactive decomposition of elements in rocks have been devised. These methods have shown that the age of the earth estimated by the old methods was less than its actual age. The application of these methods also confirmed the correctness of the relative ages used by the paleontologists.

To date a fossil bed, paleontologists look for fossils known to be restricted to a certain geological stratum, with which they establish the relative age of the bed. All the unknown fossils in the bed are then given the same age. It is of paramount importance to make sure of the location of the fossils in the bed and to know the overlying and underlying beds, since fossils are known to have been moved in and out of strata in the course of geological time. At present, when possible, samples are also collected with which ages are obtained by radioactive measurement methods.

THE MOLECULAR CLOCK OF EVOLUTION

If the neutrality theory of protein evolution (Chapter 9) were found to be correct for a sufficiently large number of proteins, then the rate of amino acid substitutions might be used to date the time of phylogenetic events. In effect, according to the neutralist theory, the rate of allelic substitution is equal to the rate at which alleles arise by mutation. If mutation rates for a given gene remain fairly constant over long periods of time, then allelic substitutions occur at a constant rate. By ascertaining the number of amino acid substitutions in a given homologous protein in two or more organisms, it would be possible to determine accurately the time when they differentiated from a common ancestor. In other words, if the assumptions are correct, proteins have a kind of built-in "clock" that dates phylogenetic events. The clock is of course a stochastic clock (like radioactive decay), but over long periods of time such a clock is fairly accurate, and furthermore, by the use of several proteins, it would be possible to refine the clock further.

We of course do not know whether the neutrality theory of evolution is correct. Recent studies on protein evolution, however, have established that the overall rate of substitution is not constant over time. Consequently, proteins apparently do not evolve at a stochastically constant rate. Nevertheless, the average rate of evolution over long periods of time and over many proteins may be used as an approximate evolutionary clock (Fig. 17.4). This information is very important and promising in the quest to date certain evolutionary events.

Table 17.1

The Geological Column.

Time Since Period Ended, in Millions of Years	Approximate Duration, in Millions of Years	Era	Period	Epoch	Some Important Evolutionary Events
1	1	Cenozoic	Quaternary	Recent, Pleistocene	Advent of man
	62		Tertiary	Pliocene, Miocene, Oligocene, Eocene, Paleocene	Development of modern families of mammals, birds, and flowering plants
63	72	Mesozoic	Cretaceous		First flowering plants
135	46		Jurassic		Age of reptiles, gymnosperms
181	59		Triassic		
230	50	Paleozoic	Permian		Radiation of amphibians; first reptiles
280	65		Carboniferous		
345	60		Devonian		First land plants and first land animals
405	20		Silurian		
425	75		Ordovician		Jawless vertebrates; first sharks
500	100		Cambrian		
600 approx. 3,500?	approx. 3,000	Precambrian			First forms of life, approx. 2,500,000,000 years ago

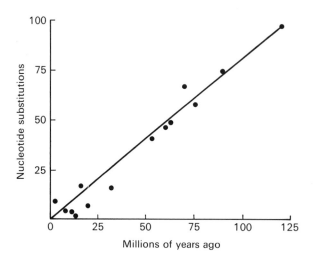

Figure 17.4

Nucleotide substitutions versus paleontological time. Nucleotide substitutions were calculated for seven proteins in pairs of mammalian species, graphed against the estimated date of the common ancestor. (The line was drawn from the origin to the outermost point.) (From W.M. Fitch, 1976. Molecular evolutionary clocks. In F.J. Ayala (ed.), *Molecular Evolution*. Sunderland, Mass.: Sinauer Associates. Reprinted from *Federation Proceedings* 35:2092–2097, 1976)

THE CONCEPT OF SPECIES IN TIME AND SPACE

We have defined species as a group of potentially interbreeding populations. We also indicated that species is the only natural nonarbitrary category, as opposed to the higher categories, which are arbitrary. But if we take into account the fourth dimension of evolution, our concept of interbreeding is no longer applicable. A present-day horse is not capable of interbreeding with one that lived in the Middle Ages, not even with one that lived in the last century, and much less so with horses of the Pleistocene. Do all these horses, then, constitute distinct species? Evidently not. But species in time lack the limits assigned to contemporary species on the basis of their breeding behavior. Furthermore, with the passage of time, the populations evolve and pass imperceptibly from one into another.

The problem is still further compounded by the scantiness of the fossil record. In very few instances does the paleontologist have a good sample of the variability of the species at a given time level. Breaks in the continuity of the fossil record are usually (all too often) the species limits. The early form has by then evolved enough so that on morphological grounds (and here again analogy with corresponding modern species is applied, so that fossil species are separated when the morphological differences are comparable to modern species) it can be considered a different species. But there is always the possibility that the intermediate forms, if they were known, would form a continuous series. A good example of a continuous series is reflected in the confused terminology of human ancestors; practically every skull has at one time been given a different generic name! The paleontologist, of course, does not know, without a continuous series through time, whether the earlier form is indeed an ancestor of the more modern one, and the evolution of the horse is a case in point.

THE EVOLUTION OF THE HORSE FAMILY (EQUIDAE)

The Equidae are represented today by one genus with six species: *Equus caballus* (the horse), *E. asinus* (African donkey), *E. hemionus* (Asiatic donkey), and three species of zebras: *E. zebra, E. grevyi,* and *E. quagga*. The genus *Equus* originated in the late Pliocene, but the horse family goes back 60 million years to the Eocene period. The Equidae developed out of the ancestral stock of a primitive and long-extinct order of mammals called condylarths that lived at the beginning of the Tertiary period. The first known member of the horse family to arise from the condylarths was *Hyracotherium*. It was somewhat doglike in size and shape, with a small head and a large tail, and it stood 25 cm high in some species and up to 50 cm in others. It had padded feet (like present-day dogs) with four toes on the forefeet and three on the hind feet, and the toes ended in small separate hoofs. As indicated by the low-crowned, simple-cusped teeth with low ridges, *Hyracotherium* fed on succulent leaves, small fruits, and soft seeds. It must therefore have been a browsing, forest-inhabiting animal. Although it lived in both New and Old Worlds, the horse family evolved mostly in North America, with repeated invasions into the Old World (Fig. 17.5).

By the Oligocene epoch, *Hyracotherium* had evolved into the genus *Mesohippus*. This animal was somewhat larger, with padded feet, only three toes on all four legs, and a more horselike head with set-back eyes and an elongated muzzle. The brain had also increased in size, and the teeth, though basically like those of *Hyracotherium*, had larger crests. In the Miocene epoch we find *Merychippus*, a considerably larger animal, which stood about 1 meter tall. The more advanced species of this genus had lost the foot pads, and although they still stood on three toes, the body was carried largely on the hoof of the central toe. The teeth had high crowns, a complex pattern of crests, and cement, which its ancestors' teeth lacked. This is indicative that the animal no longer lived in the forest but in savannalike habitats and that its diet was mainly grass. *Pliohippus* was a one-toed grazing horse, which roamed the North American plains in the Pliocene era, and from which *Equus* arose at the end of that period. Compared with its ancestors, *Equus* is large, with single-toed feet and large hoofs, and with high-crowned and elaborately crested and ridged teeth that are adapted to the diet of grasses it lives on. It

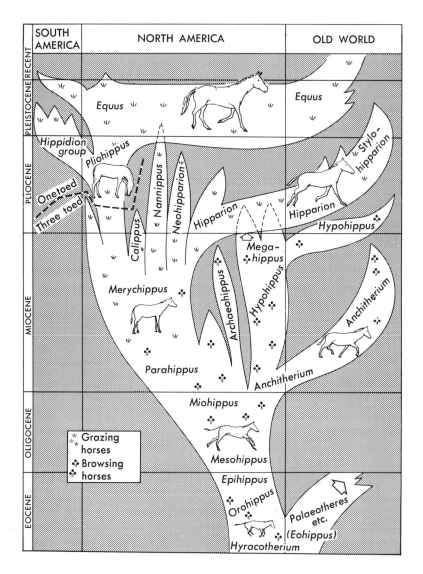

Figure 17.5
The lineages of the horse family. The restorations are to scale.

became extinct in the New World but survived in the Old World. (The so-called wild horses of western North America are recent human introductions of European domestic horses that were set free.)

In addition to this main line of descent, different other sidelines developed, such as *Anchitherium, Archaeohippus,* and *Hypohippus* in the Eocene, and *Hipparion, Nannipus,* and *Neohipparion* in the Pliocene (Fig. 17.5). The horse family is the group whose fossil record is probably best known. In regions where its members are represented by a good fossil record, the transition from one genus to the next successive one is gradual, and consequently the dividing lines between the genera are arbitrary.

The fossil record of the Equidae exemplifies the appearance and ultimate fixation of adaptive characteristics, particularly well illustrated in the evolution of the teeth. One of the cheek teeth of *Equus* has a short projection called a crochet. This projection is absent in most specimens of *Mesohippus* and other Oligocene horses. But occasional individuals, which undoubtedly represent a mutant type, had a small crochet. In *Miohippus* the crochet was present in more individuals, but there were still some without it. In the Miocene-aged *Parahippus*, the crochet had become universal in all populations, but it was still small. It became larger later in the ancestors of *Equus*. A similar situation holds for the appearance of cement in *Parahippus* and its successor *Merychippus* in the Miocene. The appearance of cement was first sporadic but eventually became general and later increased in thickness. This sequence corresponds to the time when the horses were moving from browsing to grazing. The harder grass required better-wearing teeth. Cement, which is a reinforcing material for grinding teeth, was obviously of selective advantage. So also were the development of more elaborate crests on the grinding surface of the cheek teeth and the increase in size of the crown. It is interesting to note that the development of all these characteristics proceeded with greater speed in the Miocene with the appearance of grazing horses.

RATES OF EVOLUTION

The evolution of the horse family also illustrates very nicely the fact that evolution does not proceed at a steady rate or always in the same direction. This should by now be apparent from all that has been said about the way natural selection operates. We have seen how in the Equidae the change from forest-living browsers to plains-living grazers brought about an increase in the rate of evolution of those structures that were most affected by the new situations, in particular the teeth. There were also an increase in size and a change from padded feet to the springlike action of present-day *Equus*. In other lineages there have been few or no changes during the time the horses were evolving. This is illustrated by the opossum, which has changed very little since the late Cretaceous, some 80 million years ago.

There is in fact no reason for evolution to proceed always in the same direction. Although modern-day *Equus* is considerably larger than *Hyracotherium*, it is smaller than some ancestral Pliocene species of *Equus*. In addition, some extinct Pliocene horses that were not ancestral to *Equus* (e.g., *Megahippus* and *Hypohippus*) are still larger, whereas others, such as *Calippus* and *Nannipus*, are considerably smaller.

The differences in evolutionary rates can be attributed primarily to the changes or lack of them in the environment. After a lineage has reached a high adaptive level, few changes will take place in a uniform environment. A rapidly changing environment, on the other hand, will constantly produce new selective requirements to which organisms will have to adapt. In addition to the environment, intrinsic factors also play a role. All those characteristics that increase variability (such as large populations, high number of chromosomes, short generations, and so on) are advantageous in times of environmental fluctuations, and organisms possessing them will evolve at a faster rate.

A final point to be drawn from our brief analysis of the fossil record, which is also well illustrated by the Equidae, is that the fate of most, if not all, species and genera is extinction. Some—the minority—become transformed into more evolved types, but the majority eventually can no longer adapt to the ever-present new environmental conditions or compete with better-adapted organisms, and they become extinct. This has been the fate of 17 out of 18 genera of Equidae (Fig. 17.5) and of nine-tenths of all species!

In conclusion, the fossil record provides considerable evidence for the soundness of the synthetic theory of evolution. Nevertheless, largely because of the incompleteness of the fossil record and its nonrandomness, fossils do not by themselves provide the proof of evolution, despite widely held belief that they do. Only the combined and judicious application of data taken from fossils, living organisms, and experiments can provide an approximate justification for Darwin's theory.

SUGGESTED FURTHER READING

General references

Calvin, M. 1969. *Chemical Evolution: Molecular Evolution Towards the Origin of Living Systems on the Earth and Elsewhere.* Oxford: Oxford University Press.

Kummel, B. 1961. *History of the Earth.* San Francisco: Freeman.

Simpson, G.G. 1949. *The Meaning of Evolution.* New Haven: Yale University Press.

Simpson, G.G. 1951. *Horses.* New York: Oxford University Press.

Articles and reviews

Axelrod, D.I. 1970. Mesozoic paleogeography and early angiosperm history. *Bot. Rev.* 36:277–319.

Bock, W.J. 1972. Species interactions and macroevolution. *Evol. Biol.* 5:1–24.

Fitch, W.M. 1976. Molecular evolutionary clocks. In F.J. Ayala (ed.), *Molecular Evolution.* Sunderland, Mass.: Sinauer Associates.

Simpson, G.G. 1950. History of the fauna of Latin America. *Am. Scient.* 38:361–389.

Wolfe, J.A., and E.S. Barghoorn. 1960. Generic change in Tertiary floras in relation to age. *Am. J. Sci.* 258A:388–399.

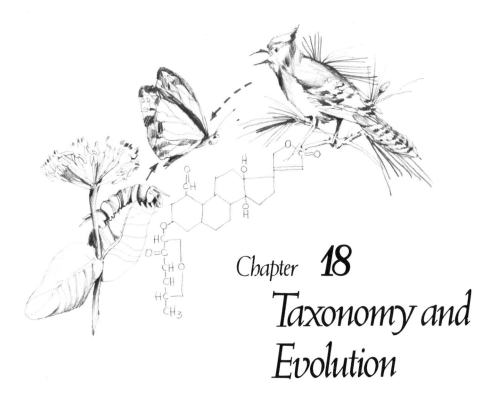

Chapter **18**

Taxonomy and Evolution

Understanding the world of organisms that surrounds us requires the ordering of the multitude of forms into some sort of rational system. Consequently, systematics is as old as the human quest for knowledge, and the effort to classify and understand the variability of animals and plants has led to the development of all the other branches of biology. The information obtained by these more specialized approaches has in turn affected and profoundly modified the field of taxonomy, transforming it from a mere effort to classify into the scientific enterprise of discovering and understanding the reasons for the apparent order of nature.

MODERN TAXONOMY

Although the rate of change has been steadily increasing, the transformation of taxonomy into a science has been slow. Obviously, not until each of the groups of organisms has been provisionally circumscribed and those groups have in turn been classed into larger groups, such as genera, families, orders, and classes, can any serious effort to study the reasons for their existence be made. So far, about 1,500,000 species have been described and classified. Of the major groups of animals and plants, only the birds, with about 8,650 species, and perhaps also the gymnosperms among plants, with about 650 species, have been more or less completely surveyed. Our knowledge of the kinds and numbers of all the other groups is incomplete, and although forecasts of the number of undescribed species are highly conjectural, it has been estimated that there are more than 100,000 undescribed species of flowering plants, 20,000 species of fishes, and more than 1,000,000 of insects! The end to the work of describing and classifying the kinds of organisms is not yet in sight.

But the work of systematists does not stop with description and classification of species. They want to know the genetic relationships and the history of the species they work with and the mechanisms that brought them into being. They also want to discover laws governing the behavior of all organisms. As biologists working in other disciplines learn more about the chemistry, physiology, anatomy, and behavior of plants and animals, taxonomists are deluged with a wealth of information to consider in attempting to formulate classificatory and evolutionary interpretations (Fig. 18.1). Probably nobody is more aware of the scientific information explosion than the small band of working taxonomists! At the same time, increased understanding of all aspects of organismic function and behavior has made systematics more interesting and exciting than ever, and this oldest of biological disciplines is experiencing a burst of activity as never before.

In this book we have analyzed aspects of some of the newer branches of biology, such as cytology, genetics, and the behavioral sciences, as well as more classical fields, such as ecology and paleontology. It must be remembered that the impulse for the development of these sciences was given by taxonomy, and that the spillover from the other disciplines that systematics is experiencing is a feedback from the early spillover of systematics. It may therefore be useful to survey briefly some historical developments and the methods and procedures of taxonomy.

THE PRINCIPLES OF CLASSIFICATION

Classification is the ordering of organisms (or any other objects) into classes. The organisms in each class are held together by some system of relationship among them. Language is a primitive form of classification in which each class of "things" is given a designation (noun). All the members of a class are held together by the possession of common characters or relationships. The first step in classification is the delimitation of the classes; the second step is the establishment of relationships between the classes and the formation of a hierarchy. Relationships can be established in two ways: first, by the overlapping or coincidence of nonidentical classes, and second, by the subordina-

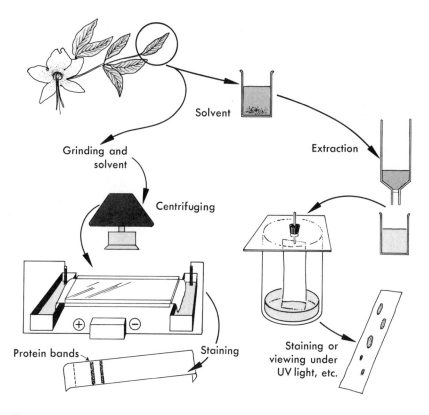

Figure 18.1

Chemosystematics. Chemistry is one of the branches of science that now provide systematic information. Groups of organisms frequently have characteristic chemical compounds or classes of compounds that can give the systematist clues to their phylogeny. (From O.T. Solbrig, 1970. *Principles and Methods of Plant Biosystematics.* New York: Macmillan)

tion of some classes to others or the inclusion of one class within another. As an example of overlapping, there can be a class of long-furred dogs and a class of brown-furred dogs, which are nonidentical. A class of long-furred, brown-colored dogs overlaps both of the given classes and is coextensive with the overlap of the two.

Classification by subordination and inclusion of classes is used in language together with overlapping. "Red foxes," "gray foxes," and "arctic foxes" all belong to the larger and inclusive class "fox." In biological classification, ordering by subordination and inclusion of classes is the usual method, and its use goes back to Aristotelian logic. All classes are included in larger ones. For example, the species *Felis felis* (the cat), *Felis concolor* (the puma), *Felis tigris* (the tiger), and *Felis onca* (the jaguar) are all included in the genus

.... In turn, the genus *Felis* is included in the family Felidae, in the order *...*arnivora, and in the class Mammalia. Such an arrangement is a hierarchy.

According to this method, a species consists of its *genus* plus its *differentiae*. The "genus" encompasses all the properties common to all the species of the next inclusive class (in this case also called the *genus*), those properties by which it is defined. The "differentiae" are the characteristics by which we differentiate one species from another within a given genus. The family is likewise defined by the characteristics common to all the genera included in the family (the "genus" of the family), and so on with each higher class. This method of classification is useful if we are interested in establishing an order based only on the logical relationships between the genus and its species, and among the species themselves. So, if we defined the genus *Felis* by the possession of tails and rigidly applied Aristotelian logic, we would have to exclude bobtailed cats from the genus (and also from the species *Felis felis*, the cat). But bobtailed cats do belong both to the species *Felis felis* and to the genus *Felis*.

In practice, we classify on the basis of the totality of the characters of the organism and not primarily on the basis of logical consistency. If an animal lacks some of the "genus" characters but has all the others, we keep it in the class to which it would belong if it had the missing characteristics. The scientific justification for this procedure is given by evolutionary theory. In adapting to different environments, populations may change in any way, and both the "genus" and the "differentia" are likely to change. We define species, whenever possible, primarily in terms of genetic and breeding relationships, and only secondarily in terms of morphological characteristics. Doing so presents a problem, however. If related classes (in a phylogenetic sense) do not have any common characteristics, how can a higher, inclusive class be defined? A partial answer to this problem is given by set theory. A genus in set theory is defined in terms of several characteristics (the more the better), and each species must have at least a minimum number of these characteristics in order to be included in the genus. This procedure would be completely satisfactory if the phylogenetic relationships were known. But they seldom are. As a matter of fact, once the classification is established, a phylogeny is deduced from it.

In taxonomic practice, an empirical, pragmatic approach is used, which combines elements of Aristotelian logic, set theory, and evolutionary theory. Field observation and experimentation, supplemented by museum and herbarium studies, provide information on variability and breeding relationships of populations. On the basis of these data, the elementary classes, the species, are established. Using the totality of the known characteristics of all the species, taxonomists then establish genera, so that the degree of difference separating the members of the same genus is less than that separating the species of different genera. This can be done in two ways. One is a strictly statistical approach, in which as many characters as possible are quantified and multiple correlations are established by the use of high-speed digital computers. This method is logically consistent in terms of set theory. Its greatest drawback is that all aspects of the phenotype are given the same importance. Yet we know that some are determined by a few genes and others by large series of genes. Furthermore, certain characteristics of the organism (those of immediate adaptive value) are very plastic, but others (such as reproductive structures) are not. In other words, the phylogenetic significance of these characters is different because the rate at which they can evolve is very different. The usual procedure is therefore to weight characters according to their potential phylogenetic significance and to establish higher classes accordingly. The great drawback of this procedure is that it introduces a subjective element (weighting of characters), but it has been upheld so far in those cases where a phylogenetic check (via the fossil record) has been possible.

TAXONOMY, SYSTEMATICS, CLASSIFICATION, AND NOMENCLATURE

The four terms combined in the heading above are often used interchangeably, leading to much confusion. We have said already that classification is the ordering of organisms into classes. *Systematics* is the comparative study of any group of organisms and of any and all relationships among them by means of the techniques of one or more branches of biology (Fig. 18.2). Systematic studies often (but not always) result in classifications. Likewise, classifications may or may not be based on systematic studies, but present-day biological classifications are almost exclusively so based. Thus these two operations tend to be synonymous. *Nomenclature,* on the other hand, is the application of distinctive names to each of the groups recognized in any classification. A set of arbitrary rules, embodied in the "codes" of botanical and zoological nomenclature, govern the application of names to plants and animals. These rules, which have been elaborated over the years, combine historical elements (like the rules of priority in naming species) with practical rulings to ensure clarity and ease in use. Nomenclature is a device to eliminate confusion and ensure one common language in biological classification.

Taxonomy has been variously defined as "the theoretical study of classification, including its bases, principles, procedures and rules" (G. G. Simpson, 1961, *Principles of Animal Taxonomy*); "the synthesis of all the facts about [organisms] into a concept and expression of the interrelationships of [organisms]" (H. L. Mason, 1950, *Madroño* 10:193–208); or the "study of the principles and practices of classification, ... in particular ... the methods, the principles, and even in part the results of biological classification" (J. Heslop-Harrison, 1953, *New Concepts in Flowering Plant Taxonomy*). The relationship of taxonomy to systematics is somewhat like that of theoretical physics to the whole field of physics. Taxonomy includes classification and nomenclature but leans heavily on systematics for its concepts. The two areas are often erroneously equated.

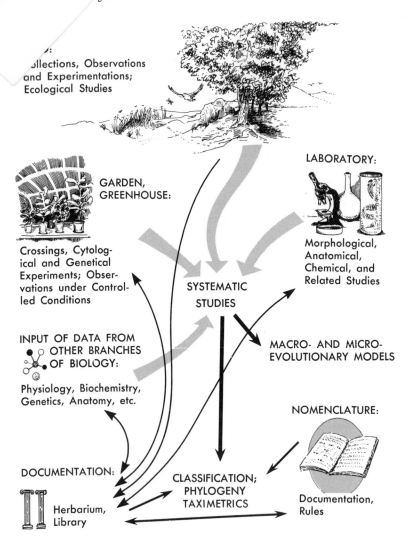

FIELD:
ɔllections, Observations
and Experimentations;
Ecological Studies

GARDEN,
GREENHOUSE:

LABORATORY:

Crossings, Cytolog-
ical and Genetical
Experiments; Obser-
vations under Control-
led Conditions

Morphological,
Anatomical,
Chemical, and
Related Studies

SYSTEMATIC
STUDIES

INPUT OF DATA FROM
OTHER BRANCHES
OF BIOLOGY:

MACRO- AND MICRO-
EVOLUTIONARY MODELS

Physiology, Biochemistry,
Genetics, Anatomy, etc.

NOMENCLATURE:

DOCUMENTATION:

Herbarium,
Library

CLASSIFICATION;
PHYLOGENY
TAXIMETRICS

Documentation,
Rules

Figure 18.2

Some of the operations involved in systematic and evolutionary studies. Arrows indicate direction of informational flow. (From O.T. Solbrig, 1966. *Evolution and Systematics*. New York: Macmillan)

ARTIFICIAL, NATURAL, AND PHYLOGENETIC CLASSIFICATIONS

The earliest classifications known to us are eminently practical ones, aiming solely at a system that allows ordering of the elements of the environment. So, for example, in Greek and Roman writing, animals were divided into those of the sea, the land, and the air, and plants into herbs, shrubs, and trees. These classifications of antiquity, known to us mainly from the works of Aristotle and Theophrastus, were slowly elaborated and perfected, particularly after the Middle Ages. The elaboration culminated in the mid-eighteenth century work of the great Swedish naturalist Carl von Linné (Linnaeus), who made the first effort to systematically classify all organisms.

Linnaeus's system had several features that made it deservedly famous in his time. For example, it allowed for the placement of newly discovered entities without disruption of the system. The basic taxa, the genera and species, were by and large fairly well delimited, and Linnaeus introduced a very practical method of binary nomenclature, which has persisted until this day. Nevertheless, Linnaeus's concept of the species was essentially the same as Aristotle's, based exclusively on morphological similarities, and his higher categories, though logically consistent, were often based on trivial characteristics. Linnaeus was a keen observer, and he had a questioning mind that led him, among other things, to conduct experiments on hybridization and inheritance in plants. This activity notwithstanding, Linnaeus believed, as did most naturalists of his time, in the immutability of species, which supposedly had been created once by a superior being in the same way and in the same number as they were known then. Consequently Linnaeus, who was a very pious man, did not attempt to discover a grand scheme of order in the universe. His practical and religious outlook probably did not permit him to do so.

It was the next generation of naturalists, in particular the French biologists, who attempted to find the grand scheme. The root of this movement goes back to the beginning of the scientific revolution. The discovery by Newton and Leibnitz of the laws of gravity and the movement of celestial bodies, together with the impact of the ideas and methods of Bacon and Descartes, produced a search for the laws of nature. That search eventually produced a new outlook about the biological world. Most still believed that species had been created by a superior being, but the creator was supposed to have had a grand scheme in mind, and the purpose of the systematists was to discover it. A search for the so-called natural affinities of organisms was initiated. This led to a great deal of good work, in the course of which true evolutionary relationships were discovered, although at the time they were not recognized as such. The ferment also gave considerable impetus to the study of comparative anatomy, and the period saw the beginnings of serious studies in physiology.

Among the earliest of the French naturalists espousing the new view were Buffon and his protégé Lamarck, who worked in the second half of the eighteenth century. Lamarck was the author of one of the first systems in which species were not considered immutable but the result of some kind of evolution. The movement to find the grand scheme flourished in France and to a somewhat lesser extent in Germany and England. Aside from Buffon and Lamarck, the names of Geoffrey, De Candolle, the German Oken, and the Englishman Erasmus Darwin should be mentioned.

The "natural systems" were not a total break with the artificial system of Linnaeus. Aristotelian logic was still, at least in theory, the rationale of the classifications. But a greater effort was made to study the totality of the characters of an organism and to try to have the classifications reflect natural affinities. This led to a more empirical approach to classification, in which logical concepts that we would associate today with set theory were applied.

The naturalists of the first half of the nineteenth century came very close to discovering the theory of evolution. In their writings we can read many modern-sounding discourses about the relationships between plants and animals. They probably did not arrive at the point of proposing an evolving biota for lack of a mechanism to explain the change. Charles Darwin's greatest contribution was therefore not the idea of evolution but his theory of natural selection, which provided a mechanism for evolution. Lamarck and others after his time had proposed systems in which species became transformed, but they were not accepted because the authors could not present a rational explanation of how the transformation might have come about.

The formulation and acceptance of the theory of natural selection did not immediately affect contemporary classifications. The principal reason was the excellent work of the systematists of the time, who in their search for "natural affinities" had discovered evolutionary affinities. Darwin himself acknowledged the influence of his four-volume work on the classification of barnacles on his formulation of the theory of evolution. Taxonomists after Darwin changed the phrase "natural" affinities to "phylogenetic" affinities, but for some time they did not drastically modify their mode of work.

The rediscovery in 1900 of Mendel's laws of inheritance at first brought a reaction against Darwin. Particulate inheritance was interpreted initially as denying natural selection. How could the accumulation of small changes account for evolution, if characters were inherited as discrete units? The reaction against Darwinism was due in part to the drastic effects of some of the first-discovered mutations. It was also due in part to the unfortunate choice of the plant genus *Oenothera*, the evening primrose, as research material by Hugo de Vries (one of the rediscoverers of Mendel's laws and a very strong opponent of natural selection). In species of this genus, morphological changes occur that were misinterpreted at the time as inheritable point mutations. Nevertheless, it was not long before Mendelism and Darwinism were united. More and better insight into genetic phenomena and the development of population genetics were most prominent in effecting this marriage. It eventually led to the formulation of the synthetic theory of evolution.

The prominent feature of the synthetic theory is a change from a typological mode of thinking to one that emphasizes populations. A feature of the Linnaean system that was still adhered to at the turn of the century was the Platonic concept of the *eidos* (idea, type, essence). This principle was embodied in the concept of the *type*—that is, the idealization of an individual as a representative of all the other individuals of the species. The idea of the type was a direct outgrowth of the period when it was believed that each species was immutable. A single specimen—the type—would be a sufficient sample of the species. Darwin recognized the importance of variability, but typological thinking did not cease at that time, partially because the dynamics of population evolution were not known. Only with the development of genetics and the realization that the population and not the individual is the unit of evolution did the thinking of systematists shift from considering individual specimens to studying series and mass samples and to moving whenever possible out of the museum into the field. The concept of "type" as the embodiment of the *eidos* of the species was abandoned, although types are still used as nomenclature devices.

The second big impact of the synthetic theory on systematics was the change in the concept of the species. Linnaeus defined species according to the dicta of Aristotelian logic. In his definition "the differentiae" were stressed, since "the genus" was common to a whole group of species. His concept emphasized differences rather than similarities and consequently obscured relationships. The lack of knowledge in Linnaeus's time about genetics and reproduction precluded the use of any characters other than strictly morphological ones. The lack of good optical equipment further restricted the use of morphological characters of an organism in their search for natural affinities, and the similarities as well as the differences between organisms were considered in their classification. The development of physiology, anatomy, embryology, and (toward the end of the century) cytology and genetics vastly increased the available criteria. Systematic studies became the basis of biological classifications in most instances—a situation that still holds. Darwinian theory provided for the first time a solid theoretical framework. This led scientists to emphasize similarities as well as differences between species. Nevertheless, the concept of the species did not change very drastically from that of Linnaeus.

It was the tremendous growth of genetics in the first half of the twentieth century and the development of the synthetic theory that brought about a major change in the concept of the species. Reproductive relationships became the principal criterion in defining species. This new species definition is closely tied to the emphasis given in the synthetic theory to breeding populations.

PHENETIC VERSUS PHYLOGENETIC CLASSIFICATIONS

The ultimate goal of the taxonomist is to produce a classification that reflects the evolutionary history of the species that are being classified. However, establishing the evolutionary history and phylogenetic affinities of a group of species is very difficult. Information is incomplete, and opinions differ among taxonomists regarding the meaning of the evidence. Therefore it is not uncommon to find more than one classification for the same group of species, each being in a sense a hypothesis regarding their possible evolutionary history.

If a classification is to reflect the evolutionary history of the group, all members of a taxon should be closely related and descended from a common ancestor. Such a classification is called *monophyletic*. But taxonomists also wish their classifications to reflect the degree of similarity of organisms, and they therefore feel that organisms that share common morphologies or special characteristics or that exploit similar environments should be included in the same taxon. For example, all "ducklike" birds are grouped in a common order (Anseriformes), which is divided into different families for the swans (Cygninae), the geese (Anserinae), the tree ducks (Dendrocygninae), the surface-feeding ducks (Anatinae), etc. In turn, each family is divided into genera. For example, the surface-feeding ducks (Anatinae) are divided into several genera (*Anas*, *Spatula*, *Aix*, etc.), and each genus is divided into species, such as the pintail duck (*Anas acuta*), the mallard (*Anas platyrhynchos*), the black duck (*Anas rubripes*), etc. All members of the Anseriformes are swimming birds that are also good flyers and may migrate long distances. They have relatively short legs, with three webbed toes and a flattened bill. This is the generalized plan that members of the order share. In turn, each family is characterized by a more specialized plan. Swans are very large with long necks; geese, which also have long necks, are smaller than swans but larger than ducks; ducks are both smaller and short-necked. Genera possess even more specialized plans, and so on. Unfortunately, monophyletic groups do not always share the same basic adaptive groundplan.

Lineages can change gradually with time (gradual speciation), or they may split, giving rise to two distinct lineages (true speciation; Chapter 11). The change of one lineage into two or more is called *cladogenesis*. Gradual change is referred to as *anagenesis*.

Anagenesis can create organisms with novel and distinct characters, such as "warm-bloodedness" (*homoiothermy*). When that happens, we say that the resultant organisms have achieved a new *grade*. Grades may be monophyletic or polyphyletic. In a monophyletic grade, only one phyletic line has achieved the character in question; in a polyphyletic grade, more than one phyletic line has achieved it. Homoiothermy is polyphyletic, since it has been evolved at least twice (birds and mammals) and possibly more times. A new phyletic line arising through cladogenesis is a *clade*, and clades are by definition monophyletic (Fig. 18.3).

The difficulty faced by taxonomists in classifying species into higher taxa is to reconcile their desire to create monophyletic taxa that also reflect anagenesis, and furthermore to do so with very incomplete knowledge (often no knowledge) regarding the actual phylogenetic history of the lineage—knowledge that can be obtained only from fossils. On the basis of knowledge derived from present organisms, intuitive feelings regarding the relative importance of characters, reconstructions of past climates, and an occasional fossil,

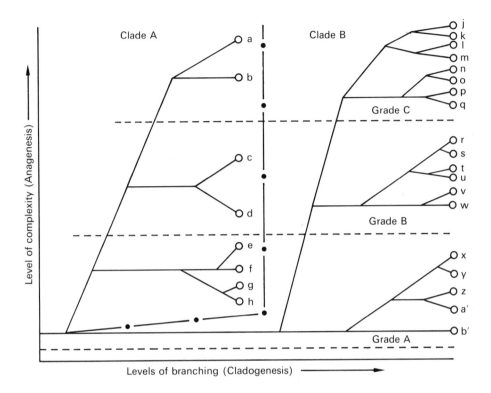

Figure 18.3

Relationship between evolutionary change in time (anagenesis) and branching of
lineages (cladogenesis); descent from single or multiple lineages (monophyly vs.
polyphyly); branches (clades) and levels of morphological or functional dissimilarity
(grades). Each circle represents an extant species. On the abscissa the level of
branching (cladogenesis) is indicated. On the ordinate the levels of complexity are
shown. Species a and b are more related phenetically to j–q since they have attained
similar levels of complexity, but they are more related phylogenetically to c and d
since they share a common ancestor. Species in the same grade may be polyphyletic,
as shown here; those in the same clade are always monophyletic.

taxonomists propose classifications that they hope reflect the phylogenetic
history of a lineage. The difficulty is that often no agreement can be reached
between taxonomists in their interpretation of the data. The result is a very
fluid situation, with different taxonomists proposing different classifications
that reflect their interpretations of the data. The consequent instability of
classifications presents a problem, for which two principal solutions have been
proposed. One solution is to classify organisms not on the basis of their phylogenetic history but on the basis of their morphological similarity. This *phenetic* approach has the advantage of operational simplicity. Morphological

characters can be easily measured or encoded numerically and compared (using computers) with the values of other organisms; such comparisons can produce an index of similarities (Fig. 18.4). Similar species can be grouped into higher taxa on the basis of a given value of the index of similarity. A major problem with a purely phenetic classification is that, depending on the criteria used, different values of the index of similarity can be produced. More important, all phylogenetic information is lost.

The other approach, known as the *cladistic* approach, is to classify taxa exclusively on the basis of the number of branchings that occurred during the

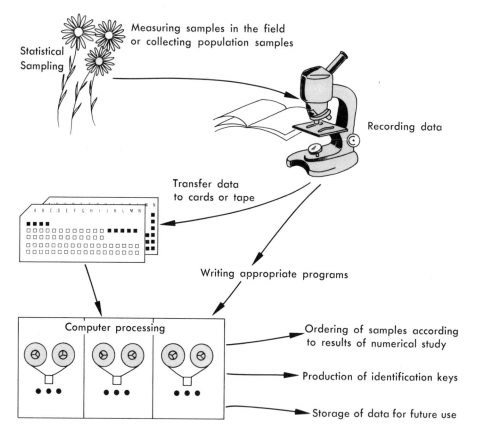

Figure 18.4

Numerical systematics. Characteristics of organisms are recorded or encoded numerically. They can then be manipulated by means of numerical techniques, often by computer. Various kinds of information can be obtained, including a systematic ordering of the organisms under study. (From O.T. Solbrig, 1970. *Principles and Methods of Plant Biosystematics.* New York: Macmillan)

evolution of a lineage. Taxonomists proposing this view argue that the branching of lineages offers the objective criterion for division into different taxa that is lacking when a phyletic line evolves by anagenesis. According to the cladistic approach, each branching episode creates sister taxa of the same rank, taxonomically separated from their ancestors (Fig. 18.5). Cladistics has the advantage of imparting a certain uniformity and simplicity to classifications, removing subjectivity from many classifications. The use of molecular clocks (Chapter 17) may help establish precisely when cladistic events took place. This simplicity, as in the case of phenetics, is gained by sacrificing evolutionary and ecological information.

It is not possible to produce a classification that reflects the evolutionary history of the species that are being classified when that history is only dimly perceived because of incomplete knowledge. Under the circumstances, a great deal of subjectivity will be involved. Strict adherence to phenetic or cladistic

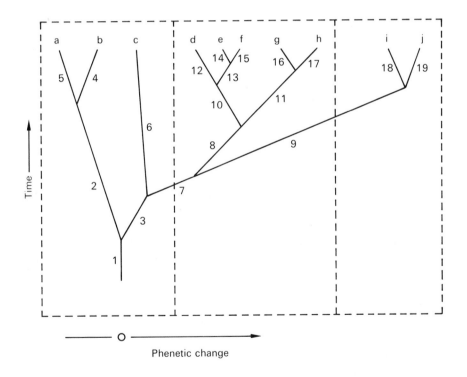

Figure 18.5

Relation between phenetic similarity and phylogenetic similarity. The numbers indicate the species according to a cladistic scheme of classification. The dashed lines include possible species according to a phenetic scheme of classification.

criteria will eliminate subjectivity. But subjective classifications are hypotheses, subject to verification and modification, and in that sense they are tools that aid the evolutionist in his quest for better knowledge.

THE TASK AHEAD

At the beginning of this chapter we pointed out the immensity of the task of naming and describing organisms that lies ahead. But taxonomy has now become a more rigorous scientific discipline than it was in the time of Linnaeus, with a set of basic tenets and a theoretical framework. Present-day taxonomists can often make predictions and design experiments to check their predictions. Resulting classifications are consequently much richer in informational content. A species name in Linnaeus's time was a means to designate a plant or an animal species but not much more than that. A name in a modern classification is the key to all that is known about it, its past history, and its relationship with other organisms, since we know so much more about the organism it designates.

Nevertheless, taxonomy has its limitations, too. The major stumbling blocks are lack of evidence concerning extinct organisms and ignorance about the direction of evolutionary trends and rates of evolution. These deficiencies create a major problem, since without that information, the weighting of characters in classification is always going to be largely subjective, and a truly phylogenetic classification will never be obtained.

Skeptics may therefore question the validity of attempting to classify organisms in such a way that the classification will reflect their past history when that history is unknown. Yet the careful study of genetics, cytology, structure, function, behavior, and ecological relationships can give us many clues to the past, as we have seen in this book. The judicious application of these data can produce a classification that approximates the past better than the artificial classifications of two centuries ago. We must always remember that science can at best present a statistical approximation to reality, and the best we can hope for is relative truth.

SUGGESTED FURTHER READING

General references

Davis, P.H., and V.H. Heywood. 1963. *Principles of Angiosperm Taxonomy.* Edinburgh: Oliver and Boyd.

Heslop-Harrison, J. 1953. *New Concepts in Flowering Plant Taxonomy.* London: W. Heineman.

Simpson, G.G. 1961. *Principles of Animal Taxonomy.* New York: Columbia University Press.

Solbrig, O.T. 1970. *Principles and Methods of Plant Biosystematics.* New York: Macmillan.

Articles and reviews

Ashlock, P.D. 1974. The uses of cladistics. *Ann. Rev. Evol. Syst.* 5:81–100.

Estabrook, G.F. 1972. Cladistic methodology: a discussion of the theoretical bases for the induction of evolutionary history. *Ann. Rev. Evol. Syst.* 3:427–456.

Hull, D.L. 1970. Contemporary systematic philosophies. *Ann. Rev. Ecol. Syst.* 1:19–54.

Mason, H.L. 1950. Taxonomy, systematic botany and biosystematics. *Madroño* 10:193–208.

William, W.T. 1971. Principles of clustering. *Ann. Rev. Evol. Syst.* 2:303–326.

Index

Page numbers in italics indicate that a definition of the term (also in italics) appears on the indicated page. For subjects discussed extensively in the book, also consult the table of contents.